Wind Power
for the World

Pan Stanford Series on Renewable Energy

Volume 2

editors
Preben Maegaard
Anna Krenz
Wolfgang Palz

The Rise of Modern Wind Energy

Wind Power
for the World

PAN STANFORD PUBLISHING

Published by

Pan Stanford Publishing Pte. Ltd.
Penthouse Level, Suntec Tower 3
8 Temasek Boulevard
Singapore 038988

Email: editorial@panstanford.com
Web: www.panstanford.com

British Library Cataloguing-in-Publication Data
A catalogue record for this book is available from the British Library.

Wind Power for the World: The Rise of Modern Wind Energy

ISBN 978-981-4364-93-5 (Hardcover)
ISBN 978-981-4364-94-2 (eBook)

Printed in the USA

The answer, my friend, is blowin' in the wind,
the answer is blowin' in the wind . . .

—Bob Dylan

Contents

Preface

We are happy to introduce *Wind Power for the World*, a book on the generation of electricity from wind. The book intends to present to the general public how wind power was able to conquer in less than 20 years a global market share of 250 GW and provide the perspective of the markets to come.

Preben Maegaard and Anna Krenz of the Nordic Folkecenter for Renewable Energy in Denmark have put this remarkable book from over 50 pioneers of wind power together. I take this opportunity to express my sincere thanks to all of them.

Pan Stanford Publishing, Singapore, had launched a book series *Pan Stanford Series on Renewable Energy* in 2010, starting with a book on photovoltaics. This book is the second in this series. Others will follow.

I am confident that the book will make its mark to the further success of wind power everywhere on our globe.

Wolfgang Palz
Series editor
Paris
Summer 2013

Acknowledgements

Bob Dylan's song was played again and again when the idea of the Folkecenter was first announced to the public more than three decades ago by Holger Sindbaek (1938–1990), a blind radio journalist. Although Holger could not see the beauty and colours of this wonderful world, he had the passion and vision that with joined forces it was still possible to have a common future, free of the fossil fuels and atomic energy. Soon the idea got general support—also from the government of Denmark.

As the fossil fuels and the atomic energy are limited resources they involve tremendous risks to the future of mankind, being the main cause of urgent worldwide climatic, environmental and economic crisis. The renewable energies are the answer, the promise to future generations. Life in harmony with nature is possible by using the abundance of perpetual sources of energy from the sun and the wind, a vague but realistic vision as our blind radio journalist saw it.

Wind Power for the World, a book of this type and size is possible only with generous financial support. It found one sponsor—the Nordic Folkecenter for Renewable Energy in Denmark. Despite its rather limited resources, the Folkecenter made it possible to edit a publication about modern wind energy, and which in itself is symbolic of the center's core activity since it was founded. The support provided the means to present to the world community the most comprehensive book ever published about the history of modern power with its all many facets.

I thank Stanford Chong that he insisted to publish a book on wind energy, a follow-up of Dr Wolfgang Palz's book from 2010, *Power for the World: The Emergence of Electricity from the Sun*. I also thank all the many authors from the five continents who so enthusiastically delivered their contributions. Without their efforts, visuals and papers, this book project would not have been possible. Someone has to coordinate and organise the many manuscripts, and sometimes even remind the authors of the deadlines as well as have their approval of changes that inevitably occur. For this, I thank Anna Krenz most heartedly. Together with Wolfgang Palz, the three of us formed a harmonious troika that maintained the focus of the book project even if it by variety and volume continued to grow and grow. For Anna Krenz, no detail was too small or too big, and in the end we included her as one of the authors. I thank my secretary, Nicolaj Stenkjær, for his steadiness in doing many of the trivial but necessary details. Finally I owe endless gratitude to my wife, Jane. She supported Wolfgang's proposal for my involvement as lead author without any delay. Patience and some sacrifices are unavoidable followers when you hold the many hundred pages of manuscripts and visuals in your hands. But Jane stood by my side and encouraged when needed till the very end.

Preben Maegaard
Denmark
May 5, 2013
Danish Day of Liberation

My personal adventure with renewable energy began more than 10 years ago, when I started my graduate studies in energy, environment and sustainable design at the Architectural Association School of Architecture (AA). It would be nothing special, if not the fact that it was my first (very) long trip abroad, really being thrown into cold water on the western side of the curtain. I was born and raised in Poland, in the times of the communist era. What can I say—it was all different back then.

When I had finished my master's studies at the AA School, I went home. I did not have any plans really—renewable energies were not so much present in the academic or popular scene in Poland. Actually, back then, nobody wanted to hear about that. And then my fellow student from AA called and said, "Why don't you go and visit my dad at the Folkecenter?" Why not? And I did. It was in 2002. Since that year I have been coming to Folkecenter every year, and I spend some weeks there, being always welcomed by warm-hearted Jane Kruse and inspiring Preben Maegaard.

Looking back, Folkecenter was a culture (and technological) shock for me. I have never met in my life people so much devoted to their cause, and this cause was a noble one. I was learning not only about renewable energy, sustainable solutions or specific technologies but about people—people are the key. Being at the Folkecenter, I came to experience the notion of "respect"—respect towards another human being (something hard to find in my homeland) and towards nature (in all its rough majesty), while bending heads towards the wind.

Starting with this book I was thrown into cold water again. When Preben asked me to make a book about wind power with him, I did not hesitate for a second. Since all these years we have known each other—I never said no to any of his challenges—and they always resulted in very hard work, but also satisfaction of learning something new and surpassing own limits.

The wind book project grew and evolved week to week. I talked and wrote to so many people—all of whom are the people of modern wind power. The more I read, the more I felt immense respect towards their achievements. It is truly unique to see people behind the development of what we today often take for granted—modern wind turbines—that originated many times from a sketch, from an experiment (even a failed one), from a stubborn urge to invent and make things work. For me, it was all history. For Preben, it is his story, his own life, his and many others' fight for the windmills. These stories needed to be told.

I am proud to be part of this project, and I am deeply grateful to Preben who asked me to join this project. Our collaboration, as always, was a great inspiring journey, challenging beyond limits. And I would like to thank all the authors, who agreed to share their stories, their thoughts and experiences.

At the same time I would like to thank all our collaborators, who engaged in the process of making of this book and helped us with their best knowledge and skills—proofreaders Jarra Hicks, Ron Ofer, JoAnna Woodruff, Natalie Rouskov, and Mike Eckhard; translators Ronny Spuer and Luise Hemmer Pihl; Nicolaj Stenkjær for support. My sincere gratitude goes to the publisher Stanford Chong for immediate responses to my technical and metaphysical queries and also to Shivani Sharma for the most brilliant cooperation towards the intensive end of the process of making of this book. And last but not least, Jane Kruse for always being there for us, my husband, Jacek Slaski, for his patience and understanding and my mom, Lonia Krenz, for travelling with me to Folkecenter and taking care of our little son, Anton (a three-year-old windmill fan), when I was working.

Anna Krenz
Lead editor
Berlin
Summer 2013

Introduction

Numerous books about wind energy have been published, addressed to professionals, investors, academia, ordinary users, environmentalists and several other potential sorts of interested circles.

However, among the many publications none had a special focus on the emergence of modern wind power seen from an international perspective; the process that resulted in well-functioning, affordable wind turbines. Basically, the story behind the scenes had to be told.

The book *Wind Power for the World* is for non-specialists and wind energy science and technology are covered to the strict minimum. The book is in two volumes: The Rise of Modern Wind Energy and International Reviews and Developments. In the first volume, we tell the story about how it became possible for wind power to develop during an almost 40-year period and emerge as a world-wide business of EUR 30 billion per year employing almost one million people by 2013. In the second volume of the book, we have collected reports and overviews of wind power status and history in various countries.

The uphill struggle that made it happen; wind energy strategies and policies that paved the way; the creative persons in politics, agencies and institutes; the industry; the world societies at large; and the challenges for which a solution was found at the end are the main topics presented in this book. The book's richly facetted stories are presented by over 50 experts from various areas of the renewable energy sector. Several of them are seniors who have made it all the way. They were first involved in their home countries and then at the international scene at the scientific, engineering, organisational, or political level, and have made valuable contributions to the successful emergence of wind power.

Without their professionalism, dedication and persistence, wind power might still have been in its infancy. The rising global climate and resource crisis, the increasing energy prices as well as the working of time, resulted in an astonishing technological development and reduced costs of equipment, which gradually made wind energy a realistic substitute for the conventional fossil fuels. The visionaries who created the way out of the feared fossil fuel and uranium trap have interesting stories to tell.

Only few countries took the initiative to proliferate contemporary wind power. We, as editors of the book, decided to focus mainly on three of these countries in the first volume: Denmark, with its absolute dominance from 1975 and for the following 15 years; Germany, which with its progressive legislation demonstrated that political visions and will, more than good wind resources, made Europe's largest economy a champion of wind energy; and China, which started from almost zero in 2005, but entered the arena with a concerted effort, and just five years later could celebrate its role as the global no. 1 in wind power sector, both in terms of installed and manufacturing capacity. A Chinese proverb says that even on the longest march, the first step has to be taken. China has shown a direction. Other countries can still use this as an example.

Some readers may find that the achievements of Denmark in the early phase of the contemporary wind energy development are overrepresented; however, most of the stories of the period have never been presented earlier to an international audience, and quite much of what happened since 1975 is still a historical mystery.

All told and with respect to every effort made to make wind energy a unique success, a search for the roots of the story points especially to the role of Denmark. Here, the first commercial, reliable and affordable wind turbines appeared as people's response to an oil crisis that had caused severe unemployment and financial problems for this small north European country's five million people. Denmark relied almost 100% on imported oil for heating, electricity and mobility. A paradigm shift within the supply of energy was absolutely necessary.

Fortunately, a large number of people were ready to spend their savings to purchase a wind turbine. At the same time many other people had visions and passion to design and manufacture the perfect windmill, that within a five-year period, through the trial-and-error method and countless experiments, resulted in what proved to become the contemporary wind turbine concept and the basis of large-scale, worldwide industrialisation.

When we decided to bring up this book on wind power, we looked 40 years back to find the roots of this exciting development and recognised that the cradle of the modern wind turbine stood in Denmark and a bottom-up development was the solution. We realised that this is the right time to get authentic stories from the Danish authors, who themselves were often part of this breakthrough, and have first-hand reports to bring forth the most important events, technologies, successes as well as mistakes. It may soon be too late.

The pioneers had to acquire fundamental knowledge on aerodynamics, technology and controls; make mistakes; learn from experience and often ruin their own economy—but every time someone would take over and experiences would not go wasted. The development of wind power was based on whatever worked best, while the negative results were left behind when they did not turn out to be any good. Thus, it is a long cavalcade of developers, inventors and manufacturers who each gave their big or small contributions to what took its beginning in 1975 and, in the course of five to eight years, became a real modern industry.

In that period the Danish concept was defined; reliable and affordable wind turbines became available. The first uncertain steps into the Fourth Revolution—the transition to renewable energies—were taken in Denmark in the same period when the heroes of the Third Revolution, people like Steve Jobs and Bill Gates, fumbled with and engineered their equipment and software for some of the first commercially successful personal computers, the backbone of the information society that later reached out to every corner of the world.

However, there is a decisive difference. Bill Gates for his part kept the copyright of his operating system for himself. He believed that other vendors would clone it. The Danish wind energy pioneers in contrast had grown up in a totally different tradition: open source.

When Denmark in 1895 got its first patent law, technologies and processes for use in agriculture were explicitly excluded from copyrights and patents. This principle of openness and sharing of knowledge by the Danish wind energy pioneers was maintained as extremely essential for the innovative culture. There was no fear of ideas being cloned and concrete solutions were never protetcted by copyrights or patents. The principle learned from Poul la Cour, the Danish wind energy pioneer 100 years earlier, was as follows: Patents, if any, should belong to the Danish people.

For decades, the universities had severely neglected research and capacity building within renewable energy, not realising that the fossil and atomic era would some day come to an end. Therefore, the Danish wind energy pioneers wanted the scarce technology know-how to be freely available to anyone who wanted to use and clone it. They believed that energy from the wind and the sun belonged to us all, and therefore, the knowledge to harvest renewable energies should be part of the heritage of mankind.

During the 1980s, wind industry in the United States, the biggest economy of the world, passed through a rapid rise and fall with later ups and downs, while the young German industry attached to the Danish supply chain with its specialised, independent suppliers of blades, control systems, etc. Emerging manufacturers made robust wind turbines long before the first German renewable energy law opened the market for large-scale investments and regular industrialisation. Wind power soon gained significant shares at the cost of conventional power sector that was dominated by big and politically influential power oligopolies.

In 1995, Spain introduced the German version of the feed-in tariffs (FITs)—the key to rapid and decentralised renewable energy development. The country did not rely on imported wind turbines and got its own advanced manufacturing sector with brands that became well known in the international market.

Ten years later in 2005, China, in search for additional electricity production capacity and with an eye for a new industrial growth sector, also joined the wind energy frontrunners. By opening the door for the best available technology, the biggest nation in the world diligently avoided mistakes that delayed the industrialisation in most other countries. Soon China got its own complete supply chain and more than 50 MW–size suppliers of wind turbines. After a five-year period of concerted effort, a completely new industry emerged that made China the absolute leader in the wind power sector, both in terms of manufacturing and installed capacity.

A Roadmap in Pursuit of the Legend of Modern Wind Power

With the rediscovery of the asynchronous generator, this book presents various sides of the origin of the contemporary wind turbine design and acknowledge people who made crucial conceptual and technological contributions in the wind power revolution, especially Johannes Juul and Christian Riisager. In 1976, Christian

Riisager was the first to commercialise a small wind turbine using J. Juul's design principles from the 1950s and like other manufactures could sell it to Danish people willing to spend their money on alternatives to the imported oil.

It was not the hybrid blade technology of Juul and Riisager, however, that lead the way to the development of the so-called Danish Concept, but the 2 MW Tvind windmill, designed and built by a group of amateurs and idealists from the Tvind School. In 1976, Tvind transferred Professor Ulrich Hütter's advanced blade technology from the Technical University of Stuttgart to Denmark and made it available for the general public. The newly founded Økær Vind Energi brought to the market Tvind's downscaled 4.5 m fibreglass blade, the basis of the emerging component wind turbine. Soon after, NIVE, divided up a wind turbine's structure into four basic elements that were manageable items within the existing specialised industries. Specifications were defined for modular wind turbine components (blades, controls) which lead to the emerging supply chain that enabled the Herborg blacksmith, members of the Danish Blacksmith's Association and other small enterprises to manufacture and assemble reliable and affordable 15 kW to 22 kW wind turbines.

This book describes how the combination of J. Juul's principles (heavy, upwind, 3-bladed, asynchronous generator, stall-regulated) with Ulrich Hütter's/Tvind's advanced blade fibreglass design and root assembly resulted in the winning wind turbine concept, called the Danish Concept.

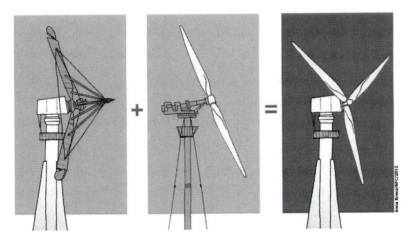

A hybrid of J. Juul's turbine with U. Hütter's blades is "the Danish Concept".

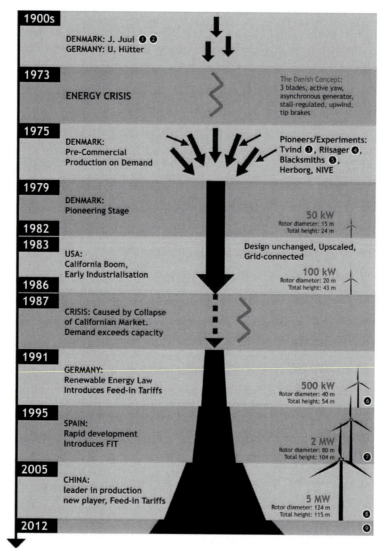

Timeline of the development of modern wind power.

This process took place between 1975 and 1979 when the later successful manufacturers Vestas, Nordtank, Bonus, etc., commercialised the concept. During the Californian wind rush from 1983, small workshops became regular factories with series production. The leading companies scaled up the concept step by step and took it to the world market that they dominated for the succeeding 15 years. When other countries like Germany, Spain and later China got into commercial wind turbine manufacturing, their designs were J. Juul–Hütter hybrids that had become the industrial standard. Some manufacturers changed to synchronous generator, however, within the same basic concept. The diagram below presents some of the milestones in the period from 1975 till 2012.

The development of modern wind turbine in pictures
(Photo 8: Hans Hillewaert).

Writing history, even the history of contemporary wind power, is a continuous process. The essence of this book is to bring forth the fact that it was up to the people to create the modern wind turbine industry with bottom-up growth, even if the early products often might appear to be technically clumsy, compared to the perfectionism characterising modern industrial products. It was the craft, design, production and operation of wind turbines that was to be learned first, and from the bottom, by people who were obsessed with belief in the new technology. The wind singing above their heads belonged to them.

Preben Maegaard
Lead author
Nordic Folkecenter for Renewable Energy
Denmark
Summer 2013

Chapter 1

The Wind Power Story

Ross Jackson

Gaia Trust, Stavnsholt Gydevej 52, 3460 Birkerod, Denmark

rossjackson@gaia.org

The development of wind power is a fascinating story that tells us a lot about how really significant change often comes about from the actions of unsung, unknown entrepreneurs with a clear vision of the future, but operating beneath the radar of the mainstream culture, and how politicians are often the last to understand the long term consequences of their actions or, unfortunately, their lack of action.

If you were to ask an American politician today if society ought to be investing far greater sums in renewable energy, many would shake their heads and mumble something about "prophets of doom" or "idealists out of touch with reality". Those that respond positively will offer as their main argument that we must reduce

Wind Power for the World: The Rise of Modern Wind Energy
Edited by Preben Maegaard, Anna Krenz and Wolfgang Palz
Copyright © 2013 Pan Stanford Publishing Pte. Ltd.
ISBN 978-981-4364-93-5 (Hardcover), 978-981-4364-94-2 (eBook)
www.panstanford.com

our dependence on unstable Middle-East sources of oil, their main concern being to continue their current lifestyles with as few disturbances as possible. Only a few would mention anything about ecological overload, climate change, peak oil, sustainability or the needs of future generations. And yet it is the latter factors that are really critical to understanding the urgent need for more renewable energy. Interestingly, these are precisely the kinds of argument you will hear from those who actually started the wind power adventure. They had an intuitive understanding of the need for clean energy if our civilisation is to survive and thrive—people like my good friend Preben Maegaard, who made the vision of a sustainable future the driving force in his life's work as far back as 1970. Today, if you could make a search engine to seek out the greatest concentration of windmills in the world, it would lead you to Preben at his Nordic Folkecenter for Renewable Energy in northwest Denmark. Let us now look at the arguments for increased investment in renewables, and particularly in wind power.

1.1   Ecological Overload

A useful quantitative measure of the degree of sustainability of a region is the so-called "ecological footprint". While it is a rough measure, it is the most useful tool we have at this time for the impact of human societies on the environment. The ecological footprint measures the amount of land that would be required by the population of a region in order to provide the renewable resources consumed and the sinks to absorb waste products. This measure is now widely used. Data are published regularly by the World Wildlife Fund for Nature (WWF) for 150 nations in their *Living Planet Report* [1].

In the period 1961–2008, the total footprint increased from approximately 53% to 150%, as shown in Fig. 1.1. The 2008 level thus corresponded to an overshoot of roughly 50%. The overshoot is probably closer to 60% in 2013. WWF figures are about three years behind for data collection reasons, but our footprint is thought to be growing by 2% per year.

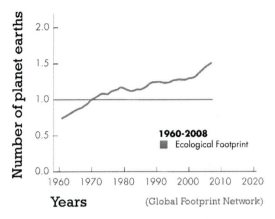

Figure 1.1   Humanity's global footprint 1961–2005.

Think about this for a moment. In just one generation we have gone from a situation where we consumed about one half of what the sun provided each year, as plant and animal resources, to a situation where we are consuming 50% *more* than the renewable resources that nature can replenish. How is this possible?

1.2   Climate Change

The explanation becomes clear if we break down the ecological footprint by component, as shown in Fig. 1.2 from the WWF 2008 report. Here we see that by far the major contributor in 2005, and the fastest growing component, is the "carbon footprint", or CO_2 emissions—now roughly 50% of the total—as opposed to only about 10% in 1961. In absolute terms the 2005 carbon footprint was roughly 13 times larger than the 1961 level!

Note that it is not so much increasing consumption, but rather the utilisation of the sinks, that is, the storage facilities for CO_2 emissions—primarily from burning fossil fuels—namely the oceans and the biosphere, that is the main problem. We will soon reach capacity in the oceans' ability to absorb CO_2. The direct results for aquatic life alone are enormous, as ocean life is ultrasensitive to small temperature increases, the increasing acidity due to dissolving massive quantities of carbon dioxide and decreasing salinity due to melting ice, e.g., coral reefs, which are a critical component of the food chain. But the greater catastrophe ahead is what happens when the ocean's CO_2 capacity is reached. The amounts entering the atmosphere will then accelerate, with the imminent danger of passing a tipping point of CO_2 concentration in the atmosphere that could trigger an irreversible temperature increase, which could be fatal for humanity.

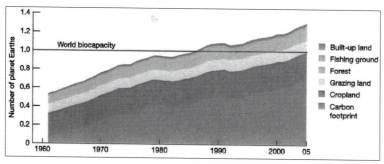

Figure 1.2   Global footprint by component, 1961–2005.

The burning of fossil fuels seemed like a reasonable thing for humans to do once oil and natural gas resources were discovered and the necessary technologies developed in the late 19th and early 20th centuries. Few, if any, guessed that this seemingly innocent act was going to cause nature to react in an unexpected and potentially disastrous way one hundred years later. Unintended consequences are typical when we mess with complex systems that we do not thoroughly understand.

It is therefore not surprising that global warming, which is a direct consequence of rising CO_2 emissions, is currently the most visible example of ecosystem breakdown. Examples of drought, floods, hurricanes, glacial melt-offs, shifting rain patterns and other consequences of what amounts to about a one-degree Celsius temperature rise are so numerous now that we all have had some personal experience of what is in store. And this is just the beginning. Clearly, we should be rationing the use of fossil fuels as if our life depended on it. But even if we had a much larger biosphere (and hence no immediate climate change problem) and much larger fossil fuel reserves (hence no immediate peak oil problem), we should still ration the use of this resource for the sake of future generations and long-term sustainability. US President George H. W. Bush once said: "The American way of life is not negotiable", but a more correct statement would be to say: "Negotiating with nature is not possible."

1.3 Peak Oil

The potential threat to humanity from global warming is in the intermediate term, that is, decades rather than centuries. There is however, a shorter term event, that is, years rather than decades, looming on the horizon that is likely to overshadow concerns about global warming and other problems caused by ecological overload.

I am referring to the coming peak in global oil production. The coming decline in oil supplies is not a threat to survival as such, but is going to cause a more immediate crisis that is a threat to our whole way of life. The basic problem is summarised in the Fig. 1.3, which shows that net additions to global oil reserves have been declining for several decades.

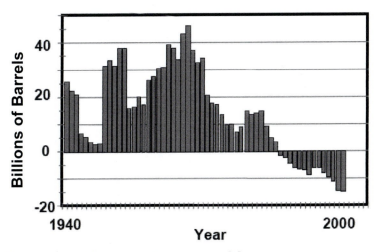

Figure 1.3   Oil discovery and consumption [2].

We are consuming more each year than we are discovering. In other words, oil reserves in the ground are decreasing for every year that goes by. Until now, supply has always exceeded demand, but just as reserves are decreasing, demand is increasing, especially from China and India. Thus, it is only a question of time before demand will exceed supply for the first time ever, and there will not be enough oil to go around. When this happens, the consequences will overshadow everything else on the political agenda. The only real doubt among the experts is *when* the peak will occur, if it hasn't already occurred.

It is the impact on economic growth that is likely to be the most immediate effect of the coming oil shortage. It will come as a great surprise because mainstream economics is fatally flawed when it comes to understanding the dependence of growth on energy. The economists' approach is to look at the national accounts and observe that energy use accounts for about 5% of GDP; no problem. But this is because their prices are all wrong. And they are wrong because mainstream economics has no concept of physical limits to growth, and the true value of natural capital like oil reserves. They, and their political masters, are in for a sobering surprise.

When physicist Robert Ayres of the European Business School, INSEAD, applied a physics approach to economic growth models, he was able to show that 75% of growth in US GDP in the 20th century was due to increased energy use, and not just 5% as suggested by the economists' accounting approach. His model was far better than the economists' traditional labour/capital model, which could only explain 25% of the actual growth. Ayres concluded: "Economic growth depends on producing continuously greater quantities of useful work" (i.e., more and more net energy) [3].

Once we enter the period of energy descent in the relatively near future, oil and natural gas production is likely to fall by about 3% per annum, according to leading oil geologist Colin Campbell, who warned of what was coming as early as 1990, and urged politicians to plan at least ten years ahead of the peak to avoid chaos. His appeal—like many others from the scientific community—was ignored [4]. The decrease in GDP for a 3% descent, according to the Ayres model, is 2.25% per year, suggesting we will enter a semi-permanent depression globally, at least for several years, if not decades.

Our whole society has been built up around the unrealistic concept of unlimited cheap oil. For example, if we take into account the oil used in chemical-based industrial agriculture for fertiliser production, farm machinery, transportation, irrigation, livestock raising (exclusive feed) and pesticide production in the United

States, over 400 gallons (1514 litres) of oil per year are used to feed each American. This figure does not include packaging, refrigeration or transportation to retailers. How we are going to feed seven billion people when oil production goes into permanent decline? How we are going to maintain a society of megacities and an oil-based automobile and aviation culture when oil prices quadruple? These are the issues that will soon be at the top of the political agenda.

Once the reality of the crisis hits home, there will be a desperate attempt to develop alternatives to oil, but probably coming much too late to avoid a long term crisis due to faulty economic models, a failure to head warnings from the scientific community and a lack of preparation.

1.4   Energy Alternatives

What about coal, nuclear, solar, wind, hydro, geothermal, biogas, wave energy, hydrogen-based fuel cells and other energy technologies? Can they not replace oil so we can continue our high consumption society? Leaving aside for the time being the problem of ecosystem overload, none of these can replace oil. They are all substitutes in the sense that they are types of energy, but none has the low cost access or concentrated energy of oil that has allowed us to build up a highly technological, centralised, automobile-based society. In the future, when considering alternative strategies, we will have to think much more in energy terms rather than economic calculations based on distorted pricing.

The key question we have to keep asking regarding useful energy production is, how much energy do we put in and how much do we get out? It costs energy to produce usable energy. The source has to be mined, the raw product refined and transported before it can be used usefully at the point of need. Table 1.1 shows roughly how much energy we can usefully get out of various energy sources for one unit of energy put into production, the so-called EROEI (Energy Return on Energy Invested): A figure less than 1 means that we use more energy on the input side than we get on the output side. Figures are based heavily on the analyses of Charles A. S. Hall, systems ecologist and professor at the State University of New York [5].

Table 1.1   Energy return on energy invested (EROEI)

Resource	Energy return
Oil	19
Natural gas	10
Coal	50
Nuclear	1.1–15
Wind	18
Solar	3.7–10
Hydrogen	< 1

Coal clearly stands out, but the above figures do not include the associated environmental costs, which are far higher for coal than any other alternative. Although some optimists claim that there are coal reserves for at least 130 years at current usage rates and quality, there are strong indications that economically available quantities are much lower. The one country that has (almost) completed the complete cycle of coal production is the United Kingdom, where it turns out that historically estimated useable reserves were grossly overestimated. Dave Rutledge of Caltech has developed a more accurate model of useable coal based on actual historical production data rather than government estimates, and concludes that traditional estimation methods, including those of Intergovernmental Panel on Climate Change (IPCC), overestimate useable reserves by 4–6 times. Rutledge estimates that 90% of all usable coal will have been burned by 2070 [6]. Others are even more pessimistic, as increasing costs of mining suggest that the EROEI of coal will slide substantially within a few decades, making coal use quite inefficient within a relatively short period [7].

Nuclear energy is only viable with huge government subsidies to cover hidden costs that are never mentioned by the promoters. Take for example the United States. The first hidden cost to taxpayers is due to a cap on possible insurance claims against private corporations at USD 10 billion per accident—as mandated by the Price-Anderson Act. The risk of a major accident, typically claimed by promoters to be negligibly small, is simply too high for private insurance companies to accept. This raises an interesting question: who is most credible on risk evaluation, the promoters of nuclear energy or the insurance companies? I vote for the insurance companies. Thus, anything above USD 10 billion must be paid by the payer of last resort—namely the taxpayer. It is impossible to put a precise cost figure on a nuclear accident because of the individual circumstances and the long-range generational effects and the human costs, but the numbers can be astronomical. The second subsidy is the cost of disposing of nuclear wastes, some of which will be around for over 100000 years. There is still no adequate solution to the disposal problem after decades of research. It is impossible to put an upper limit on what nuclear energy waste products will have cost humanity when the final record is written. The third hidden cost is public research funds. In the IEA countries in the three decades between 1974 and 2002, nuclear energy received 58% of energy research funds, or USD 175 billion, and fossil fuels 13%. Wind energy received only 1%.

Wave power has great potential but a major breakthrough still lies ahead. We cannot count on it as yet. In the meantime, we should be allocating major resources to further studies and experimentation.

The much-hyped "hydrogen economy" is not a viable solution to the energy shortage either. It takes more energy to produce hydrogen than it contains. But hydrogen does have the advantage that it can be stored as fuel for fuel cells and used to power electric cars. The problem is: where do we get the energy to produce the hydrogen? Do we use the declining supplies of non-renewable sources like natural gas, coal and nuclear power? That is certainly not a sustainable policy. Should we use the limited sources of renewable energy such as wind and solar? Firstly, that would mean diverting them from other more critical uses, and secondly, there is simply not enough energy available from these sources in the intermediate term. And besides, what is available is expensive. The best we can hope for is to maintain the automobile culture for a little while longer. Whether hydrogen production will be the best way to use the limited energy available in a post peak-oil world is highly questionable.

That leaves wind and solar as the two most promising alternatives. Considering that wind power is also the most environmentally friendly of all the alternatives (with the possible exception of wave power), and the sun is a limitless energy source (at least on the human scale), then there can be little doubt that wind and solar power is where the major investments should be made. The combination of wind and solar in decentralised combined heat and power (CHP) plants producing both electricity and heat for the local community may well be the most promising vision for the future. But additional research and experimentation is required, not least as regards energy storage, as both the wind and sun are irregular sources.

1.5   The Energy Trap

No one can foresee how the energy descent scenario will play out, other than to say it will be a traumatic and painful experience for everyone. But I would like to point out one important factor that we should bear in mind no matter what else we do in this period–the risk of falling into an "energy trap".

A fatal energy trap could condemn humanity to become a subsistence civilisation if we are not very careful with how we use the remaining fossil fuel resources. The production of alternative energy sources, such as solar power plants, windmills, and wave and geothermal energy, all require fossil fuel energy. It takes energy to mine the metals and manufacture the necessary inputs in order to build alternative energy power plants. Today this energy can only come from one source—fossil fuels. Imagine for a moment what will happen when the oil and gas runs out in a matter of decades at current usage rates. Not enough fossil fuel energy will be available to build or maintain windmills and solar power plants. Unless some unforeseen energy breakthrough occurs, humankind will be doomed to a subsistence-level low-energy future with a much smaller population. This is the energy trap we must avoid at all costs.

If we plan carefully, we should be able to fulfill our energy needs without fossil fuels in the long run. An IPCC report in May 2011 estimated that we could meet 80% of our global energy needs with renewables by 2050 while keeping the CO_2 concentration in the atmosphere under 450 ppm. The major problem, as the IPCC also points out, is political will [8].

Avoiding the energy trap may well be the greatest challenge facing humankind in the coming decades. Solar energy, supplemented by wind power, is the key to a long-term solution because of the enormous amounts that impact the planet every day and will continue to do so for at least a billion years. But in the long term, solar/wind power plants, whether based on solar thermal or photovoltaic technology, must become self-replicating, that is, produced 100% from solar and wind energy without fossil fuels. Until we reach that point, we cannot relax. From that point on, we can always produce more energy than we need. If we continue to use fossil fuel energy for running cars, heating houses, and producing non-essential products, we may not make it out of the trap. This will require sophisticated rationing and prioritising the use of the remaining fossil fuel energy.

1.6 Conclusion

The story of wind power is a never-ending story and we are still in the early days. Assuming we survive the next century, which is far from given, then the wind and the sun will be our best friends for millennia to come.

References

1. World Wildlife Fund "Living Planet Report 2008", (see www.panda. org).

2. Hirsch, R.L., Bezdek, R., Wending, R. (2005). *Peaking of World Oil Production: Impacts, Mitigation & Risk Management*, US Dept. of Energy, p. 15.

3. Ayres, R.U., Warr, B. (2009). *The Economic Growth Engine; How Energy and Work Drive Material Prosperity*, UK: Edward Elgar, p. 297.

4. Campbell, C. (1997). *The Coming Oil Crisis*, Petroconsultants, in association with Multi-Science Publishing Co. Ltd.

5. Heinberg, R. (2009). *Searching for a Miracle*, p. 55 (www.postcarbon. org).

6. Rutledge, D. (2011). "Estimating Long-Term World Coal Production with Logit and Probit Transforms", *International Journal of Coal Geology*, Vol. 85, pp. 23–33.

7. ibid., p. 3 referring to John Gever *et al.* in *Beyond Oil*, (Boulder CO: University Press, 1991).

8. Press Release "Experts Underline Significant Future Role in Cutting Greenhouse Gas Emissions and Powering Sustainable Development", *Intergovernmental Panel on Climate Change*, May, 2011 (see www. ipcc.ch).

About the author

Ross Jackson, PhD, an expert in international finance and operations research, has long been an innovative leader in both the business and NGO worlds. He is founder-chairman of Gaia Trust, a Danish-based foundation that for 25 years has supported and continues to support the Global Ecovillage Network and Gaia Education, as well as hundreds of sustainability projects in over 40 countries. He is also director and owner of Urtekram, Scandinavia's largest wholesale organic-food company, and is former chairman of Gaiacorp, foreign-exchange consultants, and hedge fund managers. This article draws heavily on his recent book *Occupy World Street: A Global Roadmap for Radical Economic and Political Reform* (Chelsea Green, 2012).

Chapter 2

Forty Years of Wind Energy Development

Jos Beurskens

ECN Wind Energy, P. O. Box 1, 1755 ZG Petten, The Netherlands

beurskens@ecn.nl

It was during the early seventies, that I started thinking about sustainability and started working on wind energy technology as one of the sustainable energy supply options. After almost 40 years of intensive involvement I realise that I am one of the few lucky ones, who experienced the entire development period of modern wind power technology. I feel it as a duty to provide a (personal view) on the developments in those 40 years, if it were only to prevent unnecessary duplications of past developments.

Below I will address some developments from my personal point of view; developments in the areas of wind turbine systems and applications in a framework which is determined by advances in science and technology, market development, integration requirements and policy.

Wind Power for the World: The Rise of Modern Wind Energy
Edited by Preben Maegaard, Anna Krenz and Wolfgang Palz
Copyright © 2013 Pan Stanford Publishing Pte. Ltd.
ISBN 978-981-4364-93-5 (Hardcover), 978-981-4364-94-2 (eBook)
www.panstanford.com

In the late sixties of the previous century a broad loosely organised political movement arose that among others strongly criticised the limitless consumption of resources by the rich industrialised nations. Limitless growth of the economies, without taking principles of sustainability into account, would lead to environmental destruction, poverty in large parts of the world, depletion of precious energy, water, food and material sources. Although these concerns were in the first instance expressed by left parties and anarchistic youth movements, the philosophies behind those protest movements were gradually embraced by industrial leaders and many politicians. In 1972, this culminated in the publication of *The Limits to Growth* [1], commissioned by the—still existing—Club of Rome, an alliance of industrial leaders and concerned scientists. When and how will exponential growth start to curve into saturation as a result of resource constraints or other boundary conditions and what will the practical consequences be?

Together with two colleague students and our scientific coach Paul Smulders at the Eindhoven University of Technology, we set out for a case study on sustainability in the area of energy supply. As we were interested in fluid dynamics, our choice was wind energy! We studied world energy demand, problems in the developing world, the history of wind energy until the early seventies, the potential role of wind energy in the future, wind statistics, wind turbine rotor models and matching load and rotor characteristics. This led to the publication of a university text book *Windenergie*, which was published in 1974. Publication was postponed for more than one year because our serious work was brutally interrupted by the energy crisis of 1973! The four of us appeared to be the first ones in the Netherlands who possessed updated documented knowledge about wind energy technology and its potential. The coincidence of "limits to growth" and the "energy crisis" led us into the world of the sustainable energy industry in general and that of the wind power industry in particular; an industrial sector which prospers until this day and will definitely determine our energy future to a great extent.

2.1   The Emergence of the Wind Energy Industry

Realising the low energy density of the wind and the vast amounts of energy that the industrialised world needs, I was not convinced that wind energy could play a significant role in the energy supply of the developed world. A commercial wind turbine from the late 1970s with a rotor diameter of 10 m to 15 m could easily double the commercial energy supply to a village of 3000 people in developing countries but could only increase the energy supply by 2% to 3% of a village of the same size in the industrialised world. So wind energy, like solar energy, in my view, could only play a significant role in developing countries, but was bound to remain marginal in the industrialised world. In developing countries the locally available wind (and solar) energy could be used to improve the quality of life by supplying water for irrigation, domestic use and cattle watering, by extending daily workable hours (lighting), conservation of food (refrigeration) and communication (powering radio, TV and much later PCs.) NGOs and groups, often associated with universities and supported by national governments, were established to utilise the wind for energy supply to rural areas in developing countries. Using a domestic energy source to meet the primary needs was considered key for development and it would hardly draw on foreign currency. Examples of these groups are CWD in the Netherlands, ITDG in the United Kingdom, the BRACE Institute in Canada, IPAT in Berlin, Folkecenter in Denmark and several groups connected to the universities in various countries among in the USA, Sri Lanka, India and Bolivia.

I started my wind energy career with Consultancy Services Wind Energy Developing Countries (CWD, formerly SWD) to develop water-pumping systems for irrigation and cattle watering (see Fig. 2.1).

To that end I lived three months in Musoma, Tanzania, at the shore of Lake Victoria and two months on the Cape Verdian Islands. Having to deal with the basic needs of life and finding energy sources to meet them, is an essential lesson. An energy supply system should never be designed without taking energy efficiency and conservation into account, a lesson which nowadays is often forgotten. Research during the CWD period was not only focused on efficient wind turbine and pump design, but also very much on system technology. By means of passive matching of rotor and load characteristics, system efficiency could be doubled or tripled, without additional cost in active control systems. System engineering and integrated design seem trivial basics for wind turbine design, but is still poorly developed and not applied in the industry to the highest standards possible.

Figure 2.1   Demonstration of a fast-running water-pumping wind turbine developed by CWD, the Netherlands, manufactured and tested by ECN, the Netherlands, at the Silsoe (agricultural) College, UK (Photo: Jos Beurskens ©).

Wind and solar energy technology, however, developed completely different from what I expected in the early days. Actually we have seen a dual development track from the start: decentralised applications, later extended to grid-connected wind turbines, on the one hand and centralised grid-connected large-scale wind farms on the other.

Even before the oil crisis, pioneers, both individuals and small companies, developed small wind turbines with a generating capacity approximately equal to the average electricity demand of a household, farm or small enterprise.

In a way, these pioneers followed upon a craft which was developed during the Second World War, when many people built wind power generators for electricity supply as the regular electricity supply was interrupted frequently. Some of the self-made designs of the pioneers of the early 1970s made it to the industrial phase. Examples are the Lagerwey turbine (the

Netherlands), Enercon (Germany), Carter (USA), Enertech (USA). Concept wise these designs were very innovative (passive blade control, variable rotation speed drive systems, flexible elements in the blade hub connection). In Denmark members of the Smedemesterforeningen (Danish Blacksmiths' Association), among many other pioneers, built small grid-connected turbines mostly according to the proven "Danish design", which featured fixed pitched blades, constant speed induction generators and (passive) stall control. Companies like Vestas, Bonus, NEG and Micon emerged from this pioneering work. In the early 1980s in Denmark alone some 30 companies were active on the market and in the Netherlands more than 20.

In Denmark, the Netherlands, the USA and Germany test stations for small and medium-size wind turbines were built to test industrial turbine types under field conditions. The intention was to support manufacturers in marketing their products by providing evidence of reliability and energy efficiency of their machines.

After consultation of our colleagues at Risø (DK) and Pellworm (D), at the Energy Research Centre (ECN) we launched the Informal Meetings of Test Stations (IMTS) network. The objective was to improve testing procedures, develop codes of practice and improve exchangeability of test results between test stations. The industry, which was more and more acting on the international market, required one internationally recognised test certificate. The IMTS network quickly grew by memberships from Canada, France, the United Kingdom, the United States, Germany, Sweden, Russia and other countries.

The IMTS network was very fruitful and, primarily with the support of the European Commission, the work among others resulted in recommended practices for measuring procedures [published by the International Energy Association (IEA) and inputs for standards which were issued by the IEC. Another important result of the collaboration of test stations was the acceptance of MEASNET (Network of European Measuring Institutes), a self-imposed set of quality measures that test stations are obliged to meet in order to use the MEASNET quality brand. This system, created with support from the EC, is still being used in the present market and both public and private testing laboratories seek MEASNET accreditation before entering the market.Essentially MEASNET enables a number of distributed test stations to act as one entity as an alternative for one centralised accredited testing laboratory.

The demand for small capacity wind turbines vanished slowly in favour of medium-size grid-connected turbines. At the same time the wind turbine market developed from a domestic to a continental and intercontinental one. It was the Danish manufacturers who started selling their machines to different European countries and were very successful in exporting them to the United States.

2.2   National Programs

In parallel to the developments described above, as of the late seventies, national governments commissioned large technology firms like MBB, MAN, Boeing and Stork and research establishments like Risø (DK), ECN (NL) and SERI (USA) to develop large experimental wind turbines. These projects were elements of various national R&D programs, which also incorporated resource assessment, grid integration, technology assessment (vertical versus horizontal wind turbines), aerodynamic rotor modelling, aero-elastics, structural dynamics, design tools, component testing and standards.

The first large-scale turbines were commissioned in 1979 (NIBE 1 and 2 in Denmark) and the last experimental, non-commercial, turbine was completed in 1993. In total some 30 non-commercial experimental machines were built with substantial support from national governments. The technical layouts of these wind turbines were very diverse. Many of the innovations dating from before 1960 were improved and implemented.

The spectrum of concepts included:

- Rotors with 3, 2 and 1 blade(s) for horizontal-axis turbines and 3 and 2 blades for vertical-axis machines;
- Fixed, teetering and flexible hubs;
- Fixed blades, stall control and full and partial pitch control;
- (Nearly) constant and variable rotational speed transmission systems, incorporating multiple control options.

Also spectacular installation techniques were realised. The methods varied from conventional installation by means of cranes to using the tower as a hoisting device to platforms which climbed along the tower to put the nacelle and blades on top of it.

Considerable improvements were achieved by applying fibre reinforced plastics in the blade structures and by new electrical conversion systems, incorporating power electronics. During the biannual executive committee meetings of the IEA Wind Energy Programme country representatives reported about progress of the realisation of the largest turbines ever built at that time. The largest among the giants of that time was the German GROWIAN turbine with a rotor diameter of 100.4 m and an installed generator capacity of 3 MW. It felt like a race for the first-man-on-the-moon.

Finite element methods (FEM), although not as refined as today's, were used to more critically design the sensitive components of the wind turbines, notably the hub structures. The design base for comprehensive design methods was far from complete. In aerodynamics proper modelling of stall, three-dimensional effects, aero-elastic phenomena, etc. were missing or were inaccurate. The same applied to wind field descriptions in the rotor plane, the effects of turbulence on performance and loading and wake modelling and wake interaction.

At ECN, one of the experimental turbines, the relatively small but versatile two bladed 25mHAT made its first rotation on the first of March 1981. The wind turbine, which later became the turbine with the longest operational lifetime among the experimental machines in the world (1981–1995), had a variable rotation DC generator with static inverter, by which the load characteristics of various electrical conversion systems could be simulated. The versatility of the facility appeared from the experiments which were carried out for various national and European research programs:

- Verification of various control strategies
- Boundary layer measurements
- Determination of the radar cross-section
- Determination of acoustic noise emission as a function of blade section by means of an acoustic telescope
- Duration test of the operation of tip brakes of Stork blades
- Testing of the FLEXTEETER concept (Figs. 2.2 and 2.3)
- Field testing of blades with 150 pressure holes and sensors as part of an IEA Annex.

Figure 2.2   The 25mHAT experimental wind turbine at the premises of ECN, Petten, the Netherlands, equipped with the FLEXTEETER rotor, 1992 (Photo: Jos Beurskens ©).

Figure 2.3   The FLEXTEETER experimental rotor and hub structure, 1992 (Photo: Jos Beurskens ©).

The tests on the FLEXTEETER showed promising results with respect to the decrease of the fatigue load spectrum by introducing flexibility in the rotor concept (Fig. 2.4).

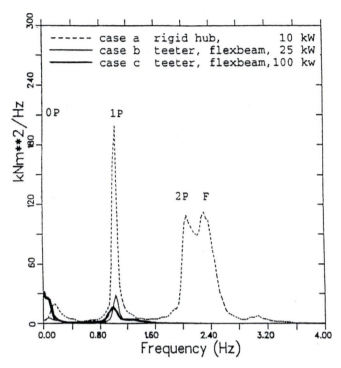

Figure 2.4   Blade bending moment load spectrum of the FLEXTEETER rotor, with flexible- and rigid-blade configuration. Note the dramatic reduction of the fatigue load spectrum due to flexibility (Image: Dekker, c.s. ECN).

Results however were not implemented immediately by the industry because it would require a complete redesign of their commercial product line and risks were considered too high. Time for long-term development strategies was lacking.

2.3   European Wind Turbine Development

In the early 1990s the small wind turbines had reached the same size as the smaller government-sponsored machines of the 1980s. The manufacturers of small and medium-size machines, partly benefitting from the results of national and European research programs, were consequently upscaling their designs and quickly reached the dimensions of the government-sponsored machines (Fig. 2.5).

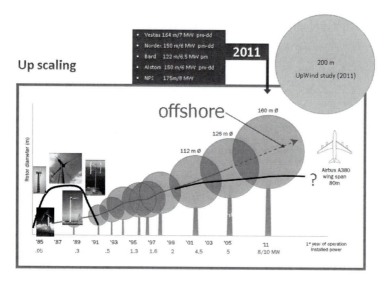

Figure 2.5   The evolution of the capacity of wind turbines since 1985. In the left corner the emergence and decline of the government-sponsored experimental large-scale wind turbines are indicated.

The European Commission played a crucial role in the long lasting success of the manufacturers by providing financial support for new designs, for associated research and for building and verifying prototypes and 0-series. With the support through the WEGA and THERMIE programs manufacturers were able to build up a track record, which was a necessary condition in order to enter the market successfully. At this point in history the two initially separated development lines merged into one. Upscaling continues as a result of the demand for very large wind turbines, which appear to be potentially more competitive for offshore applications.

I remember well that the start of the WEGA program was preceded by fierce discussions between representatives of the wind turbine manufacturers, scientists and EC officials about the optimum turbine size and market potential. David Milborrow published an article in which he showed that the optimum capacity of a wind turbine was about 1 MW. If this was true it would have placed a bomb under the EC's plan to launch a tender for the design of large commercial (multi-MW) turbines. To shed more light on this issue, the EC issued a study which was carried out by Bob Harrison, Erich Hau and my colleague Herman Snel. The report "Large Wind Turbines" [2], published as a book by Wiley, showed that there was no strict physical limit to the maximum size of wind turbines, provided flexible elements and materials with a higher strength-to-weight ratio are being applied.

Table 2.1   First European large wind turbine development and testing programs

Wind turbine (Country)	Rotor diameter [m]	Rated power [MW]	Year of first rotation	Commercial follow-up
European Program WEGA I				
Tjaereborg (Esbjerg, DK)	61	2	1989	No
Richborough (UK)	55	1	1989	No
AWEC-60 (Cabo Villano, E) (Fig. 2.6)	60	1.2	1989	No
European Program WEGA II				
Bonus (Esbjerg, DK)	54	1	1996	Yes
ENERCON E-66 (D)	66	1.5	1996	Yes
Nordic (S)	53	1	1996	No
Vestas V63	63	1.5	1996	Yes
WEG MS4	41	0.6	1996	No
European Demonstration Program THERMIE				
Aeolus II (D and S)	80	3	1993	No
Monoptoros	56	0.64	1990	No
NEWECS 45 (Stork, NL) (Fig. 2.7)	45	1	1991	No
WKA-60 (MAN, Helgoland, D)	60	1	1989	No
NEG-MICON (DK)	60	1.5	1995	Yes
NedWind (NL)	53	1	1994	Yes

Figure 2.6   The AWEC-60 at Cabo Villano, Spain, 1990 (Photo: Jos Beurskens ©).

Figure 2.7   The NEWECS 45 at Medemblik, Netherlands, 1990 (Photo: Jos Beurskens ©).

As a result EC launched the call for proposals and the manufacturers who earlier were involved in manufacturing smaller machines and who had a genuine interest in commercialising large wind turbine systems responded to the EC calls and won contracts to design and build commercial prototypes of MW-size wind turbines (Table 2.1).

The involvement of the commercial industry in these programs changed the scene considerably. The testing and evaluation programs, sponsored by individual governments and realised by large engineering and aircraft companies came gradually to an end. Many wind turbines were decommissioned. The physical end of some of the wind turbines was quite spectacular as they were dynamited, like the MOD 2s in the USA and the GROWIAN and Aeolus II in Germany. The technology firms were not interested in commercialising their products, either because of disbelief in a future for wind energy or they considered themselves too expensive. Many of these companies were not used to cost constraints dictated by the wind energy market.

The WEGA and THERMIE programs mark the beginning of consequent upscaling of wind turbines. The developments can be characterised by step-by-step upscaling of smaller, successful commercial wind turbines. Many of the advanced technical concepts, like teetering hubs, downwind rotors, fast-running 1- and 2-bladed rotors, were abandoned by the commercial companies. Their first prototypes were on the conservative side, because reliability was the most important selling point rather than the future promise of cost reduction of advanced systems. Later innovations were introduced which were different from the first generation machines. The most spectacular improvements included the introduction power electronic convertors which increased the controllability of turbines significantly. These conversion systems combined with blade pitch control and advanced multi-parameter control strategies made the modern machines compatible with the grid requirements. Critical design procedures and the application of new materials resulted in weight reductions and consequently to the reduction of electricity generation cost.

2.4   Research and Development

Apart national programs, notably in Denmark, Germany, the Netherlands, the United Kingdom, Sweden, Canada, USA, Italy and

Spain), research was carried out within the EC and IEA programs. The added value of the EC program in particular was pooling of expertise and resources, so projects could be carried out that exceeded the capability of individual countries. Without the EC programs, which had their own financial resources, and those of the IEA we would never have seen such an intensive exchange of scientists, engineers and views around Europe and the world.

Among the striking examples of high-impact European projects are the European Wind Atlas, the WEGA and THERMIE wind turbine development and test projects and the development, improvement and verification of design tools to which also ECN contributed considerably over a very long time span. I personally had the honour to initiate and co-manage the Integrated Project "UpWind", run by our colleagues of Risø, Denmark, which has explored the design limits of very large wind turbines. It was the largest single EC project in the area of Renewable Energy and involved more than 30 participants.

Advances in design related knowledge were impressive, in particular in the areas of aerodynamics, wake modelling in wind farms, aero-elasticity, FEM, structural dynamics, measuring techniques, system modelling and control technology. Analytical results had to be verified and to that end dedicated experimental open air turbines were realised for field aerodynamic experiments. Laboratory facilities mainly comprised blade test rigs, material test benches, drive train test beds and dedicated wind energy wind tunnels. Most of the research hardware facilities were national initiatives and were used with occasional support from the European Commission for European projects. These facilities all had some unique features. The most important open air wind turbine test beds include:

- Uniwecs (16 m diameter, 2-bladed, downwind): hub configuration could be changed by computer controlled hydraulics (individually hinged blades, teetering hub, and fixed hub) and the damping and stiffness parameters were adjustable.

- 25mHAT (25 m diameter, 2-bladed, upwind): Fully adjustable generator load characteristics, using DC generator and 12 pulse inverter).

- NREL, Phase II, III, IV turbines.
- Risø, Tellus turbine.
- TUD Open Air Facility (10 m diameter, 2-bladed, upwind): the very same fully instrumented blade for pressure distribution measurements could be placed in a wind tunnel for comparison with well defined flow conditions in a wind tunnel.

Along with the growth of the capacity of wind turbines, also the size of wind energy projects (wind farms) grew from farms with a total capacity of 1 MW to wind farms of several hundreds of MWs. In order to be able to design wind farms efficiently, know how was needed about wake interaction and flow phenomena inside wind farms. Research into those aspects started in the early 1980s with analysis of the physics of wakes and experimental research in wind tunnels. Since the first investigations in wind farm flow, this research has become more and more important. External conditions like wind shear, turbulence intensity and the stability of the atmosphere play a large role in the propagation of rotor wakes. As there is a big difference between those conditions on land and offshore, economic planning of offshore wind farms depends to a large extent on results on wind farm flow research.

Traditionally research was focused on wind turbines and wind farms. How to keep them operational at low cost during the lifetime never has been an issue until the first wind turbines installed offshore in 1991. At ECN we realised that the single focus on wind turbine and farm design could make the research community less relevant to the industry if no more attention were paid to operation and maintenance (O&M) issues. For the Netherlands this was of particular importance as the main manufacturers were moving outside the country and the offshore contractors and operators showed interest in developing O&M services. Corrective maintenance has to be substituted by preventive maintenance in order to decrease O&M cost. To this end conditioning monitoring techniques have to be developed to determine the life time consumption of critical components for condition based maintenance. Other topics like a "flight leader system", access technology and dedicated logistics all need to be improved or developed in association with efficient O&M systems.

2.5   International Networks

The fact that wind energy technology knowledge base is still coherent, up to date and accessible, is not so much due to direct government actions but more due to the degree of self-organisation of the scientific world and to the intense contacts between the research community and the industry, facilitated by the EC. Though the European Commission actively facilitated cooperation, it was the wind energy sector itself which developed many initiatives for coordination, task sharing and developing strategic research programs. Examples of those initiatives in Europe include the IMTS, European Wind Energy Association, the European Renewable Energy Centres (EUREC) agency, the European Academy for Wind Energy (EAWE), the measurement quality assurance network MEASNET and the European Technology Platform Wind Energy.

Apart from the scientific and technical discussion also strategic implementation issues were discussed during the numerous meetings of the associations and networks. For renewable energy to make a significant impact on the world's energy supply, systems have to be produced and installed in enormous numbers. It was Wolfgang Palz, the EC official responsible for R&D on renewable energy, who again and again stressed this important point and stimulated me to make the transition from thinking on a laboratory scale to world market scale.

2.6   Concluding Remarks

All joint efforts of industrial pioneers, researchers, government and EC officials resulted in wind energy not being an alternative energy technology anymore, but an established energy option. The evidence is provided by the following facts as it stood at the end of 2011:

- Total installed wind power was 238 GW, of which 93.9 GW in Europe.
- Wind energy covered 6.3% of Europe's electricity demand.
- Europe is leading the offshore wind power system technology and harbours almost all the world's offshore turbines, totalling 3.7 GW out of 4.1 GW globally.

The capacity of offshore wind farms is growing fast, which appears from the following figures:

- Average capacity of all wind farms (2011): 43 MW/project
- Average capacity of the 10 smallest older wind farms: 8 MW/project
- Average capacity of the 10 biggest recent wind farms: 198 MW/project

Europe is heading for 280 GW by the year 2030 according to the EC scenario and for 400 GW according to the EWEA by the year 2030.

Considering the capacity of wind power European countries intend to realise offshore, an area without any electrical infrastructure, the need for a renewed approach of the electrical grid becomes obvious. This need becomes even more urgent if the future levels of variable output renewable energy sources are taken into account. It specifically concerns wind, concentrated solar power and hydro plants. As the lead times for extending the transport grid are very long—10 years or even more—the construction of the grid extension should have started 10 years ago in order to meet the present needs. This however is not the case, which is a perfect illustration of the most important bottleneck for further extension of the offshore wind energy capacity: the grid.

As the best onshore wind sites are being used up, offshore wind energy offers the opportunity to fully exploit the European wind energy resource. The offshore potential is larger than all electricity we need in Europe. The future developments in wind energy will be dominated by offshore developments. So we will see larger wind turbines than the ones on the present market (see also Fig. 2.5). In the further future we will see dedicated radical, extremely reliable offshore concepts which feature full integration of wind turbine, support structure and installation technology. Despite the fact that present wind energy R&D is driven by offshore applications, that research results will be applied to onshore turbines and to small and medium-size turbines.

There will be a new market for the small wind turbine applications which I described in the beginning of this chapter and where I started my career with. The needs are still there. The estimated number of people in rural areas in developing countries who will never be connected to an electricity grid has not decreased since the 1970s and still stands at 1.4 billion. A revival of the development of wind systems for rural areas is likely to be much more successful than in the seventies as technical innovations in power electronics, materials, design methods have made enormous progress. This will create the possibility to realise extremely reliable small and medium wind power systems for both stand alone and grid-connected applications. The only missing ingredient is capital. My sincere hope is that we have to draw a third line in the right corner of Fig. 2.5, indicating the (re)emergence of wind systems for those 1.4 billion people who are kept from access to electricity.

References

1. Meadows, D.H., Meadows, D.L., Randers, J., Behrens III, W.W. (1972). *The Limits to Growth*, Universe Books, New York.
2. Harrison, R., Hau, E., Snel, H., (2000). *Large Wind Turbines*, John Wiley & Sons, Chicester, UK.

About the author

 h.c.Ir. Jos (H.J.M) Beurskens is the scientific director of the We@Sea Foundation, which carried out R&D on the application of wind energy plants offshore. From 1989 to 2005, he was head of the Renewable Energy Unit and the Wind Energy Unit of the Energy Research Centre of the Netherlands (ECN), and he is still attached to ECNs wind energy unit as a senior scientist.

Dr Beurskens is member of the board of trustees of the Windunie, a co-operative enterprise of Dutch farmers operating about 440 MW of wind power on a private basis. He was the technical and scientific advisor to the European Wind Energy Association (EWEA) and was a member of the scientific advisory board of the former Institute for Solar Energy Technology (ISET) of Kassel, Germany (at present part of the Fraunhofer Gesellschaft). He is chairman of the scientific advisory board of For Wind (joint wind energy research institute of the universities of Oldenburg, Bremen and Hannover, Germany) and a member of the Scientific Advisory Council of the Hanse Wissenschaftskolleg, Delmenhorst, Germany. Dr Beurskens was recently reappointed to serve on the Steering Committee of the European Wind Energy Technology Platform and act as the chairman of WG 2 (Wind Power Systems). He received the honour of Wind Energy Pioneer of the British Wind Energy Association. In 2008 he was awarded the Poul la Cour's prize, which was presented to him by Mr Janez Potočnik, EU commissioner for Science and Research. In November 2009, Dr Beurskens received a honorary doctorate degree at the University of Oldenburg, Germany, for his work on initiating European research in the field of wind energy.

Throughout his career, Dr Beurskens has been extensively involved with various industry associations and policy groups involved in renewable energy, particularly those active at the European level. He was a founding member of the Netherlands Wind Energy Association (NWEA, former NEWIN) and the European Wind Energy Association (EWEA) and currently serves them as a board member. He was one of the founders of the International Meeting of Test Stations (IMTS) and the European Academy of Wind Energy (EAWE). Dr Beurskens represented ECN at the

College of Members of the European Renewable Energy Centres Agency (EUREC) and was also one of its founders. He has chaired the Executive Committee of the Wind Energy program of the IEA and has been retained as an advisor to the European Commission, and several national governments on R&D programs in the renewable energy field.

Chapter 3

History of Danish Wind Power

Benny Christensen

*Danish Wind Historical Collection (DVS), Smed Hansens Vej 11,
DK 6940 Lem St., Denmark*

bemctim@gmail.com

To understand the worldwide success of Danish wind turbines over the last decades, it is essential to look back at the last 150 years of wind power history in Denmark. With more than 7 000 km of coastlines that are located close to the North Sea, Denmark has extraordinary wind resources. But until the middle of the 19th century, the level of wind energy use in Denmark was very similar to other European countries like Germany, the Netherlands, France and Great Britain. The establishment of new windmills had during centuries been strongly regulated by the crown, but from 1852 to 1862 these restrictions were gradually removed. One result was a growth in the total number of big, commercial grain-grinding windmills from about 800 to more than 2 000 during the rest of the 19th century.

Wind Power for the World: The Rise of Modern Wind Energy
Edited by Preben Maegaard, Anna Krenz and Wolfgang Palz
Copyright © 2013 Pan Stanford Publishing Pte. Ltd.
ISBN 978-981-4364-93-5 (Hardcover), 978-981-4364-94-2 (eBook)
www.panstanford.com

An even more important effect was that Danish farmers were now allowed to have their own windmill and during the next decades, thousands of small "farm windmills" were installed. Usually they had four blades with wooden structure, equipped with canvas sails. The windmills were placed on the top of the barn roof and were turned by hand according to wind direction. The wind power was used for fodder grain grinders and other kinds of farm machinery. Most of these farm windmills were built by the same local millwrights as the big commercial windmills.

3.1 Industrial Windmill Production: 1876–1900

The earliest development of Danish industrially manufactured windmills was inspired by the American multi-bladed "pumping windmills", introduced in the second half of the 19th century by Daniel Halladay and others.

Figure 3.1 The front page of a price list (ca. 1886) shows the place for the first Danish industrial windmill production. A big Halladay-type windmill, based on N. J. Poulsen's 1883 patent, is being assembled and a similar windmill is mounted on the roof of the factory (Photo: Esbjerg Town Historical Archives).

In 1876, a young Danish blacksmith and millwright, N. J. Poulsen, left his father's workshop on the island of Samsø, and arrived to Esbjerg in the southwestern coast of Jutland. Here, the first big Danish North Sea harbour was under construction, and a local blacksmith had established a workshop servicing the construction companies. N. J. Poulsen joined the harbour blacksmith Franz Møller and the small workshop soon grew to a machine factory and iron foundry. During the following years, the Esbjerg factory was the first to introduce industrially produced multi-bladed wind motors on the Danish market in competition with imported US products [1]. Poulsen quickly adopted new ideas from abroad. In 1877, he made a wind motor with 7.5 m rotor diameter and fixed curved metal blades for a land reclamation project in southwestern Jutland and during the years 1883–1886 he delivered 11 Halladay-type wind motors for the water supply at the Danish State Railways stations. In the years 1881, 1883 and 1888 he obtained Danish patents for various types of multi-bladed windmills.

While the US wind motors were primarily used for water pumping, bigger types of new Danish windmills were often used as a power source in, for example, brickyards, spinning mills and wood workshops. The more effective industrially produced "self-regulating" power windmills were also competition to the traditional farm mills in Danish agriculture during the 1880s.

For more than ten years the Esbjerg factory was the only Danish industrial "wind motor" producer, but in 1889, the products got competition from inventor and millwright Christian Sørensen, who established the Skanderborg Wind Motor Factory. Whereas the Esbjerg wind motors were inspired by American technology, Sørensen based his products on innovative use of the European windmill tradition. During the 1870s he had built several big windmills of the "Dutch-type" in many places in Jutland and made experiments with different blade types and windmills having more than four blades.

In 1896, he made a patent application for a special type of a conical wind motor, where the blades (five to eight or more) with adjustable vanes had a combination of convex and concave profiles, supposed to "catch the wind" in a more effective way. In a few years more than 100 windmills of this type were produced. A license agreement was also made with the firm Theodor Reuter & and Schumann in Kiel, Germany, where the conical wind motors were produced for the German market. Here, they were often simply named "Sorensen Motors".

Figure 3.2   Sørensen's conical wind motor was an early example of the industrial windmill type with adjustable vanes, which later on were one of the Danish "standard types" (Photo: The Danish Energy Museum).

During the 1890s, several other producers entered the market. In these years wind power was not only attractive for farm use, but was also still used in many workshops in towns, where electricity was not yet introduced. It is difficult to find valid statistics covering this sector, but from national agricultural statistics it is known that 4 600 Danish farms had windmills in 1907. The major part of these was surely built locally by traditional millwrights, but at least a few hundred of them must have been industrially produced "wind motors" [2].

3.2 The Pioneer Work of Poul la Cour: 1891–1902

The internationally most renowned Danish wind power pioneer from this period, Poul la Cour (1846–1908), had a university degree in physics and meteorology. In 1872, he became employed at the new Danish Institute of Meteorology (DMI), but also worked with inventions in telegraphy technology. In 1878, he left his job as deputy director at DMI and became teacher at the Folk High School at Askov in southern Jutland [3]. There, far away from the university and established research facilities, Poul la Cour continued his work as an inventor while teaching physics and mathematics. His experiments were primarily concentrated on the fundamentals of wind energy and the practical use of wind power for electricity production. From 1891, his work was supported financially by the Danish government, and that year, the first experimental windmill was built in Askov. It was delivered by N. J. Poulsen and its 4-bladed construction combined elements from the traditional Danish windmills and the new technology used in the industrially produced "wind motors".

Figure 3.3 Poul la Cour's first experimental windmill from 1891 (Photo: The Poul la Cour Museum).

La Cour used the windmill for electricity production.[1] The direct current (DC) produced by the dynamo was used for electrolysis, splitting water into oxygen and hydrogen. These gasses were stored and from 1895 used to light up the school. La Cour also used the gasses for autogenous welding and made experiments with the use of hydrogen in a gas engine. In 1897, la Cour had got funding for a new bigger experimental windmill. He had at that time been contacted by the other wind power pioneer Christian Sørensen, who was keen to prove the superiority of his patented conical rotor by wind tunnel tests. Sørensen also delivered the 6-bladed rotor for the new big windmill and during the following years, la Cour made systematic wind tunnel tests of different blades and rotor types, getting deeper knowledge of the basic aerodynamic laws [4].

Figure 3.4　The laboratory windmills in 1900. The big windmill now has an ordinary 4-blade rotor. On the right is the original small windmill from 1891—now equipped with adjustable vanes (Photo: The Poul la Cour Museum).

One of the results of these experiments was that though Sørensen's conical rotor was superior to more traditional blade types, even higher efficiencies could be obtained by a carefully designed

---

[1] For more information about la Cour, see chapter *The Aerodynamic Research on Windmill Sails of Poul la Cour, 1896–1900* by Povl-Otto Nissen.

rotor with fewer blades and without the characteristic conical form. At the same time, there were continuous problems with the heavy 6-bladed rotor on the new windmill and at last, la Cour decided to replace it by a more traditional construction with four blades. The cooperation between the two wind power pioneers then finished in a long and bitter public dispute.

3.3 Wind Electricity for Rural Areas: 1903–1919

After the turn of the century, Poul la Cour's interest was concentrated on the use of wind power for electrification of rural areas and small towns. In 1903, he was the leading force in the establishment of the Danish Society for Wind Electricity (DVES). Two years later, DVES had already helped to establish 25 small local power stations. Twenty-two of these used wind power with battery storage and kerosene or gas engine as a supplement for periods without wind. The power plants delivered electricity for one or more farms or small communities [5].

Beginning in 1904, yearly courses for "rural electricians" or "wind electricians" took place at the test mill at Askov Folk High School. After three months of theoretical training, the students performed practical installation work on one of the small power plants planned by DVES.

The youngest student on the very first course, Johannes Juul, was in fact too young for the course. When it started he was only 17, but he was so eager to take part in the course, that he had manipulated the year of birth in his papers. As this was revealed, he had to wait for his diploma until his 18th birthday in May 1905. During this waiting period he was an apprentice at DVES, assisting with installation work at power stations. For his stay, he carried a camera along and thank to him the Danish Museum of Energy now owns a unique collection of photos showing the "rural electricians" at work.

More than 40 years later, this enthusiastic young man, after a long career as an electrician, producer of electric equipment and employee in a Danish utility, took up the legacy of Poul la Cour and became a key person in the development towards modern wind turbines.

Figure 3.5   Participants of the first class of "rural electricians" in 1904. Poul la Cour is sitting at the left end of the middle row. Number three from the right in the back row is the young Johannes Juul (Photo: The Danish Energy Museum).

DVES played an important role for the use of electricity in rural areas and small towns. In 1906, the first two cooperatively owned power plants using wind power were built in towns in western Jutland. However it was soon evident that wind power could not economically compete with combustion engines. Already in 1908, power stations without wind power dominated the list of 19 plants, planned by DVES that year. The activities at DVES continued until 1916, but wind power was only an important part of them during the first few years. With the scarcity of imported fuels during the First World War (1914–1918), many small power stations again invested in wind power. Around 120 of them used wind power as an additional power source during the war.

Figure 3.6   A 14 m diameter windmill at a farm (1905). The DC dynamo produced electricity for lamps and two electro motors used by the farmer and his neighbour. The windmill was fabricated by the company Lykkegaard and the electric plant was planned and installed by DVES (Photo: The Danish Energy Museum).

Figure 3.7   Photo (left) and drawing (right) of the local electric power plant in Vemb, western Jutland (1906). The building contains a dynamo, a kerosene engine and (in the right part) a battery storage (Photo: The Danish Energy Museum).

3.4 The First "Golden Age" of Danish Wind Power: 1900–1920

During the last decades of the 19th century, three outstanding personalities have had a deep influence on the development of wind power in Denmark. One of them, N. J. Poulsen, by a creative interpretation of the new US technologies; the second one, Christian Sørensen, by innovative use of the European windmill tradition and the third one, Poul la Cour, by a systematic investigation in the basic fundamentals of wind power [1].

Around 1900, several new windmill producers with different backgrounds entered the market. A few of them had originated as millwrights, but as traditional mill building was mostly wood work, while iron and steel were the primary materials for the new types of windmills, many of the new producers had a background as blacksmiths. Iron foundries also played an important role in the development, both as suppliers of parts for the windmills or in some cases also as windmill producers.

Three different types of the new windmills were produced:

(1) Multi-bladed "power windmills" with pitch-regulated wooden blades—one of the types originally produced by the Esbjerg factory.

(2) Windmills with 4, 5 or 6 (in few cases 8) blades, each with a row of adjustable vanes (also named "shuttered sails")—as introduced by Christian Sørensen and further developed by Poul la Cour.

(3) Small multi-bladed windmills with fixed iron blades, used for water pumping. They were in principle very similar to the US Aermotor, introduced by Thomas O. Perry, and later copied and produced all over the world.

The first two types were produced with diameters up to 18 m or 20 m, and usually named "wind motors", while the third type had a diameter of 3 m to 5 m.

The total number of windmills produced during the first decades of the 20th century can be estimated from the second agricultural statistics for the year 1923 [6]. In that year, 16 600 Danish farms had their own windmill. The statistics were made at a time when the use of wind power was already in decline, and a few years before, the number may have been even higher. Some of the oldest windmills in the earlier statistics from 1907 had surely also been scrapped during the period and replaced by new industrially produced "wind motors". Wind power was also used for electricity production, waterworks and land reclamation projects and the numbers of windmills in these sectors were of course not included in the statistics. A fair estimate could then be that during the first two decades of the 20th century, at least 15 000 to 18 000 windmills (or a yearly average of 750 to 900) were produced and erected in Denmark.

Figure 3.8 The exhibitions of machinery in connection with many local and regional cattle shows were used for marketing of the windmills. On this photo there are 5-bladed windmills with adjustable vanes from two different producers and a number of small pumping windmills with fixed blades from various producers (Photo in the Danish weekly *Illustreret Tidende* 1908).

Figure 3.9 Multi-bladed power windmill at a Danish farm in northern Jutland (Postcard in Sæby Museum Archives).

Statistics from the Danish factory inspection indicates that this development culminated in the years before and during the First World War. In 1916, the number of new installed windmills in the statistics reached a maximum of more than 1 300.

3.5 Denmark on a Different Path of Windmill Production

These numbers could be compared with data from Germany from the same period. Here statistics from 1925 show that 6 700 farms used wind power at that time and Matthias Heymann [7] has estimated that a maximum of around 8 300 windmills were used in Germany in the years before the First World War. This means that a maximum of 1% of the German farms used wind power—while in Denmark as a whole it was more than 8% in 1923.

Already when the first simple "farm windmills" were introduced in the 1860s, there was a high degree of regional diversification in the use of wind power in Denmark. The strongest concentrations of windmills were found in the northern and northwestern part of Jutland, where the wind resources were most abundant. Later, the same pattern of wind power usage was found for the new "wind motors". In 1923, one out of four farms in northern Jutland used wind power to drive the grain, threshing machine, chaff cutter and other farming machinery. In southeastern Jutland and on the islands Funen and Zealand only 5%–10% of the farms used wind power and there it was primarily used for water pumping.

Also on the manufacturing side the development in Denmark was different from Germany. When it culminated around 1912–1913, the German market was (still according to Heymann) divided between a few rather large manufacturers located in the big cities, with one single factory producing more than half of the 500–600 units delivered per year. In Denmark, there was a multitude of small and middle-size producers—many of them located in small communities—and 10 to 15 enterprises reached a yearly maximum production of 50-100 units or more. The major part of new manufacturers of wind motors were–like the windmills–located in the northwestern part of the country. Around 1910, more than 10 the factories with wind motors as an important product were placed in this area.

Some of the wind motor producers had their own iron foundry, but others bought their cast iron parts from one of the many local foundries. One of such iron foundries in western Jutland (Holstebro Jernstøberi A/S) also supplied their own complete "construction kits" with all necessary mechanical parts for different sizes of wind motors.

Figure 3.10 Construction kit "Type 23" with mechanical parts for a 5-bladed windmill for small local windmill producers–delivered from Holstebro Iron Foundry (Illustration from a catalogue in the Local Historical Archives, Holstebro Municipality) (above); "Type 23" in the only surviving windmill with one of the "construction kit" from Holstebro Iron Foundry, produced by a local craftsman—and today standing at the Hjerl Hede open air museum (Photo: Benny Christensen, DVS) (below).

The 70–80 single parts were delivered fully machined and ready to mount. The rest of the process of making a wind motor (blades and tower) was then a traditional blacksmiths job. This concept enabled many local blacksmiths and millwrights to get a part in the flourishing Danish wind power market during the first decades of the 20th century. Between 1910 and 1920, the Holstebro Iron Foundry delivered nearly 200 of these "kits" to more than 40 local craftsmen.

During the two decades, here called "the first golden age" of Danish wind power, Denmark had a windmill-industry with high capacity in relation to the size of the country. The knowledge of windmill-technology was widespread to a great number of small- and medium-sized producers—most of them far away from the big industrial centres, but close to the potential market. This unique background could probably partly give an explanation for the successful development of a new Danish wind turbine industry several decades later.

Figure 3.11 The windmills from "the first golden age" were strong survivors. This 6-bladed windmill was delivered in 1933 to the big farm Søe Hovedgaard at the island of Mors by the producer D. M. Heide. Placed close to a modern wind turbine, in the early 1990s it was still in use, driving a grinder (Photo: Benny Christensen, DVS).

3.6 New Technology and Survival Strategies in the 1920s

When imported fuels were available again after the war, the windmill market experienced a steep decline. Many of the windmill factories were shut down or forced to rely on other products. However, the renewed interest in wind power during the war had initiated new technical developments [8].

In 1918, the last year of the war, the Danish engineers Povl Vinding and R. Johs. Jensen began developing a new type of windmill—in fact an important step against the contemporary wind turbines. In 1919, the aerodynamic blade profiles on the "Agricco" windmill, which were inspired by the development of aeroplane propellers and wings during the war, were patented internationally [9]. From the very beginning of the project, the intention was to use the Agricco to produce alternating current (AC) for the public grid and in 1921 an Agricco with a 40 kW asynchronous generator produced electricity for the 10 kV grid of the North Zealand utility NESA. The Agricco went into production, but most of the windmills produced were used for drainage purposes driving big water-screws.

Figure 3.12 The Agricco windmill from 1919 used the new aerodynamic knowledge from aeroplanes—but was introduced at a time, when the interest in wind power was on a decline (Photo: The Danish Energy Museum).

Tests made in 1921–1924 by the Danish State Test Authority showed that the Agricco used the wind energy with a much higher efficiency than la Cour's "ideal" windmill. Several developers in other European countries introduced windmills with aerodynamic blades after the war, but the Agricco's position was also internationally documented with tests by the Institute of Agricultural Engineering at the Oxford University [10]. In 1925, the manufacturing rights for the Netherlands, Argentine and Brazil were sold to the firm Werkspoor in Amsterdam, which produced windmills for the use with drainage pumps. In the same year a proposal for production of the turbine was made in the United States [11]. However, as the market for windmills was in decline, the Danish production closed in 1926 and it must be concluded, that the Agricco was "the right windmill, but at the wrong time".

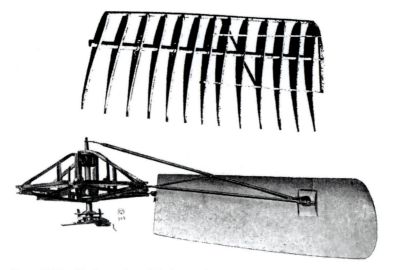

Figure 3.13 Pitch-regulated blades on the Agricco windmill had a steel tube as the main structural element, wooden profiles and covering of metal plate (Illustrations from a catalogue in The Danish Energy Museum).

For the next two decades, the few remaining Danish windmill producers used different survival strategies. From a complete collection of notebooks from one of the medium-sized Danish windmill producers, D.M. Heide on the island of Mors, the yearly production during the period 1905–1934 can be followed. It culminated from 1909 until 1915, when around one windmill per week was delivered, but in 1920, the yearly production had declined to less than 10 windmills.

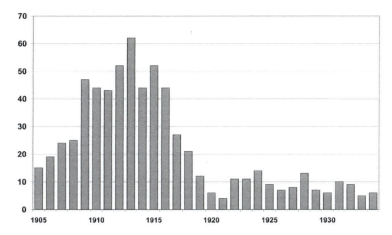

Figure 3.14   The annual number of windmills during the period 1905–1934 from the producer D. M. Heide reflects the development of the wind power market. Heide maintained production through the 1920s and 1930s, but at a reduced level (Data from copies of Heide's notebooks in the Danish Wind Historical Collection, DVS).

During the following years, Heide and other producers found a new market offering a technical upgrade for the traditional Dutch windmills, which were in close competition with industrial flour and grain producers using steam power and electricity. The cap, the original four blades and part of the wood-based transmission were removed. Instead an iron tower with a 5- or 6-bladed rotor was mounted at the top of the mill house, giving more wind power and higher production capacity.

Another survival strategy was used by the Lykkegaard Machine Factory of Funen. The founder, Niels Hansen, was one of the two winners in Poul la Cour's contest for practical construction of the "ideal windmill" in 1901. The name of the family farm "Lykkegaard" was adopted both for the firm and for the next generation. During more than 50 years (until 1957), the Lykkegaard factory produced 4-bladed windmills true to the la Cour tradition.

Figure 3.15   In 1922, the turnable cap and big 4-bladed wooden rotor on this traditional windmill was replaced by a steel lattice tower and a 6-bladed rotor similar to the type used on the popular farm windmills. The local manufacturer D.M. Heide made this kind of modification on 20 of the 31 commercial grain windmills on the island Mors (Photo: Morsø Local Historical Archives) (left). The Lykkegaard Machine Factory maintained production between the two world wars by export and by "leasing-arrangements" with small local power stations. The 4-bladed windmills for these projects were true to the Poul la Cour tradition (Picture: The Danish Energy Museum) (right).

During the difficult decades after the First World War, Lykkegaard offered small local power stations to install a windmill and ensure its operation. The local power plant had to guarantee for a loan of 80% of the value of the windmill and be responsible for the daily operation. Lykkegaard did service, repair work and replacement of parts and demanded a fixed price for each kWh the windmill generated. In a good windy location, this arrangement could be a good deal for both parts.

3.7   Technological Development during the Second World War

When the breakout of the Second World War in 1939 limited the supply of imported fossil fuels, the Danish knowledge of wind power was not far away. A few windmill producers had survived since the "golden age" 20 years ago and statistics from 1935 showed that more than 15 000 windmills were still installed in Danish farms. Some of them were out of service, but many farmers still considered wind power as useful means to reduce electricity bills in times of crisis. In most cases the windmills were used in the traditional way with a mechanical drive to the farm machines, but more than 1 400 farmers established their own electricity production during the war, either by connecting a dynamo to their old windmill or buying one of the new small (<1 kW) propeller-type windmills produced by a number of Danish companies [8].

Among the commercial electricity producers, the war also gave wind a new chance. The number of local power stations using wind power grew from 16 in 1940 to around 90 at the end of the war. Most of the windmills were delivered by the Lykkegaard Machine Factory and followed the la Cour tradition using four blades with wooden shuttered sails. The "standard model" for small power stations had an 18 m diameter, a 20 m steel lattice tower and a 30 kW generator.

Though, the basic principles were 40 years old, details in the technology had been modified throughout the years. Light running ball bearings were used instead of the traditional white metal bearings, and at the bottom of the tower, the combined pinion- and belt-drive to the dynamo was replaced by a sturdy gearbox. On windy locations along the west coast of Jutland or at one of the small Danish islands, this type of windmill could have had a yearly production of 60 000–70 000 kWh.

However, the problems caused by the war also initiated development of quite a new type of "wind turbine" with aerodynamic blade profiles, taking up the legacy from the ill-fated Agricco 20 years before. In May 1940, the big Danish industrial company F. L. Smidth & Co., the world-known producer of machinery for the cement industry, decided to enter the wind power market. The FLS industrial group included a small factory in Aalborg, Skandinavisk Aero Industri, which started production of private aeroplanes a few years before the war. This company played an important role in the development of the new product.

Figure 3.16 When the Second World War caused problems with oil supply, Lykkegaard was ready with a wind power solution. During the years 1940–1942, 60 of the factory's 18 m windmills were supplied to small power stations (Photo: The Danish Energy Museum).

In August 1940, after only a few months the decision to go ahead was taken, a prototype wind turbine was tested at one of the company's cement factories in Aalborg. The two blades on the upwind-placed rotor had aerodynamic profiles, developed by K. G. Zeuthen, engineer and co-founder of the small aeroplane factory. The 17.5 m diameter rotor was driving a 60 kW generator through a gearbox, developed on basis of the high-precision gears, produced by FLS for the cement industry. The gearbox casing with bearings for the main shaft also acted as an important structural part of the unit, placed on the top of the 24 m steel lattice tower.

A similar turbine was erected in March 1941 at the B&W shipyard in Copenhagen, but for the rest of the FLS Aeromotors, as the new turbines were named, the steel lattice tower were replaced by tubular concrete towers. This choice was partly made because of the lack of steel during the war, but also because this type of tower could be delivered by the company Danalith—a member of the FLS group, specialised in reinforced concrete constructions. In May 1941, the first Aeromotor with a concrete tower was erected in Nexø on the island of Bornholm. Standing a few metres from the shore, it was a possible candidate as the world's first offshore wind turbine.

Figure 3.17   The 2-bladed FLS Aeromotor (Photo in the FLS files of the Danish Wind Historical Collection, DVS) (left); the 3-bladed FLS Aeromotor (right).

Later during the same year, seven more 2-bladed Aeromotors were placed at Danish power stations and early in 1942, a 3-bladed version with a 24 m rotor and a 70 kW generator was presented. During the war, thirteen 2-bladed and seven 3-bladed Aeromotors were built. Despite the very short development time, rather few technical problems occurred. The worst problem came very soon after the testing of the first turbines with concrete towers. At maximum speed (90 rpm for the 2-bladed and 60 rpm for the 3-bladed version) the frequency for the blade passage in front of the tower became close to the natural frequency of the tower. This caused heavy tower vibrations and cracks in the concrete, but the problems were eliminated by stiffening the towers with three outside "pilasters" and increasing the natural frequency above the critical level. Another weak point showed to be the durability of the blades. They had a main beam made from laminated wood and wooden ribs. On the prototype the covering was canvas, as used on aeroplanes, but later a changeover to water-resistant plywood was made. But even that had to be treated often with a special lacquer to withstand the rather harsh Danish climate. The two longest-living Aeromotors, in service for 16 and 19 years, both had to have their blades changed.

Figure 3.18   Vital parts from the two longest-living Aeromotors have been preserved. The Danish Wind Historical Collection has the complete gear and generator unit from a two-bladed turbine— illustrated here—and the two new metal-covered blades from 1954. At the Nordic Folkecenter for Renewable Energy, the gear and blade shaft unit from a three-wind Aeromotor has been preserved (Photo: Benny Christensen, DVS).

Though the Aeromotor was an important step in the development towards modern wind turbines, it was still producing direct current (DC). And the performance was not only dependent on local wind resources, but also on the capacity of the local grid. Some of the 2-bladed Aeromotors had a yearly production of 80 000 kWh or more, while others were producing less than 30 000 kWh/year. The 3-bladed version, mostly placed on locations with good wind resources, produced from 90 000 kWh/year to more than 130 000 kWh/year.

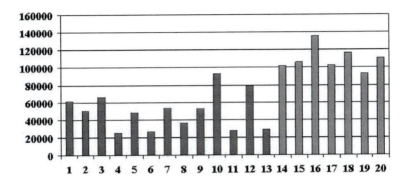

Figure 3.19 The maximal annual production for the 20 Aeromotors. Numbers 1–13 are of the 2-bladed type, 14–20 are the 3-bladed version.

3.8 From DC to AC: Juul's First Experiments after the War

In order to get the full benefits of wind power, it was obvious that wind turbines should be connected to a high voltage AC-grid—as already demonstrated with the Agricco turbine in 1921. After the war, this development was taken up again by Johannes Juul. As mentioned earlier in this chapter, he had as a young man attended the electricity course of Poul la Cour in 1904. Since then, he had worked with electricity and in 1926 was employed in the utility SEAS in Southern Zealand. Beside his work, he was allowed to make independent research and development work and it resulted in several inventions and products for the Danish industry [12].

In 1947, at the age of 60, he decided to take up again the wind tunnel experiments made by Poul la Cour in 1891–1903, but this time concentrated on different types of aerodynamic blade profiles. In 1948, a wind tunnel was built from plywood at the SEAS facilities in Haslev and in 1950, the first experimental wind turbine was built by SEAS to make practical tests, based on Juul's research. Originally it had a downwind rotor with two 4 m blades, but it was soon changed to an upwind rotor, where the blades were supported by wires and stays. The 15 kW AC-generator was an ordinary asynchronous motor, coupled to the grid. It started as a motor, driving the rotor (at under-synchronous speed), but when the wind speed was high enough, it operated at over-synchronous speed and acted as a generator, delivering electricity into the grid. By a careful combination of generator size and blade profile, the turbine could be "self-regulating". When the wind speed got too high, the force on the blade was reduced because of the "stalling phenomena", known from aeroplane wings. As en extra safety feature, the blades were equipped with revolving tip brakes—a feature that Juul got patented in 1952—and many years later became "standard equipment" on many modern wind turbines.

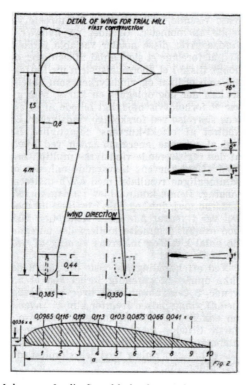

Figure 3.20 Johannes Juul's first blade design for the 2-bladed turbine. Here, a downwind concept was chosen. The design includes Juul's patented revolving tip brakes (Picture: The Danish Energy Museum).

The testing of the first experimental turbine gave good results, but in fact, the choice of two blades was a low cost solution. A rotor with three blades could have been more stable. When in 1952 SEAS took over the FLS Aeromotor from a small utility on the island of Bogø, Juul got a chance to test a 3-bladed rotor. The original 18 m 2-bladed rotor from the Aeromotor was replaced by a 13 m 3-bladed rotor and the turbine was connected to the AC grid.

Figure 3.21   The final design for the 2-bladed 15 kW test turbine (left); The 65 kW Bogø wind turbine with Juul's 3-bladed rotor was mounted on a concrete tower from a 2-bladed Aeromotor (right) (Photos: The Danish Energy Museum).

The modified turbine had been in service for 10 years until 1962. Despite of the smaller rotor diameter, the average yearly production of 80 000 kWh was more than three times the production of the original two-bladed turbine. It was partly because of the 3-bladed rotor with the more efficient aerodynamic profiles, developed by Juul—but first of all it showed the effect of the connection to the high capacity AC grid.

3.9   Post-War Development Blocked by Cheap Oil and Coal

Juul's experiments were remarked outside Denmark and in 1950 he was invited to take part in international meetings on wind power in Paris and London. In Denmark however the attitude towards wind power was less positive. Juul's experiments were supported by his employers at SEAS, but the majority of Danish utilities relied on a bright future based on coal and oil being still cheaper during the post-war years, and with nuclear power as a future option. By a government initiative, however, a "wind power committee" was established in 1950 and finally in 1954 decided that a new, bigger experimental wind turbine should be built based on Juul's experiments. The 200 kW wind turbine, built at Gedser, near the south most point of Denmark, had a 24 m 3-bladed rotor and a 200 kW AC-generator. The design also included tip brakes used on the first experimental wind turbines. The transmission, however, was something of a compromise.

As the project budget could not afford a specially designed gearbox, a double chain drive was chosen, giving some problems with oil spill. But apart from this detail, the Gedser wind turbine was successful. It was in service for ten years (1957–1967) and had a maximum yearly production of more than 360 000 kWh. The official Danish authorities and the utilities were not in favour of wind energy, and in the final report from the committee in 1962, it was concluded that wind energy could not compete with fossil fuels like coal and oil, which at that time were still cheaper. Juul was as the only member of the committee voting against this conclusion. He had five points in favour of wind energy:

- Possible savings in the import of fuels;
- Wind power was a valuable supplement in periods with cold weather and high energy consumption for heating;
- Building of wind turbines could give extra employment in crisis times;
- Danish wind power could be exchanged with hydropower from Norway and Sweden;
- Denmark could export wind turbines to other European countries and wind power could be an important energy source for the developing countries.

After publishing the report, many enquiries from countries in Europe, South America and Asia asked for possible deliveries of Danish wind turbines. But wind power was officially rejected by the Danish authorities. Johannes Juul died in 1969—before being able to see the future results of his pioneering work.

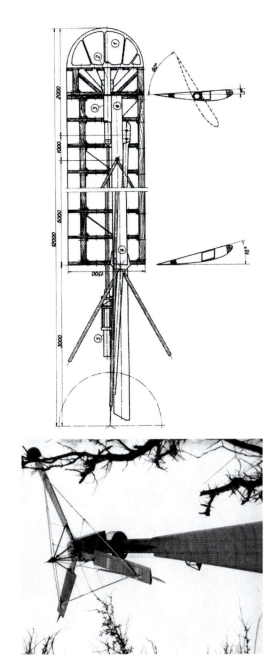

Figure 3.22 The Gedser wind turbine (left); the blade construction for the Gedser wind turbine with steel main beam, tree profiles and aluminium covering (right) (Images: The Danish Energy Museum).

Figure 3.23 Mounting of the rotor on the Gedser wind turbine. The photo shows the big chain wheel on the rotor for the first stage of the chain gear (above). The Gedser wind turbine was taken down and restored in 1990. It is now displayed at The Danish Museum of Energy (below) (Photos: The Danish Energy Museum).

3.10 Re-Invention of Wind Power: 1975–1976

When the first oil crisis hit Denmark in 1973–1974, wind power was nearly forgotten or considered as a technology of the past. With an abundance of cheap oil for more than a decade, Denmark had built up a high energy consumption and a high degree (92%) dependency of imported oil.

In the first Danish energy plan from 1976, wind power and other renewable energy sources were nearly absent. Coal, nuclear power and natural gas were considered as the only possible substitutes for expensive oil. This attitude was supported by Danish utilities, big industries and big political majority in the Danish parliament [13]. But wind power was re-invented by local blacksmiths, students, independent researchers and other innovative persons. The inspiration could be found—not in the official plans, but in the history and it was visible in the Danish landscape. At Gedser, near the southern-most point of Denmark, Juul's 200 kW experimental wind turbine could still be seen—though out of service since 1967. Close to it an even older 3-bladed FLS Aeromotor from the last World War was also still standing. And in many other places in the landscape—not at least in northwestern Jutland at the other end of the country—some of the more than 15 000 windmills once used on Danish farms were still visible in the landscape. A few of them were still in use for grinding grain and many local craftsmen had the experience of doing service and repair work on these types of windmills.

Figure 3.24   There were no suppliers of wind turbine technology on the market and the new Danish wind power pioneers started from scratch with parts from scrap dealers. A rear axle from a car or a truck was a favourite part of the transmission in many of the new wind turbines (Photo: Benny Christensen, DVS).

In the years 1975–1976, a number of new more or less home-built wind turbines appeared in the Danish landscape. Many of them were built with parts from the older windmills or material from scrap dealers. Studies were also made in the technology of the historical Danish windmills and new technologies from abroad. A popular handbook *Sol og Vind*[2] (*Sun and Wind*) published in the autumn of 1976 by Carl Herforth and Claus Nybroe presented some of the new home-built wind turbines and supplemented with knowledge of wind energy and basics of aerodynamics, blade design, heat production and electricity technology. It soon became a sort of "bible" for the new generation of wind power pioneers.

Figure 3.25 Christian Riisagers first 7 kW wind turbine, which in 1975 was connected to the public grid (left); One of Riisager's first two 22 kW turbines. The blade and rotor design was inspired by Juul's Gedser turbine (right) (Photos: The Danish Energy Museum).

---

[2]Herforth, C., Nybroe, C. (1976) *Sol og Vind*, Information Publishers, Copenhagen.

In private households, a substitute for the expensive heating-oil was the most urgent need. As a consequence most new wind turbines were designed for heat production with a DC generator producing power for a heating element or using a mechanical churn system. But some of the new pioneers wanted to produce AC and connect the wind turbines to the public grid (like Agricco did in 1919 and Johannes Juul in 1950). One of them was the carpenter Christian Riisager. He built a small 7 kW wind turbine in his garden at Skærbæk, near Herning in central Jutland and in 1975 he tried—without permission—to connect the wind turbine to the grid. Inside the house his wife registered that the electric meter was running backwards. Later he got the formal permission and official rules were made for the sale of electricity to the public grid.

3.11   The NASA Connection: 1974–1978

After the oil crisis in 1973–1974, a new worldwide interest in wind power resulted in several international workshops and conferences. At a workshop in Stockholm, August 1974, three key-persons in Danish wind power development took part. Niels I. Meyer[3], professor at the Technical University of Copenhagen and an early proponent of renewable energy, gave a summary of the Danish experience in the field and was supplemented by the former project manager for the FLS Aeromotor project, Helge Claudi Westh. The third person was Jean Fischer, a visionary high level employee at the FLS-group, who at that time tried to revive the interest for wind power in Danish industry. He presented a number of innovative design sketches for vertical-axis Darrieus turbines. Also present at the workshop were Louis Divone and Joseph Savino from NASA, both of them involved in the design work on the US 100 kW "Mod-0" experimental wind turbine. They had already showed keen interest in Danish wind turbine design and NASA published translations of French and German articles on Danish use of wind power during the Second World War [14,15]. After the Stockholm workshop, the two experts from NASA visited Gedser together with Meyer, Fischer and Westh in order to have a closer

[3]See chapter *Danish Pioneering of Modern Wind Power* by Niels I. Meyer.

look at the FLS Aeromotor from 1943 and Johannes Juul's 200 kW
AC turbine from 1957.

Figure 3.26 With Jean Fischer and Helge Claudi Westh as guides, key
persons from the US "Mod-0" project were visiting the
historical Danish wind turbines at Gedser in August 1974—
here looking at the FLS Aeromotor from 1943 (Photo in the
FLS files of the Danish Wind Historical Collection).

During the following months, NASA published a translation
of the technical description and servicing instructions for the FLS
Aeromotor [16]. Fischer and Westh were invited to a workshop
in Washington in June 1975. Only Jean Fischer took part [17], but
Claudi Westh delivered a paper comparing the technology of the
Danish Aeromotor and Juul's Gedser turbine with two new US
100 kW turbines. After the workshop, one of Jean Fisher's Darrieus
models was on the cover of the *Science* magazine in July 1975.

From August 1974 until November 1976, there was an intensive correspondence between the NASA team and Claudi Westh. The design of NASA's 100 kW "Mod-0" was in many ways inspired by German Ulrich Hütter's 100 kW turbine from 1957 and had a 2-blade downwind rotor. Claudi Westh tried in vain to convince Divone and Savino that the Danish 3-blade upwind solution should be preferred in order to avoid vibration problems due to the blades passing behind the tower [18]. He referred to the experience with these problems on the Aeromotors in spite of its less critical upwind solution.

Figure 3.27   Jean Fischer's model of a Darrieus-type wind turbine with six rotors and a concrete tower on the cover of "Science" in July 1975.

The discussion on the best choice of configuration continued across the Atlantic during 1977, when NASA had experienced heavy vibrations and blade stresses during the first tests of "Mod 0" [18]. The vibration problems also stopped discussions about using FLS-type concrete towers on future US downwind turbines. Nevertheless, for the next NASA turbine, the 2 MW "Mod 1", the downwind solution was once more the choice. Later after more vibration problems an upwind rotor was finally chosen for the Boeing-designed 2.5 MW "Mod 2" in 1982. At that time, Danish wind turbines with upwind rotors had already entered the US market. A more positive result of the dialogue between NASA and Denmark was US funding for repair of Juul's 200 kW Gedser turbine during the years 1978–1979 in order to put it in service again and get more experience from it. Parts of this experience were used in two 630 kW test wind turbines, built by the Danish utilities in 1979.

3.12   The Tvind Wind Turbine: 1975–1978

In the Danish energy plan from 1976, nuclear power was planned to cover 6% of the Danish electricity consumption in 1985 and more than 25% in 1995. The introduction of nuclear energy was supported by a massive political majority in the Danish parliament. But in spite of an intensive campaign from the Danish utilities, started already in 1973, the public opinion in Denmark was sceptical. And this scepticism was intensified after the selection of nine locations as suitable for the first nuclear stations in 1974. In January of that year, a grassroots movement against nuclear power (OOA) was established and in September 1975 another grassroots movement supporting renewable energy (OVE) was introduced. The scene was set for an intensive energy debate during the next 10 years [14].

Another important and spectacular contribution to this debate was the building of a big wind turbine at the Tvind schools near Ulfborg in western Jutland. The first steps were taken in May 1975 and the wind turbine was running for the first time nearly three years later, in March 1978. With a 54 m rotor diameter and a 50 m concrete tower, it was at that time the world's biggest wind turbine in operation.

The project demonstrated the possibilities of wind energy as an alternative to both fossil fuels and nuclear power. Moreover, it made the schools and colleges self-sufficient with cheap and clean energy. The self-financed project, which was realised with the help from several hundred volunteers, inspired many new wind pioneers and the big wind turbine had both symbolic and technical influence on the further development of wind power in Denmark.

Mechanical parts were of high quality; some were bought second hand or constructed on site. The main shaft was found at a ship graveyard in Rotterdam (it had been used as a propeller shaft in an oil tanker), the 1 725 kW ASEA synchronous generator was from 1954 and the gearbox with a weight of 20 tons, also produced by ASEA in 1958, had never been used, but had served as a spare unit in a Swedish copper mine. The most important technological feature of the Tvind turbine was the self-supporting fibreglass blades, which were based on the principles developed by Ulrich Hütter at the Technical University in Stuttgart on a 100 kW wind turbine in 1959. His blade design was characterised by the way in which the blades were fastened around the hub bolts at the blade root with glass fibre strands. This technology was introduced in Denmark by Tvind and it became the "standard solution" for the next generation of Danish wind turbines.

Figure 3.28   The main shaft for the Tvind turbine was originally used as propeller shaft in an oil tanker. Here it is coupled to the 16 tons welded rotor hub, produced at the site during the spring of 1977 (left); One of the finished 54 m blades is carried out of the workshop tent—by many hands (Photos: Tvindkraft).

The Tvind people were innovative and courageous, but also seeking the best technical solutions and searched for inspiration and assistance from experts in Denmark and abroad. At one single point, however, the inspiration from Hütter showed to be problematic. Like NASA and others in the US and Germany, they adopted his downwind layout with the rotor placed behind the tower. And like many others they experienced vibration problems due to the blades passing in the "wind shadow" behind the tower. This was a contributing cause to the fact that the wind turbine was not able to reach the projected maximum of 40 rpm. In this case, vibration problems were avoided by limiting the speed of the turbine to a lower level. As the turbine was running at variable speeds and the electricity from the AC-generator had to pass a converter to obtain grid frequency of 50 Hz, it gave no extra technical problems. But the production was of course reduced, because only about half of the generator capacity could be used. However, the goal was never to obtain a maximum production of electricity for export to the grid—but to get enough power and heating for the schools. And here, the 900 kW that could be obtained was more than needed. Reduced speed gave an additional advantage in a lower stress and a longer lifetime for mechanical components.

Figure 3.29 The Tvind turbine with its red and white decoration, applied for the 25th birthday in 2003 and designed by the Danish architect Jan Utzon, son and partner of the late Jørn Utzon, who designed the Sydney Opera House (Photo: Benny Christensen, DVS).

The Tvind turbine proved to have a long life. After more than 30 years it is still in operation in 2012 with all the original parts, except for the blades and blade bearings, which had to be changed in 1993–1994. Compared with the fate of international projects for big wind turbines from this early period, supported with heavy budgets from national research funding and big industries, it is a remarkable achievement for which Tvind was awarded with the European Solar Prize 2008.

3.13   Blades for a New Wind Turbine Generation

While NASA learned from Danish wartime and post-war experience with wind power and the big Tvind turbine was growing, the new Danish pioneers gradually developed the first home-built wind turbines into a simple and efficient industrial product. The first generation of commercial turbines followed Riisager's concept using mechanical yawing and stayed blades with wooden main beams.

An important inspiration for the next steps in the development came from the Tvind project. One of the basic ideas of the Tvind people was expressed in the slogan "Let 100 windmills bloom!" (which in a short time showed to be an understatement). In order to boost this development, all the technology and the experience from the project was made available for everybody. As a consequence, a team of students built a small wind turbine (PTG turbine) at the same time as construction of the big wind turbine took place. Its 4.5 m self-supporting fibreglass blades, downscaled from the full size blades for the big turbine, became an important help for the new Danish wind turbine industry [17].

The small blade mould was lent out to anyone who would like to build their own wind turbine. A mould set, copying the PTG-blades, was used by a group of wind power enthusiasts in southern Jutland in 1977. As they gave up the project at an early stage, the mould was bought by Erik Grove-Nielsen, who after some years of study at the Technical University in Copenhagen and experience with aerodynamics from work with model planes and sailplanes, started producing blades in a remote farm in Økær, Middle-Jutland.[4]

---

[4]See chapter *Økær Vind Energi—Standard Blades for the Early Wind Industry* by Erik Grove-Nielsen.

Figure 3.30   The PTG turbine, built at Tvind in 1977 and using the same
blade design as the big Tvind turbine (Photo: Tvindkraft).

The Økær blades soon became the standard solution for
most of the new Danish wind turbines. Only four sets of the original
4.5 m blades were produced. They were used on wind turbines
constructed by electro-mechanic Svend Adolphsen and two other
members of the southern Jutland group. With its downwind rotor
and slender lattice tower stabilised by guy wires, the Adolphsen
turbine was in many ways inspired by Tvind's PTG turbine. It
was later modified and produced in Ulfborg under the name
"Kuriant" and was the first important commercial competitor to the
Riisager turbines.

The Adolphsen/Kuriant turbine in 1979 became the first
wind turbine to obtain the "system approval" from Risø, the new
national Danish test centre for wind turbines. It was established
in 1978, at the nuclear research centre at Risø. The placing of this
"ugly duckling" (to quote Andersen's famous fairytale) there amidst
some of the most active nuclear lobbyists was considered strange
by many on both sides of the intense nuclear debate. Although the

leader of the new wind facilities, Helge Petersen, came from Risø's reactor department, he had as a private person been involved in the development of the blades for the Tvind turbine. Young engineers employed at the new centre had backgrounds in renewable energy projects and grassroots movements, and during the following years, the test centre became a valuable partner for the Danish wind turbine industry.

Figure 3.31 Erik Grove-Nielsen with the final version of the Økær blade with revolving tip brakes (Photo: Erik Grove-Nielsen, www. windsofchange.dk) (above). Three stages in the development of blades for modern Danish wind turbines: To the left is a Riisager blade with wooden main beam, to the right of the first are 4.5 m Økær blades based on the original Tvind design and used for 11 kW Adolphsen turbine. In the middle is a late 5 m Økær blade, produced by Alternegy for 30 kW Bonus (Photo: Benny Christensen, DVS) (below).

3.14   The Birth of the "Danish Concept": 1978–1979

Since 1974, a group of engineers, blacksmiths, technical school teachers and others in northwestern Jutland—including the founder of the Folkecenter for Renewable Energy a few years later, Preben Maegaard—had studied and worked on different kinds of renewable energy. In 1977, this group (NIVE) designed a 22 kW wind turbine, which was to be produced locally.[5] It was a 3-bladed upwind turbine with an asynchronous generator, an industrial gear, electrical yawing and locally developed control system [14]. Maegaard asked Erik Grove-Nielsen to develop a new 5 m blade, and ordered 15 blades still before it was even designed and tested. During the process, an important choice was made. All traditional Danish windmill turned counter-clockwise (seen from the wind side) and it was also the case for the Riisager turbines and Tvind's PTG turbine. But Grove-Nielsen designed the new blades for clockwise rotation. After that, it became the "normal" direction for wind turbines, both in Denmark and abroad. The first set of Økær blades for the NIVE turbine was delivered in June 1978.

At the same time near the small town Vildbjerg in the Middle Jutland a young student had got his first experience with wind power. After finishing high school in 1976, Henrik Stiesdal made experiments at his home farm near Vildbjerg between Herning and Holstebro in western Jutland.[6] He placed his first 2-bladed sail-wing turbine at an agricultural trailer and towed it to a place on the fields, where the wind was the strongest. Together with his father he also visited Tvind several times and they decided to build a 15 kW sail-wing turbine to produce electricity for the farm. In February 1978, he got assistance with production of some parts for this turbine by the blacksmith Karl-Erik Jørgensen, who had a small mechanical workshop in nearby Herborg. Jørgensen was at that time working on his own small wind turbine, and it became the start of a close cooperation during the next couple of years. Jørgensen's practical skills, extreme willpower and determination along with Stiesdal's theoretical knowledge showed

---

[5] See chapter *From Energy Crisis to Industrial Adventure—A Chronicle* by Preben Maegaard.
[6] See chapter *From Herborg Blacksmith to Vestas* by Henrik Stiesdal.

to be a lucky combination and together they brought Danish wind power technology important steps forward.

Figure 3.32 The NIVE wind turbine—first to use the new 5 m Økær blades that turned clockwise (Photo: Folkecenter of Renewable Energy).

Figure 3.33 Karl Erik Jørgensen's and Henrik Stiesdal's 22 kW turbine at Herborg in the summer of 1978 (Photo: Henrik Stiesdal).

Their first prototype from June 1978, financed with DKK 50 000 from a public fund for inventors, was a 22 kW turbine with three 5 m Økær blades and electric yawing. The rotor hub was mounted directly on an industrial standard gearmotor with reinforced bearings. On the basis of the experience with this turbine, they decided to make a 30 kW version. One of the hard lessons from the test of the prototype was that the blades needed an automatic braking, preventing runaway at high wind speeds. After that, the Økær blades got revolving tip brakes after the same principles as on Juul's Gedser turbine from 1957. The 30 kW wind turbine with the designation "HVK-10" (HVK for Herborg Wind Power—10 for the rotor diameter) was finished in the spring of 1979.

Inspired by the work of the first pioneers—Riisager and others—the NIVE turbine and the HVK-10 added new features for the next generation of Danish wind turbines: Upwind rotor with three self-supporting, fixed fibreglass blades, electrical yawing, industrial gear and disc brake between rotor and generator. The HVK introduced tip brakes and two generators (a 30 kW for higher wind speeds and a 5.5 kW for lower wind). What later had been named the "Danish Concept" for modern wind turbines was born.

3.15   Establishing the New Industry: 1979–1981

In the 1970s, the company Vestas in Lem, near Ringkoebing, was a small machine factory, mainly producing machinery for the agricultural sector. As this market was in a deep crisis, Vestas was motivated to look for new products and turned their interest to wind power. Experiments were made in 1978 with several types of vertical-axis Darrieus-type turbines, designed by the innovative Danish engineer Leon Bjervig, but in the end the choice was the more conventional HVK design. In the autumn 1979, a contract was signed with Herborg Wind Power (Karl-Erik Jørgensen and Henrik Stiesdal) and the 30 kW HVK wind turbine with the designation "V-10" (with reference to the rotor diameter—as all later Vestas wind turbines) became Vestas' entrance on the wind turbine market.

In 1979, a new Vestas 55 kW was already under development at HVK using the same basic construction as the 30 kW. Økær developed 7.5 m blades for the new turbine, named V-15, and in March 1980 a prototype was erected at the factory in Lem. At the same time wind turbines got their own department at the Vestas factory, managed by Birger T. Madsen, who together with Finn M. Hansen (son of the owner Peter Hansen) had started the first experiments two years earlier. In the summer of 1981, a new version of the 55 kW V-15 was introduced. It was considered during the following years to be the "standard product" from the factory.

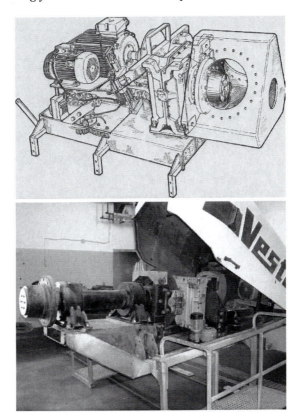

Figure 3.34 The first 55 kW Vestas V-15 from 1980 had the same compact design as the V-10 (illustration: Vestas). In 1981, the second version shown below was introduced with the disc brake in front of the gearbox, two bearings for the main axle—and more space around the components in the long nacelle (Photo: Benny Christensen, DVS).

In the meantime, a pair of other Danish firms entered the wind turbine market. The first one—Nordtank—originally produced mostly road tankers for the oil companies, but the oil crisis has nearly eliminated this market. In the search for a new product, wind turbines were the choice. Here, Nordtank could also use its knowledge of rolling and welding steel tank sections and introduced the tubular steel tower—built by cylindrical and conical sections— which became the trade mark of Nordtank and soon were adopted by other producers.

Figure 3.35 Nordtank—with their background in oil tanker vehicles—was the first to introduce. in 1980, the tubular steel tower, which in various designs substituted the traditional steel lattice tower on most Danish wind turbines during the following years (Photo: Jan Vium Nielsen, DVS) (left); Bonus adapted the tubular tower from the start. The first Bonus 30 kW turbine, sold in September 1981, was still in use, as seen in the photo, in the summer 2010 (Photo: Erik Grove-Nielsen, www. windsofchange.dk) (right).

Danregn was the other Danish firm entering the market in 1981.[7] It was specialised in mobile water irrigation machines and pipes for farms. As the market for this product had an obvious seasonal character, there was a need for a product to "fill up" the

---

[7]See chapter *From Danregn to Bonus* by Egon Kristensen.

production facilities in the off-season periods. On a visit to an exhibition, the Danregn noticed new wind turbines from Vestas and Nordtank—and tried to get a licence agreement with Nordtank. As this attempt failed, they decided to develop their own design and consulted Preben Maegaard among others. The wind turbine was named "Bonus", which after some years also became the company name. Apart from the tubular towers, which also were adopted by Danregn/Bonus—the basic constructions from the three factories were very similar—all using Økær blades and following the "Danish Concept". Several other producers of wind turbines also entered the market—some of them with designs, inspired by Riisager, others adopting the new "Danish Concept" and a few were still working with other ideas.

The development in these years was characterised by a free exchange of ideas and experience in open gatherings once or twice a year, first organised by the Organisation for Renewable Energy (OVE) in 1976, and from 1978 arranged in cooperation with the wind turbine owner's organisation (DV). A manufacturer's organisation (FDV) was established in 1981 and after one year it had 19 wind turbine manufacturers and 4 producers of blades as members [20]. An effort to encourage and support Danish small and medium-sized enterprises to start wind turbine production was made in 1977 by the Danish Blacksmiths Association (DS) in cooperation with the NIVE-group and the Folkecenter for Renewable Energy (established in 1983). Detailed "construction manuals" for the "Danish Blacksmith Wind Turbine" and assistance with search for and development of suitable components gave many new producers (Vind-Syssel, Dencon, Lolland, Hanstholm and others) the possibility to enter the market. Several others used the Folkecenter design for local production—Reymo, Ribe Vind, Codan and Meonia, producing, however, just a single or a few units. The Wind World was a design spin-off of the Folkecenter concept, with integrated gear, which was also applied in early Nordex designs of 150 kW and 200 kW windmills.

3.16 Danish Export to California: 1982–1986

The new Danish wind turbine industry got its big chance in 1982 when the state government in California had launched ambitious plans for renewable energy supported by adding tax credits for investors to an existing more moderate federal tax credit. It created a flourishing market for wind turbines, and the Danish products soon got a good reputation for reliability compared with most of their American and European competitors.

In 1982, the first 35 Danish wind turbines were in operation in California. The 55 kW turbines, established as the "standard product" from most producers, were upgraded to 65 kW for the US market due to the 60 Hz grid-frequency. There was rapid growth in the export, and in the year 1985, the sale culminated with nearly 3 500 units. It was 90% of the total Danish production this year [20]. Many Danish wind turbine producers took part in the export. Vestas, Nordtank and Bonus were of course there, along with Micon, which was started in 1983 as a breakaway from Nordtank. These four factories covered in this period around 70% of the Danish home market and played an important role in California. But also Christian Riisager, who now had moved his wind turbine production to the Faroe Islands, and many new manufacturers, including producers of the "blacksmith wind turbine" all got a share of the US market.

Much has been written, both in scientific and more popular literature, about the Danish role in the "Californian adventure", and many analyses have been done to explain the Danish success at the international wind turbine market [21, 22]. A common explanation has been the difference between the "development strategy" in Denmark and in other countries. In the United States and Germany a "top down" approach was chosen, including big high-tech industries and building large experimental turbines, based on advanced aerospace technology. In Denmark a "bottom-up" development was followed, starting with small wind turbines and low technology solutions, gradually up-scaling along with learning from practical experience. This may be part of the explanation. The Danish bottom-up development was not a "strategy" chosen by the government. It was the result of a close and open cooperation and exchange of ideas between the actors—inventors, researchers and innovative industries—in the first creative development phase.

The "flying start" at the Californian market had in several ways deep consequences for the further Danish wind power development. On the one hand, the new industry got a lot of technological experience, as well as practical experience with sale and service at the export market. On the other hand—when the "adventure" finished in the end of 1986 as the tax subsidies disappeared—the Danish wind turbine industry had a production capacity based on 85% export and a surplus production, which could by no means be absorbed by the home market. Most of the factories went bankrupt and some of them were re-organised, while other disappeared. After the first 10 years of development, the wind turbine industry was during the next decade moving into a new phase of big business, export and international competition.

3.17 Public Support, Political Scepticism and Local Participation

At the political level, the attitude to wind power and renewable energy during the first 10 years of development had been ambiguous. Already in 1979, a political majority had agreed on giving a 30% subsidy on private investments in wind turbines, but still renewable energy was only considered as a small supplement in relation to fossil fuels and nuclear energy. In 1983, a leaflet from the Danish Ministry of Energy concluded that "it is considered as unrealistic, that big parts of the energy consumption in the modern society could be covered by renewable sources".

Nevertheless, the public interest in wind power was growing, stimulated by the 30% investment grant. In 1983, more than 900 private wind turbines with a total capacity of around 20 MW have been erected. In 1986, the number had grown to more than 1 700 and the capacity even more to 80 MW, due to the bigger size turbines.

The first investors in Denmark were families investing in a wind turbine to make the household self-sufficient with heat and power. The surplus electricity could be delivered to the public grid, though at a lower rate, than was paid for purchasing power from the grid. When turbine sizes grew to 55 kW or more, families were motivated to share the investments with neighbours. Cooperative ownership became an option and already in 1980, the first cooperative "wind turbine guild" was established. It followed a good Danish democratic practice, which also had been used for the start of dairies in the end of the 19th century and for Poul la Cour's rural power stations some decades later. During the following decades, the legal framework regulating membership of the cooperatives, depending on consumption and geographical criteria as well of the taxation rules, were changed several times.

Figure 3.36   150 kW Micon wind turbine. Micon was established in 1983 as a breakout from Nordtank and soon became an important player at the Californian market. In 1997, Micon was again merged with Nordtank in NEG-Micon (left); Despite the common trend for bigger wind turbines using "the Danish concept", experiments were still made with other designs and smaller wind turbines. Here the wind pioneer Claus Nybroe is showing his small "Windflower" beside big wind turbines from German producer Enercon at an international exhibition at Husum, Germany. Unfortunately, he was ahead of his time. This size of wind turbine for a single household got a revival more than 10 years later (right) (Photos: Benny Christensen, DVS).

Already in 1978, wind turbine owners had been organised in the organisation Danske Vindkraftværker (Danish Wind Power Stations). DV should take care of the owners' mutual interest in relation to utilities, authorities and manufacturers. With cooperative ownership dominating during the 1980s, DV soon became an important player in the debate and the development. Their monthly production statistics covering more than 2 000 wind turbines became a good reference for customers and DV's local members formed strong support for wind power. It is estimated, that at the maximum, DV membership involved more than 150 000 Danish families.

8. årgang nr. 8 April 1986 ISSN 0106-1127 Kr. 25,00 i.m.

Figure 3.37   On this cover of the magazine *Naturlig Energi* (*Natural Energy*) from 1986, published by the Danish wind turbine owner's organisation DV, Danish cartoonist Klaus Albrechtsen visualises the cooperative ownership, which was an important factor for the wind power development in Denmark during the 1980s.

In 1985, the political majority in the Danish parliament decided to cancel the plans for use of nuclear power. This decision forced the Danish utilities to revise their attitude in relation to renewable energy—and especially wind power. In order to speed up this process, the parliament also required the utilities to build 100 MW wind power during a five-year period.

3.18   New Ambitious Targets for Wind Energy: 1990–1996

In April 1990, the Minister of Energy Jens Bilgrav-Nielsen, presented the third Danish energy plan, "Energy 2000". It was very different from the two previous plans from 1976 and 1981.

Figure 3.38 The Danish energy plan from 1990, "Energi 2000" with the subtitle "Action Plan for a Sustainable Development", was inspired by the Brundtland report and introduced new ambitious goals for the use of wind energy.

Inspired by the UN Brundtland Report from 1987 and its focus on sustainable development, the new plan had an ambitious goal of a 20% reduction of Danish CO_2 emissions by 2005. To reach this goal, the plan adopted recommendations of the Brundtland report on reduction of the total energy consumption. So whereas the two previous plans envisaged future growth, the new plan headed for an energy reduction of 15% by 2005 and drafted three scenarios for 2030 with a reduction of 20–40% from the 1988 level. For CO_2 the scenarios for 2030 would give a reduction of 38–65%[8].

---

[8]The plan did not include the transport sector. Here, less ambitious goals with stabilization on 1988 level in 2005 and 25% reduction in 2030 for both energy and CO_2 were included in a separate plan in May 1990.

Wind power should play an important role. The goal was that wind turbines should cover 10% of the electricity consumption in 2005. A new deal with the utilities covered another 100 MW installed capacity before 1994. However, the private investments in wind power during the early 1990s were delayed by uncertainty around payment for the electricity produced and local planning for placement of wind turbines.

In 1996, with new government since 1992 and Svend Auken as the Minister for Energy and Environment, Denmark got its fourth energy plan. The wind power goal for 2005 was the same as in the 1990-plan, but was supplemented with a long-term target for 2030. In this year, 5500 MW wind power should be installed— 4000 MW of the capacity should be placed offshore. This was twice the capacity of the most ambitious scenario in the 1990 plan.

At the same time, the regional and local planning for wind turbine sites were nearly finished and new legislation in 1998 favoured ownership by investors and companies without the former geographical restrictions. This gave a strong growth in the wind capacity. From 1995 to 2000, the total Danish wind turbine capacity grew from 400 MW to 2 400 MW and the share of wind energy grew from 3.5% to 12% of the electricity consumption. Wind power was in that way already several years ahead of the schedule in the energy plans from 1990 and 1996.

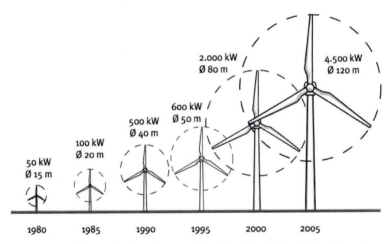

Figure 3.39 The gradual evolution of the size of Danish wind turbines from 1980 until 2005—with capacity (kW) and rotor diameter (m) indicated (Picture: The Danish Energy Museum).

During the 1990s, wind turbines grew to MW-size. It meant that the new growth in capacity could be ensured by fewer turbines. But bigger wind turbines and absence of local ownership also created more conflicts with neighbours around the sites, where wind turbines were to be erected, disturbing some of the good-will, the "green" renewable energy originally had. In some cases, there were obvious reasons for the local protests, but they were often heavily stimulated by outside lobbyists from the long battle on "nuclear contra renewables" more than 10 years before.

3.19   Wind Power Goes Offshore

One of the solutions for the neighbour-problem was to place wind turbines offshore. This is because offshore areas offer plenty of space with no neighbours to disturb. It was also possible to find vast areas for wind parks with a large number of big turbines—hard to find in a densely populated country like Denmark. It was also an expensive solution, giving new challenges.

The first wind parks—some of them with more than 30 moderate-sized wind turbines (55–95 kW)—were erected in 1983–1986 by cooperatives, utilities, municipalities or private investors. They were all land-based, though a municipality owned park with 16 turbines at Ebeltoft installed in 1985 on a wave braker connected to the shore, could claim to be "nearly offshore". The first "true" offshore wind park had eleven 450 kW Bonus turbines and was placed 4 km from the north shore of the island of Lolland in 1991. The next one came in 1995 and had ten 500 kW Vestas turbines, placed between Jutland and the island Tunø. Both these wind parks were built by utilities.

With ambitious goals for offshore capacity in the new energy plans, the utilities were also expected to be important players in this field. But even then, cooperatives were still driving forces for the development. In 2000, an offshore wind park—at that time the world's largest—was built at Middelgrunden, close to Copenhagen.[9] The long curved row of 20 wind offshore turbines became a spectacular sight for passengers arriving to or leaving the Copenhagen Airport. Ten of the twenty 2 MW Bonus wind turbines were owned by a cooperative with 8 650 members, the

[9]About Middelgrunden, see chapter *Cooperative Energy Movement in Copenhagen* by Jens Larsen.

other ten by the local utility. The wind farm now delivers more than 3% of the electricity used in the Danish capital.

In 2002–2003, two big offshore wind parks were built by Danish utilities. They were the first of five planned offshore wind parks that ensured the further development towards the long-term target in the 1996 energy plan. The first one was placed at Horns Rev at the North Sea and had 80 Vestas 2 MW turbines. The following year, a wind park of similar size with 72 Bonus 2.3 MW turbines was built in the Baltic Sea. Also in 2003, an offshore wind park with 10 Bonus 2.3 MW turbines was established on the island Samsø, which in 1998 was nominated as the "Danish Renewable Energy Island", made self-sufficient with energy.

Figure 3.40 Middelgrunden offshore wind park by Copenhagen was established in 2000 and was at that time the world's biggest offshore installation (Photo: Siemens Wind Power) (left); Installation of one of the eighty 2 MW Vestas wind turbines in the Horns Rev offshore wind farm in 2002 (Photo: Jan Vium Nielsen, DVS) (right).

3.20 Political Changes, and Changes of Mind: 2001–2011

These projects placed the Danish wind turbine industry and utilities in the forefront of the promising offshore market. But a sudden break in the development occurred after a change of the government in the autumn of 2001. The new right wing government abandoned more than 15 years of political consensus on Danish energy policy. Public support for renewable energy development and demonstration projects was reduced and the plans for three more offshore wind parks were cancelled. There would be no more national energy plans and the development would rely on the market forces alone. The consequences of these changes were reflected in the development of Danish wind turbine capacity. Until 2003, the growth from the 1990s continued, as the first two big offshore wind parks, that could not be cancelled, were connected to the grid. During the following five years, the capacity stagnated at a level of 3000 MW, corresponding to around 16% of the Danish electricity consumption in 2003. For the wind turbine industry, the loss of the home market came at a critical period with growing international competition. The changes also had damaging effect on the Danish image as a "green frontrunner".

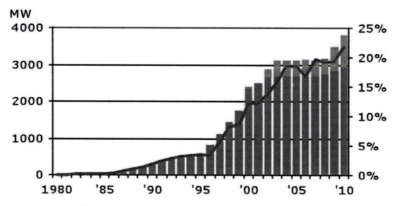

Figure 3.41 The curve on the graph shows the share of wind power in the Danish electricity supply (scale to the right) and the columns indicate the total installed capacity (scale to the left). During the period 1995–2003, there was an average growth of more than 300 MW per year. After that followed five years of stagnation (Annual statistics from 2010, The Danish Ministry of Climate and Energy).

As the government continued after a general election in November 2007, Connie Hedegaard, who had been Minister for

Environment since 2004, was appointed Minister of Climate and Energy. Hedegaard worked hard to re-establish Denmark's international reputation in the field. One of her achievements in 2008 was a Danish energy policy including two new 200 MW offshore wind farms scheduled to start energy production in 2012. And already in 2009 and 2010, growth in the amount of Denmark's wind turbines resumed. A new left-wing government after the election in 2011 is supposed to continue and develop a strategy with wind power delivering 50% of the Danish electricity consumption by 2020.

3.21   Denmark Is Still Ahead: 2012

Thirty-five years after the "re-invention" of wind power, Denmark is still in a leading position. The size of wind turbines has constantly grown since then. The first Vestas had the designation "V-10", indicating the 10 m rotor diameter. Today, Vestas has already delivered more than 1 000 of the 3 MW "V-112" and even bigger turbines are planned. The "Danish concept" has been modified along the way. Already in 1986 the first designs with pitch-regulated blades instead of the fixed blades with stall-regulation were developed—and from the 200 kW "V-25" in 1989, it was the standard solution for Vestas turbines. Later, it became the new standard for big wind turbines.

The newest technical trend—especially for offshore turbines—is gearless direct drive wind turbines using generators with permanent magnets (PMGs). This technology has been tried before. In 1999, the Folkecenter made tests with a 100 kW prototype PMG, following the trend in other countries, where gearless wind turbines were already produced. Today's version—a compact construction, enabled by new advanced materials for permanent magnets—was introduced on a Siemens 3 MW prototype in 2008.

During the last two decades, the Danish wind turbine factories have become fewer but bigger. Nordtank merged with Micon in 1997 to NEG-Micon which was finally merged with Vestas in 2002. And Bonus was sold to Siemens in 2004—and became Siemens Windpower–but is still based in Denmark.

Figure 3.42 The erection of a 3.6 MW Siemens test wind turbine with direct drive and PMG near Ringkøbing, Denmark, in July 2008 (on the photo). Three years later, a 6 MW direct drive Siemens turbine was introduced (Photo: Siemens AG) (left); In 2011, the biggest Vestas turbine produced was the 3 MW V-140 (140 m rotor diameter). It uses PMG, but still combined with a gearbox. In December 2010, three Vestas V-140 were placed at the surf of the North Sea in Hvide Sande, Jutland—an optimal onshore location (Photo: Preben Maegaard).

In 1996, the Danish wind turbine industry covered nearly 60% of the global market for wind turbines. Of course, this position could not be maintained in a rapidly growing global market. But in 2011, Vestas was still the leading global producer with a market share of 12.9%, while Siemens was number 9 with 6.3%. In 2012 the US company GE Wind took over the leader position with Vestas on a second place. But both the Denmark-based producers has increased their market shares. Vestas had now 14.0%, while Siemens had moved a third place with 9.5% [23]. And on the emerging offshore market the two Danish-based producers had delivered 85% of the capacity installed until the end of 2011. Among the blade producers, LM Glasfiber, entered the market in 1978 by producing blades for the Riisager turbines, and was in 2011 still the world's biggest independent producer, delivering blades for wind turbine industries in many countries. And the wind power share of the

Danish electricity consumption (30% in 2012) can truly not be challenged by other countries in the nearest future.

The development of modern Danish wind power has sometimes been described as the "Danish wind power adventure". In my Oxford Dictionary, an adventure is described as "an unusual, exciting and daring experience". This is surely a true description of what happened in Denmark in this period. But parts of it also come close to the famous fairytales of Hans Christian Andersen. What happened for instance with "the ugly duckling"—the small wind power test station, that in 1978 was manned with a few long-haired young enthusiasts and placed at Risø among all the nuclear lobbyists? Today, the nuclear test reactors at Risø have been decommissioned, and in 2012 DTU (Danish Technical University) Risø was reorganised—there are two new departments established—DTU Wind Energy and DTU Energy Conversion as well as a new centre—DTU NUTECH, with wind power research as one of the main activities. This could be the happy end of one of Andersen's fairytales.

References

1. Benny Christensen. "Danish Windmill Production before and after Poul la Cour—and the First "Golden Age" of Wind Power 1900-1920" in "Wind Power—The Danish Way", The Poul la Cour Foundation 2009, pp. 24–31.

2. Danmarks Statistik. Statistiske Meddelelser, 4. Række, 34. Bind, 2. Hæfte. København 1910.

3. Povl-Otto Nissen. "The Scientist, Inventor and Teacher Poul la Cour" in "Wind Power—The Danish Way", The Poul la Cour Foundation 2009, pp. 6–11.

4. Therese Quistgaard. "The Experimental Windmills at Askov 1891–1903" in "Wind Power—The Danish Way", The Poul la Cour Foundation 2009, pp. 12–17.

5. Jytte Thorndahl. Electricity and Wind Power for the Rural Areas 1903–1915" in "Wind Power—The Danish Way", The Poul la Cour Foundation 2009, pp. 18–23.

6. Danmarks Statistik. Statistiske Meddelelser, 4. Række, 72. Bind, 3. Hæfte. København 1925.

7. Matthias Heymann. Die Geschichte der Windenergienutzung 1890–1990, Campus Verlag, Frankfurt 1995, p. 92.

8. Jytte Thorndahl and Benny Christensen. "Time for Survival and Development 1920–1945" in "Wind Power—The Danish Way", The Poul la Cour Foundation 2009, pp. 32–39.

9. Patent 1,467,699 (Application filed June 25, 1919) United States Patent Office, Sept. 23, 1923.

10. A Report on the Use of Windmills for the Generation of Electricity. Oxford University, Institute of Agricultural Engineering, Bulletin No. 1, Oxford, England, The Clarendon Press, 1926, p. 63 (from the Research files of T. Lindsay Baker).

11. Document in the Kregel Windmill Company papers, Nebraska State Historical Society, Lincoln, Nebraska (from the research files of T. Lindsay Baker).

12. Jytte Thorndahl. "Johannes Juul and the Birth of Modern Wind Turbines" in "Wind Power—The Danish Way", The Poul la Cour Foundation 2009, pp. 40–45

13. Preben Maegaard. "The New Wind Power Pioneers and the Emergence of the Modern Wind Industry 1975–1979" in "Wind Power—The Danish Way", The Poul la Cour Foundation 2009, pp. 46–51.

14. Statistical Summary and Evaluation on Electric Power Generation from Wind Power Stations (translation of articles by Dimitri R. Stein in "Electrizitätswirtschaft, Vol. 50, No. 10, October 11, 1951, pp. 279–285, and No. 11, November 1951, pp. 325–329.) NASA Technical Translation F-15, pp. 651–652, June 1974.

15. Utilization of Wind Energy in Denmark (translation of "Utilisation de l'energie du vent au Danemark", La Technologie Moderne, Vol. 35, No. 13–14, July 1–15, 1943 pp. 106–109.) NASA Technical Translation F-15, p. 868, September 1974.

16. Instruction for the FLS Aeromotor (translation of FLS Internal Memo 7050, April 2, 1942). NASA Technical Translation F-16, p. 138, January 1975.

17. Jean Fischer. F.L.S. Overseas Innovation, "The Past and the Future of Wind Energy in Denmark", Proceedings of the Second Workshop on Wind Energy Conversion Systems, Washington D.C., June 9–11, 1975 pp. 162–172.

18. Private correspondence of Helge Claudi Westh in the Research Files of the Danish Wind Historical Collection (DVS).

19. More details on the blade development can be found in chapter *Økær Vind Energi—Standard Blades for the Early Wind Industry* by Erik Grove-Nielsen.

20. Birger T. Madsen. "Public Initiatives and Industrial Development after 1979" in "Wind Power—The Danish Way", The Poul la Cour Foundation 2009, pp. 52–59.

21. Matthias Heymann. Signs of Hubris—The shaping of wind technology styles in Germany, Denmark and the United States 1940–1990. Technology and Culture, 1998, 39(4), pp. 641–670.

22. Kristian Hvidtfelt Hansen. "International Perspectives on the History of Danish Wind Power" in "Wind Power —The Danish Way", The Poul la Cour Foundation 2009, pp. 60–65.

23. World Market Update 2012, BTM Consult—Navigant Research, March 2013.

About the author

Benny Christensen (1937), M.Sc. in Mechanical Engineering, was active professionally and on grassroots level in the field of renewable energy and environment since the 1970s. Board member in the Organisation for Renewable Energy (OVE) in the 1980s. He has written several articles and books and presented lectures on energy technology and development. He is editor of the book *Wind Power—the Danish Way* (The Poul la Cour Foundation, 2009). He has supported the education sector, development work in private industries, environmental regulation and regional wind energy planning through his work and has also worked as project manager for hydrogen development projects. He is also a board member of the Danish Wind Historical Collection (DVS).

Chapter 4

The Aerodynamic Research on Windmill Sails of Poul la Cour, 1896–1900

Povl-Otto Nissen

The Poul la Cour Museum, Moellevej 21, Askov, DK-6600 Vejen, Denmark

pon@povlonis.dk

The discovery of the vacuum force on windmill sails super-seded the idea of making the total front sail area of a rotor as large as possible. The Danish meteorologist, inventor and college Professor, Poul la Cour (1846–1908), made a breakthrough in the understanding of the forces on windmill sails. His research on mill models in front of two wind tunnels took place over the period 1896–1900. He developed a concept called the "ideal" sail. How did he do that? What were the improvements? What were the consequences? Before his aerodynamic research, la Cour received state support to build the first electricity-producing windmill in Denmark in 1891, which included the storage of wind energy by separating water into hydrogen and oxygen.

Wind Power for the World: The Rise of Modern Wind Energy
Edited by Preben Maegaard, Anna Krenz and Wolfgang Palz
Copyright © 2013 Pan Stanford Publishing Pte. Ltd.
ISBN 978-981-4364-93-5 (Hardcover), 978-981-4364-94-2 (eBook)
www.panstanford.com

4.1   Introduction

This chapter is based on studies of the surviving documents from the research carried out by Professor Poul la Cour over the period 1891 to 1900, especially the report "The Research Mill I and II".[1] About 8 000 documents remain in the archive at the Poul la Cour Museum in Askov, Denmark. The Poul la Cour Museum researches and distributes the history of the technology of electricity-producing windmills—nowadays called wind turbines.

After H. C. Oersted discovered the interaction between electric current and magnetism in 1820, Michael Faraday took over and developed the laws of electromagnetic induction in 1831. Then the field was open for inventions of both telegraphy and electric generators. In the area of using and distributing electricity Thomas Edison developed the electric bulb and established the first electric power stations from 1881. This technology reached Denmark in 1891. The first Danish electric power generating stations were installed in the three cities of Køge, Copenhagen and Odense in 1891. The fuel used was coal or oil, which are energy resources not occurring in Denmark. La Cour wondered if it would be possible to use the almost ever-present wind over Denmark to drive turbines for making electricity instead of having to buy fuel from abroad.

4.2   The First Danish Electricity-Producing Windmill

Everybody told him that it could not be done. The wind does not blow all the time, so it was necessary to find a way to store the energy for use during periods with no wind. The technology of lead/acid batteries was known, but was too expensive. La Cour then had the idea of storing the power of the wind using electrolysis to separate water into hydrogen and oxygen so that the two gases could be used later for heating and lighting. Also it was known, from a firm in Italy, that these gases could be used in autogen welding, which could be useful for small industries in rural areas. With the help of his older brother, who was the president of the Royal Danish Farm Household Society, the

---

[1]*Forsøgsmøllen I og II*, Copenhagen, 1900

Minister of the Interior (Home Secretary) became interested in the scheme. La Cour was then granted the financial support of DKK 4000 for research into the subject of wind energy. No doubt it helped that he was already known as a highly respected inventor of telegraph equipment. A new experimental windmill was built and ready to work in May 1891 in the southern part of the small village where the Folk High School of Askov was founded after the war with the Prussians in 1864.

Figure 4.1   The first experimental mill in Denmark with wooden shutters.

The very first known electricity-producing windmill in the world was built by Charles F. Brush in 1888 in Cleveland, Ohio. It was huge—it had many blades and it delivered power to 240 lamps in the owner's house. This system worked for twenty years. There is no evidence that Poul la Cour knew about the Brush mill in Cleveland, but its existence must be acknowledged.

Poul la Cour chose the millwright N. J. Poulsen in Esbjerg, Denmark, to build his new mill. This firm was one of the bigger millwrighting companies in Denmark. They even built the very special type called Halladay turbines, where the blades could be turned parallel to the turbine axis.

Figure 4.2   The Brush wind turbine (above); Drawing of a Halladay turbine (below).

The American tradition of windmills utilised many blades. This type of windmill is very suited for pumping water. These windmills also influenced the Danish windmill design during the 19th century. Building of the railways provided an opportunity to use this type of windmill in conjunction with water towers on railway stations for pumping water for the steam powered loco-motives.

Figure 4.3   Drawing of a typical railway station with a Halladay turbine (above), and photograph from Hjørring railway station in Denmark (below).

However, it is obvious that la Cour's Askov windmill was of quite a different design compared to what the factory was used to build. Just to prove that his concept was correct la Cour only needed his windmill to be as cheap as possible. So he chose the well-known four-bladed rotor from the standard Dutch windmill design, but of course he had no need for the interior machinery used for flour production.

Figure 4.4   La Cour's windmill in Askov.

Instead he built a slim wooden tower, sufficient to carry the turbine and a vertical spindle down to the DC dynamo in the building. To begin with it was used with canvas sails, but very soon these were changed to wooden shutters, which were a well-known technology. But la Cour is known for the improvement of the automatic regulating system of the shutters.

Figure 4.5   Details of the sails.

The zoomed sections of the photos show, that the middle of the shutters are hinged on the beam and are connected with a rod parallel to the beam. Earlier the sails were to one side of the beams. The rods again are connected with strings to a spider-like device at the rotor centre. From there a string goes through the hollow rotor shaft to a weight to which gravity decide the tension of the shutters against the wind.

4.3 How to Store Wind Energy

The well-known way of storing electricity in lead/acid batteries was too expensive at that time. Therefore Poul la Cour chose to let the direct current separate water into hydrogen and oxygen by electrolysis. This will produce hydrogen gas at the minus electrode and oxygen at the positive electrode.

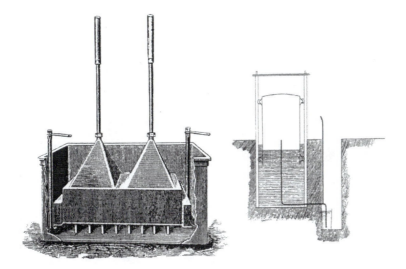

Figure 4.6   La Cour's electrolysis system (left); Hydrogen tank (right).

The electrolysing vessels were placed in the larger building besides the mill building as well as two big gas containers in the ground. From here the two gases were transported through lead pipes 300 m to the college buildings where they were used as lighting gas in lamps made especially for this application.

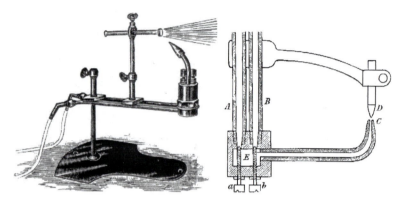

Figure 4.7   Drawing of a gas lamp (left); section (right).

The two gases go through pipes to a little chamber, where they are mixed and pressed through a pipe and ignited. The flame itself does not give much light (pale blue/white), but is very hot. The flame is heating a small ceramic object, which contains zirconium. It gives a very strong bright white glowing light. Hydrogen, however, did not find general application for lighting in the following years. The electric light bulb was preferred.

The installation functioned for seven years from 1895 to 1902. It is important that this success released more financial support and made it possible to continue the research as the State Experimental Wind Laboratory with Poul la Cour as director.

4.4   An Aerodynamic Surprise

Millwrights were excited because a new market had appeared: the production of electricity from windmills. One millwright, Christian Sørensen from Skanderborg, contacted Poul la Cour in 1896. Sørensen had invented a new design of windmill rotor, the conical wind catcher (Keglevindfanget), and he asked la Cour for help to produce documented test results in order to seek a patent. Sørensen himself had started research on models with different rotor types—all with same diameter—in front of a barrel with a blowing ventilator in one of the open ends. It was called a blowing cylinder (blæsecylinder) and possibly it was the first experimental wind tunnel. Until then windmill rotors most often were tested on a rotating merry-go-round (carousel) arm.

Using this method it was difficult to create a constant air flow long enough to make reliable measurements, but la Cour now had suitable equipment. In 1892, he invented the Kratostate, a device which could smooth out the changing rotation of the mill to a constant rotation and driving force on a dynamo. Now it could be used to deliver a very constant air flow through the wind tunnel.

Figure 4.8   "Keglevindfang" on the right with twelve sails (left); The conical wind catcher rotor as a roof mill with only six sails (right).

The first model Sørensen designed had twelve sails. La Cour found almost immediately that performance improved when some of the sails were removed. That was a surprise because the most common opinion at that time—even among millwrights—was that a mill should have as many sails as possible. This was supported by the hitherto accepted scientific formula

$$L = 0.0338 \, F \, v^3$$

where L is power, F the sum of the frontal area of the sails, and v the velocity of the incoming wind.

The formula says that if the total front area of the sails is increased, then the power output will increase proportionally. La Cour proved that this was totally wrong. Only the term for the cube of the wind velocity is still valid from this formula.

La Cour presented his findings in a paper "Experiments with small mill models" at a meeting in the Danish Engineering

Association.[2] A quote from this paper, translated as precisely as possible, is as follows:

> "So it is clear, that the rotor must collect energy from other wind streams than the ones hitting the sails. The rotor must also collect energy from the wind streams passing between the sails, and this collection must in certain cases, especially with open rotors, play an important role. The contribution coming from the space between the sails appears to increase with increasing velocity. The experimental findings show that the wind passing between the sails is highly active in driving the sails. It can easily be understood, when it is considered that the air particles shortly after passing through the spaces, take up a position behind the sails where they cause a vacuum by their inertia, which accelerates the sails and at the same time decelerates the air particles."

Thus la Cour had discovered the importance of the vacuum lift on windmill sails.

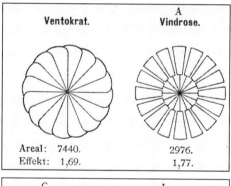

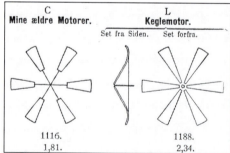

Figure 4.9   Drawing showing various models and measurements.

---

[2]The paper was published in the magazine *The Engineer* (*Ingeniøren*), issue no. 10, March 1897.

The first row of numbers is the frontal area of the blades. The second row is the corresponding power. Millwright Sørensen received his patent, which was the best known at the time. This encouraged him to build the new mill on top of a new stone building. The building itself was designed by a young architect P. V. Jensen-Klint, who later became famous church architect. Today the stone building houses the Poul la Cour Museum.

Figure 4.10   The new laboratory mill building from 1897 with millwright Sørensen's conical rotor.

4.5   Further Aerodynamic Research

Poul la Cour aimed to continue his research in order to find what he called the ideal sail. He wondered why in earlier times mathematicians and physicists such as Bernoulli, Maclaurin, d'Alembert, Euler, Lambert and Smeaton had all propagated a completely incorrect formula. An example of what will go wrong, he claimed, when "the starting point is rational thinking followed by trying to force the real world to fit into a formula with the best choice of constants".[3] Their excuse of course was that they did not have the unique test facilities that Poul la Cour had developed.

In the machinery hall la Cour installed two wind tunnels of his own design and invented a unique apparatus for measuring the

---

[3]For example, 0.0338.

wind pressure on different sail profiles. With his Kratostate he was able to produce a constant airflow through the wind tunnels with a deviation of only one per mille.

Figure 4.11 The wind tunnels (left); Drawing showing the Kratostate (right).

Both wind tunnels were 2.2 m in length. The big tunnel was 1 m in diameter, and the small one 0.5 m. The mill model for testing was placed in front of the big wind tunnel with a friction dynamometer. The pressure measure instrument was placed in front of the small tunnel.

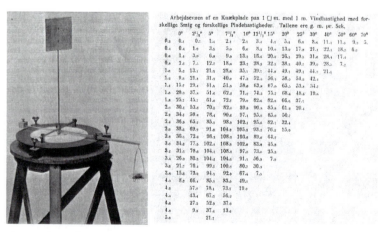

Arbejdsevnen af en Knækplade paa 1 ☐ m. med 1 m. Vindhastighed med forskellige Smig og forskellige Pladehastigheder. Tallene ere g. m. pr. Sek.

	0°	2½°	5°	7½°	10°	12½°	15°	20°	25°	30°	40°	50°	60°	70°
0.0	0.1	0.2	1.4	2.1	2.2	3.1	4.1	5.1	6.2	8.4	11.1	11.5	9.5	5.
0.1	0.4	1.0	3.1	5.0	6.5	8.2	10.0	13.2	17.5	21.1	22.1	18.5	6.0	
0.3	1.1	3.0	6.5	9.5	13.2	16.4	20.0	26.1	29.5	31.6	28.1	17.1		
0.6	2.2	7.3	12.7	18.5	23.2	28.3	32.5	38.1	40.3	39.9	28.3	7.3		
1.0	5.2	13.1	21.1	28.5	35.1	39.7	44.0	49.1	49.1	44.5	21.1			
1.3	9.5	21.4	31.2	40.5	47.3	52.5	56.1	58.5	54.5	42.4				
1.4	15.7	29.4	41.5	51.5	58.6	63.5	67.5	65.5	53.5	34.7				
1.6	20.0	37.5	51.6	62.5	71.0	74.1	75.7	68.1	48.5	19.5				
1.8	25.3	45.7	61.0	72.5	79.0	82.6	82.5	66.5	37.7					
2.0	30.1	53.2	70.5	82.5	89.5	90.5	85.5	61.2	20.4					
2.3	34.1	59.5	78.4	90.4	97.1	95.5	85.5	50.7						
2.4	36.5	65.2	85.5	98.5	102.1	95.5	82.1	32.4						
2.5	38.5	69.2	91.5	104.5	105.1	93.7	76.1	15.9						
2.8	36.1	72.5	96.3	108.5	103.5	89.5	64.7							
3.0	34.5	77.5	102.1	108.3	102.5	83.4	45.5							
3.3	31.1	78.5	104.1	108.5	97.5	73.5	25.5							
3.4	26.5	80.3	104.5	104.5	91.1	56.5	7.5							
3.5	21.7	76.1	99.5	100.5	80.7	30.5								
3.6	15.5	73.0	94.1	92.5	67.4	7.5								
4.0	8.5	66.4	85.5	83.5	49.1									
4.2		57.0	78.1	73.7	19.2									
4.4		43.4	67.5	56.7										
4.5		27.5	52.5	37.0										
4.8		9.5	37.4	13.4										
5.0			21.7											

Figure 4.12 The pressure measurer (left); Table with Poul la Cour's measurements with a bend profile (right).

Poul la Cour did not develop a new formula, but he developed a diagram to determine the best design of a sail based on very careful and precise measurements. It is called the "mussel" diagram, because the shape of it reminds about the mussel, which is very common along the Danish coastlines. With this diagram available a millwright should be able to complete the practical work to an optimum of efficiency. With the wind flow of 1 m/s determined from the small wind tunnel, he measured the influence successively on a plane, a curved and a bent profile with dimensions 10 cm × 10 cm.

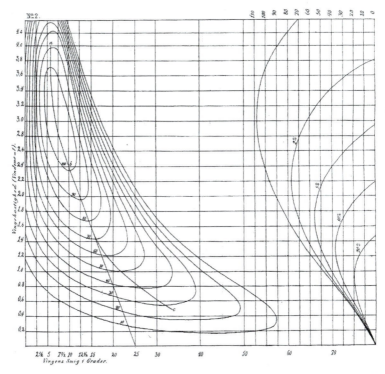

Figure 4.13   Graphical representation of the measurements, the "mussel" diagram.

The origin of the diagram is the rotational centre of the sail. The numbers on the horizontal axis are the bevel angles of the sail to the rotation plane. The numbers on the vertical axis are the velocity factors for that particular point of the sail compared to the velocity of the incoming wind. The diagram gives the impression of a hill on a map with height contour curves. In this case it has a maximum power (gram-metres/second) at the velocity factor three. Reading the corresponding angle on the horizontal axis gives 7½ degrees. To make the best possible sail, take the least steep path from the top of the hill, which gives the path a–c. Obviously it means that the sail should be slightly twisted with large angles near the centre and smaller ones further out. The curved path a–c could not be followed at that time using materials such as wood and iron. Today with glass fibre this is possible. Consequently, from about the velocity factor 1.4 it is necessary to compromise with a straight line hitting the horizontal diagram axis near the rotation centre at 25 degrees.

The research procedures for these findings were as follows:

(1) Initially it is assumed that the flow pressure perpendicular to the plane is proportional to the area of the plane.

(2) Secondly, it is determined that the pressure perpendicular to the plane is proportional to the velocity of the wind flow squared.

(3) After that, the pressure perpendicular to a plane profile from a flow directed from different angles of 1 to 90 degrees on the plane is measured. The surprise was that these results did not fit with the values calculated in advance using Newtonian physics. They were not proportional with the width of the impacting air stream. It must be concluded that the resulting pressure mysteriously turned out to have other directions than perpendicular to the plane.

To determine the directions of the resulting pressure on the plane from the air streams coming from different angles, the rotational axis of the plane is turned in every case until the plane or profile is balancing back and forth in a regular manner. Then the resultant pressure difference would be lying right over the shaft, and the optimum bevel angle could be measured.

Repeating the same with a curved profile and a bent profile it was concluded that the pressure resultant under certain circumstances (bevel and shape) lies to the "favourable" side of the perpendicular direction to the chord of the sail, sometimes to the "unfavourable" side. The term "favourable" refers to a case where the resulting pressure has a component, which causes the sail to rotate in the right direction. The task was then to find the right shape and give it the right bevel angle.

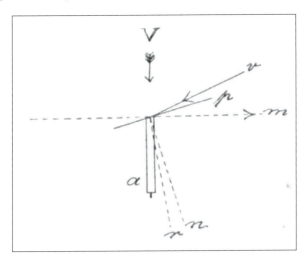

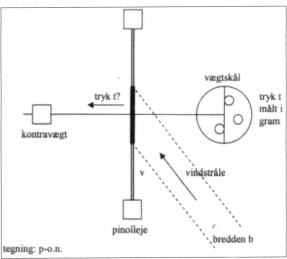

Figure 4.14   Two drawings showing the pressure of a flow from different angles on a plane surface: (above) la Cour's own description from his research report; (below) the authors interpretation.

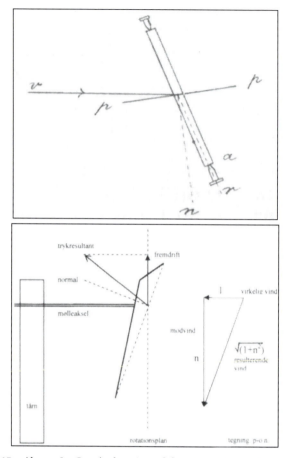

Figure 4.15 Above: La Cour's drawing of the pressure resultant lying over the axis from his research report. Below: The author's drawing showing the pressure resultant lying to the favourable side of the chord normal of a bent wing profile.

4.6 The "Ideal" Windmill

According to la Cour, the specifications of the ideal windmill are as follows:

(1) The resistant surfaces in the four sails should be as few as possible, especially near the tip of the sail.

(2) The width of the sail should be nearly 1/4 or 1/5 of the length of the sail, and almost same all over.

(3) The surface of the sail starts from the axis at a distance of about 1/4 of the sail length, so the surface is 3/4 of the sail length.

(4) The cross-section of the sail profile is not a straight line, but bent at a point, which is 1/4 or 1/6 from the fore edge and the fold is so strong that it is 3 to 4% of the chord of the profile, especially at the tip of the sail, however, near the axis it can be straight.

(5) The bevel angle is calculated from the profile chord, not from the large sail surface, nor from the small sail surface. It is 10 degrees at the sail tip and increases regularly so it is 15 degrees at 2/3 of the distance from the axis and 20 degrees 1/3 from the axis and would be 25 degrees at the axis.

(6) The mill should be arranged so that the tip of the sail runs with a velocity factor of 2.4 the velocity of the incoming wind, out of which it is desirable to extract the greatest possible power (it will most commonly be near to 6 m/s or maybe only 5 m/s).

(7) The work that such a mill produces can be calculated as the research shows, by the fact that there is 60 g/m^2 sail surface at a wind velocity of 1 m/s. The work will be a factor 60 of the total sail area in square metres multiplied by the wind velocity cubed and will be expressed in gram-metres. By dividing this number with 1 000 the value in kilogram-metre per second will be determined. Divided again by 75 will give the answer in horse power.

Example: If there are four sails, each of length 8 m and width 2 m, and the power at wind velocity of 6 m/s is to be determined, then first the tip velocity should be calculated, which will be 6 × 2.4 = 14.4 m/s. The circumference of the circle made by the wing tips will be about 50 m, so this mill will make (14.4 × 60)/50 = 17.3 turns/min. Its sail surface will be 4 × 6 × 2 = 48 m^2, and its power with the aforementioned wind velocity will be 60 × 48 × 6^3 = 622080 g-m = 622 kg-m = 8.3 hp.

After publishing his research report "Forsøgsmøllen I og II" (*The Research Mill I and II*) in 1900, la Cour initiated a contest urging millwrights to build the ideal mill. Some of them became very angry as they were convinced they were already doing that! Millwright Chr. Sørensen was especially angry, because his attractive conical rotor on the laboratory building had, in the meantime, been exchanged for an ordinary four-sail rotor. The problem was that the conical rotor in this large size had been too heavy and unstable, but Sorensen would not accept the removal and caused a lot of trouble in public, which finally involved the Danish parliament.

The end result was that in future the State Wind Laboratory should be a centre for information about the use and development of electricity in the rural areas of Denmark. A Danish Wind Electricity Society was formed and a consultant office was established. A magazine *Tidsskrift for Vind-Elektrisitet* was published four times a year. And finally, the training of the so-called "Country Electricians" (Landlige Elektrikere) began in 1904 and was repeated every year until 1916.

Figure 4.16 The design of an electricity-producing mill for farms recommended by la Cour.

One of the students, Johannes Juul, was only 17. But the number seven of his birth year 1887 looked very much like a four (1884) on his application, so the impression was given that he was old enough to participate. Of course this was discovered, and after an extra year of practice he also received his diploma. He carried out further studies and became high-voltage engineer, invented electric household equipment, and finally, in the 1950s, he developed the first three-sailed AC wind turbine, first the Bogø wind turbine in 1953 and later the Gedser turbine. These turbines became "mothers" of what is seen all over the world today, showing the influence of Denmark on the rest of the world.

Figure 4.17   Johannes Juul (left); The Gedser wind turbine (right).

4.7   Conclusions

Poul la Cour published "Forsøgsmøllen III and IV" in 1903, as a supplement to his report of 1900. He admitted that his expression "the ideal sail" was unfortunate, because it could be misunderstood as a claim to its invention, which would upset some people. The phrase was only used to point out the importance of the right shape and the right bevel angle. He also admitted that hundreds of years of millwrighting experiences provided sails that were close to the ideal, but seldom 100% correct. He invited mill owners to send him descriptions and dimensions of their mills and offered to advise them on any improvements. He also discussed the ratio between the length and width of the sails compared to the space between them. However, it was not until 1919 that a new formula was presented by Albert Betz. He showed that the power of a mill depends on the proportion of the circle area covered by the rotor—not the frontal area of the sails. Betz also calculated that the theoretical and practical limit would be 59% of the wind power available.

The contribution of Poul la Cour was an important step on the road to understand the force of vacuum lift on mill sails. This force was discovered on windmill sails 5–6 years before the Wright brothers started flying, although it is the same force that keeps an aeroplane in the air. As far as we know, the "mussel" diagram was constructed by Poul la Cour and thereby introduced the new concept of the tip velocity ratio, which is commonly used in descriptions of modern wind turbines.

Poul la Cour died from a lung infection on 24 April 1908. His contribution to knowledge in many fields was outstanding. He covered every step from fundamental theoretical to practical research, making inventions and implementing his ideas for the good of society. In addition he was the author of several textbooks such as *Historical Mathematics* and *Historical Physics*. Besides being a scientist, he was also a very religious man and felt no contradiction in that. His epitaph, carved in a memorial stone a few kilometres south of Askov, expresses that:

As Light from Heaven gives Light to our Eye.
Thoughts are inspired by Light from on High

About the author

 **Povl-Otto Nissen** has a master's degree in science education (cand.pæd. fysik) and was formerly senior lecturer at Ribe State College of Education, Denmark, from 1979 to 2000. He was principal of a Danish College for Adult Education (højskoleforstander) from 1975 to 1979. Since 2000 he has been chairman of The Friends of the Poul la Cour Museum, and heavily engaged in developing the content and activities at the museum. He has authored several papers on the history of technology and biographies of Poul la Cour and Conrad Wilhelm Röntgen.

Chapter 5

Networks of Wind Energy Enthusiasts and the Development of the "Danish Concept"

Katherine Dykes

Massachusetts Institute of Technology (MIT), 77 Massachusetts Avenue, Cambridge, MA 02139, USA

dykesk@alum.mit.edu

5.1 Introduction

Energy from the wind has played an important role in the productive activities of civilisations for nearly as long as historical records exist, perhaps even longer. Despite this long history, wind turbine technology used for the production of electricity still receives a great deal of attention as a "novel" energy source. In the most recent era of wind energy expansion beginning in the wake of the 1970s oil embargo, wind electricity–conversion systems (WECs)—modern wind turbines, became one of several technologies that provided hope for the development of a society powered by energy sources free from pollution.

However, support for the advance of the technology varied considerably among the different stakeholders from the public

Wind Power for the World: The Rise of Modern Wind Energy
Edited by Preben Maegaard, Anna Krenz and Wolfgang Palz
Copyright © 2013 Pan Stanford Publishing Pte. Ltd.
ISBN 978-981-4364-93-5 (Hardcover), 978-981-4364-94-2 (eBook)
www.panstanford.com

and entrepreneurs to the politicians, industry and utility companies. Even among the proponents and the "new" wind industry activists, there were competing interests concerning the direction of technology development and the associated political implications of those directions. Grassroots and counter-culture enthusiasts saw the development and adoption of small-scale wind turbines as a way of promoting their agenda of energy independence from large centrally controlled utility systems. On the other hand, political support for the technology was realised on a grander scale *via* industrial giants and associated scientific communities *via* the advance of large-scale wind turbines that could be co-located to create electricity generation facilities on the same scale in size as contemporary power plants. While the technological developments of such efforts were not mutually exclusive, they were sufficiently dissimilar to create a divide between the two groups. Rhetoric from both sides sought to undermine the initiatives of the other. Large-wind proponents made knowledge claims that small-wind could not achieve the performance needed for wind to be a significant contributor to various countries' energy independence. Small-wind proponents attacked large-wind programs as being overrun by hubris and wasting both monetary and human capital.

Several decades have passed and various accounts have now been written about both the large-scale federal and small-scale entrepreneurial wind development efforts from the 1970s to today. Such accounts have almost universally determined that the small-scale, predominantly Danish, wind energy entrepreneurs provided a superior avenue for innovation and proliferation of the technology to that provided by large-scale industrial efforts which were largely funded by federal government programs in the United States and elsewhere.[1] Such claims highlight that the big

---

[1]The development of WECs—modern wind turbines—have been treated with many analytical lenses—some popular accounts (Gipe, 1995; Maegaard 2009a), some interpretations by political theory (Van Est, 1999), several within the vein of innovation theory (Garud and Karnoe, 2003; Christensen, 2009), and still others within the field of history of technology (Heymann, 1995; Heymann, 1998; Serchuk, 1996). These latter accounts from the history of technology tradition focus on the implications of different social and cultural contexts on the development of the technology. The work of Matthias Heymann is a comprehensive work looking at the cross-country development of the technology with emphasis on federal programs in Germany, the United States

multi-million dollar federal wind programs had not achieved a significant commercial impact by the time that most such programs were cancelled when the oil crisis ended in the mid-1980s. At the same time, the bottom-up technology-development efforts of the small-scale wind energy entrepreneurs with an emphasis on the development of small and reliable machines succeeded in capturing the market which grew into the modern wind industry. These arguments use social and political causal relationships to explain the development of the technology and the industry. The arguments also tend to highlight the relative value of different instances of the technology itself as influential in the success or failure of the different efforts to develop a stable wind industry. Without explicitly recognising it, these analyses are thus contributing to a standing academic debate on the "nature" of technology and more specifically, the level of autonomy it has and the degree of influence it exerts on society.

This chapter[2] brings to the fore the role that assumptions about the nature of the technology play in the treatment of its history and also the intricacies of the relationships that exist across the different development efforts. To do this, a theoretical reference frame known as actor-network theory (ANT) is employed. Core to this theory is the idea of a dual-shaping of socio-political and technological worlds.[3] Thus, *via* ANT, a voice is

---

and Denmark, while the work of Adam Serchuk provides a detailed contrast of the activities of two groups in the United States: the federal wind program and small entrepreneurs. Similarly, the work of Garud and Karnoe also highlight the social and cultural aspects of wind energy development by contrasting the federal US programs (emphasising breakthrough) with the Danish entrepreneurial efforts (emphasising bricolage—improvisation and the search for "modest yet steady gains" in performance).

[2]This work is a condensed treatment of the history of wind energy technology using ANT methods. The full version is contained in a 2013 doctoral thesis by the author on the same subject. For more detail regarding the history, methods and related topics, the reader is referred to that body of work.

[3]While theories on society and technology are beyond the scope of this work, some treatment of critical concepts is useful to introduce the reader to the underlying motivation of the work and justify the use of the particular analytical lens of actor-network theory (ANT). Since Lewis Mumford published his work on *Technics and Civilization* in 1934, many scholars in the area of technology studies became involved in analysis of and debate over the nature of the relationship between society and technology. However, many still disagree on the extent and direction of causality between the technical and the social.

given to the wind turbine technology itself as a mediator of its own history. It becomes an "actor" within the various interwoven networks that embodies a sense of agency—the capacity to exert power or influence. ANT, then, seeks to be fair to all the different actors and networks in a history of technology development by providing to the extent possible a voice for each— for those who have traditionally been considered "successful", for those who have traditionally been considered "failures", and most importantly to the very technology itself.

Given the number of actor-networks relevant to the history of wind energy technology, this work will focus in particular on Danish actor-networks that formed during the mid-1970s and how the interactions within and between these networks with a particular focus on emergence of the "Danish Concept" for wind energy technology.[4]

5.2 Historical Wind Technology Development

In terms of the technological context, the form of current wind energy technology builds upon a history that began nearly as

---

On the one extreme are the "technological determinists" who see technology as an independent and autonomous entity inducing pressure to bring about changes in society. On the other, social constructionists view technology as a product constructed entirely by social processes. Between these two extremes fall theories such as "soft" technological determinism and systems theory, co-construction of technological systems, and ANT. The latter framework provides both human actors and technological actors with a voice and defines their relationships to one another through a network or even a network of networks. More importantly, in ANT, the "social" is not pre-supposed and the technology neither wholly determines nor is wholly determined by human or social forces. The main difference of this approach to other intermediate approaches is the denial of any stable social or technological structure that can serve to define a simplistic set of cause-and-effect relationships. Using such a lens, it can be argued that a richer history of the technology may be developed than one which is constantly trying to build and test hypothesis. An approach using ANT methodology would seek the social aspect from the human and non-human actor-networks that leave their historical traces as they progress through time. In essence, it is an inductive rather than a deductive approach, and as such, the identification of important relationships and directions of influence emerge out of the work and allow for a myriad of such relationships to be identified one might argue is more closely aligned with the actual complexity of the history.
[4]See the 2013 doctoral thesis by the author for the full treatment. This chapter cannot do justice to the entire landscape of Danish wind actor-networks. Other chapters in this book highlight additional networks not captured here.

early as historical records of human civilisation exist. The exact origin of the windmill, a wind-powered system used to grind agricultural products such as grain, is a subject of debate. The oldest windmills are thought to have been developed concurrently with the large agricultural systems of the ancient civilisations in the Near East.

Figure 5.1   Model of a "Persian Windmill" in the Deutsches Museum, Munich.

Evidence from records of the era as well as more modern instances of these windmills indicate that they were vertical-axis in orientation with sails spinning around a central shaft in order to turn a stone at the base that was used for grinding. Windmills with a horizontal-axis in orientation first appeared in historical records in 12th century England and nearby regions. No direct link between vertical-axis windmills of the Near East and horizontal-axis windmills of Western Europe has been proven though various theories have been proposed.

The most popular account is that soldiers earlier in the century observed windmills during the Crusades. Other accounts suggest the technology was developed independently as an extension of watermill technology that already existed in the regions. By whatever source of invention, the windmills of Western Europe took on their own distinct character in the form of the "post-mill", the horizontal-axis windmill attached to a post that encases the gears and axles for translating the horizontal to vertical motion for grinding. Initially, the entire structure of the post-mill had to be rotated into the wind for use.

Figure 5.2   Early image of a post-mill from "Schembartbuch", 1590–1640.

The technology evolved into the "tower-mill" and other variations which still used a horizontal-axis but only required rotation of the sails, wind-shaft and brake-wheel rather than the entire tower structure. Moving into the 18th and 19th centuries, various innovations in particular surrounding the control of these windmills were developed including mechanisms for orienting the mill with respect to the incident wind direction, braking mechanisms and regulation of power production. For example, the "yaw" mechanism for orienting the rotor in and out of the wind was automated using a fantail on the backside of the tower, oriented perpendicular to the rotor axis. For braking, governors dependent on centrifugal forces building up at higher wind speeds would engage automatic mechanical braking mechanisms or furling of the rotor blades. Having achieved such advancements, many elements of modern wind turbines already existed by the 19th century: upwind orientation, a small number of blades, a geared driveshaft connecting to a work machine, and automated control mechanisms as mentioned above.

Figure 5.3   Post-mill near Papenhorst, Hessenpark, Germany (Photo: Quartl) (left); Tissen tower-mill in Straelen-Herongen, Germany, converted to a house (Photo: nicKäm).

In adapting windmills to supply water across the Great Plains of the United States during the mid-19th century, various additional innovations were developed including sectional wheels (or wind rose configurations) as well as the all-metal windmill and a variety of additional control techniques. The prevalence of horizontal-axis wind rose windmills in the United States influenced the design of the first wind turbines in the country.

Figure 5.4   Wind wheel and water tank wagon at the National Ranching Heritage Center at Texas Technical University in Lubbock, USA.

The introduction of electricity-generation systems in the late 19th century along with increased use of electric appliances for the home and the workplace likely stimulated a number of independent attempts to convert windmills for use as electricity generators. However, none of these attempts in the Unites States were as large in scale, both in size and generation output, as that of inventor Charles Brush's who built his 12 kW "wind dynamo" in Cleveland, Ohio in 1888. The machine included a 60 ft. (18 m) tower with a wind rose rotor configuration of 56 ft (17 m) in diameter with 144 wooden blades. The DC electric generator designed specifically for the wind turbine was capable of providing 12 kW of power at a full load of 500 rpm. Word of Brush's dynamo spread across the country and even received coverage in the magazine *Scientific American*.

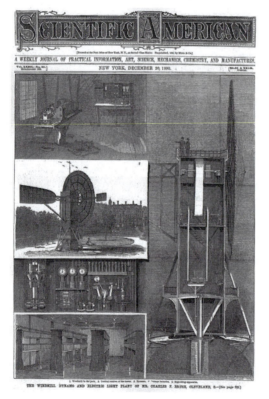

Figure 5.5   "Mr Brush's Windmill Dynamo" published in *Scientific American* (20.12.1890, 63, p. 54).

Smaller dynamos that could be integrated into existing windmill systems were developed in the early 1890s by the Lewis Electric Company of New York, but it was not until the invention of the propeller-based airplane and its development during WWI that the propeller-rotor horizontal-axis wind turbines was developed and inspired growth of a considerable-sized market. Perkins Corp., HEBCO, Jacobs WEC, and Wincharger Corp. were some of the more prominent manufacturers of these 2- to 4-bladed propeller rotor turbines in the United States.[5] Turbines descendant from American wind-pumps as well as new propeller–rotor types were developed in countries in Europe as well. Germany in particular reacted to the "coal crisis" that began with World War I with both state and independent initiatives to promote the adoption of wind energy. German Wind Turbine Works in Dresden produced American style wind machines for electricity production while the fundamental science of wind energy was pioneered by Albert Betz and colleagues at the Aerodynamic Research Institute (AVA) in Göttingen. Similar efforts involving both state-funded science programs and industrial marketing of small wind turbines existed in other European countries as well through the first half of the century including the United Kingdom and Denmark.[6] Despite the global efforts for the development and promotion of the technology, however, the expansion of centralised electricity networks into rural areas all but eliminated the markets for the technology by the mid-20th century.

Still, motivated in large part by World Wars I and II as well as general developments in the science and technology of aviation, there were a few notable attempts at large-scale wind turbine development in the first half of the 20th century. Such attempts sought to integrate "fuel-free" wind energy with modern large-scale electricity networks. The Smith-Putnam 1 250 kW wind turbine was developed as a collaborative and experimental initiative privately financed by the S. Morgan Smith Company. Palmer Putnam, an MIT graduate and engineer, originally conceived of the idea to build a wind turbine as a way of reducing his electricity bill and as the idea was developed, the scale of

---

[5]For a thorough treatment on the development of wind turbine science and technology during the first half of the 20th century in the United States, the reader is referred to R. Righter's 1996 work.

[6]See chapter *The Aerodynamic Research on Windmill Sails of Poul la Cour, 1896–1900* by Povl-Otto Nissen.

electricity needed to become independent from the grid led to the design size of a 53 m rotor diameter 2-bladed horizontal-axis turbine with a rated power of 1 250 kW. The Putnam turbine was built from 1940 to 1941 and operated for four years until a blade sheared off during operation in March 1945.[7]

Figure 5.6   Smith–Putnam wind turbine, the world's first MW-size wind turbine in 1941, installed in Grandpa's Knob in Castleton, Vermont, USA (Photos: NREL/DOE).

Other nations similarly invested in the development of large-scale experimental turbines. Turbines of sizes greater than 100 kW were developed in the 1950s and 1960s in England, Germany, Denmark, France and Russia. In particular, Ulrich Hütter, a member of the Stuttgart University's German Air and Space Laboratory, became involved with both theoretical and experimental work to pursue the development of wind energy technology based on science. His notable practical experience included a 10 kW turbine that was built and ran from 1950 to 1960 and a subsequent 100 kW turbine which ran intermittently from 1961 to 1966. His work on the aerodynamics and blade design for wind energy would become an important aspect of the worldwide wind energy technology development in the 1970s including the Danish history to be discussed.

Despite the efforts of Hütter, Putnam and others, in particular Johannes Juul who will be discussed later, the use of wind turbines for commercial electricity generation was not fully

---

[7]The account of Putnam's turbine is provided in detail in his book *Power from the Wind* (1948).

realised and it would take another crisis, in the form of the OPEC oil crisis, to spur the modern era of wind energy development and the creation of a global market for the technology that would persist for several decades.

5.3 Wind Actor-Networks in Denmark

Moving to the modern era of wind turbine development, the scope of this work will be limited more-or-less to Danish actor networks. Extending the above discussion of 20th century wind energy technology history to Denmark, several notable achievements deserve mention.

Wind turbine history in early 20th century Denmark is often tied to a social movement associated with larger political developments in the country stemming from the mid-19th century. Following the Napoleonic wars and the two wars of Schleswig, Denmark had to cede control of large areas of land including Norway, Schleswig and Holstein. The impact of the wars resulted not just in a loss of fertile lands within its control, but also resulted in a larger national crisis in terms of political stability, economic growth and the sense of national identity. This crisis sparked a national movement that included the formation of various "folkehoejskole", or folk high schools. The schools were inspired by Danish philosopher Nikolaj Grundtvig and were meant to serve in contrast to more conservative ideals of education. The folk high schools, found throughout rural Denmark, emphasised practical knowledge and enlightenment especially for rural agrarian citizens and gave origin to three very important developments for Danish wind energy technology.

Figure 5.7   Portrait of Nikolaj F. S. Grundtvig (1783–1872) by Constantin Hansen, 1862.

In the late 1800s, Poul la Cour,[8] a folk high school teacher of mathematics and science, began to consider that the rural population would benefit from electric power generation. He began experimenting with wind energy as a way to drive an electric generator. These first Danish WECs typically featured four or more blades, sometimes slatted, and typically in a "smock-mill" configuration where the driveshaft was used to drive a DC generator rather than pumps or milling equipment. The Askov folk high school of Poul la Cour would serve as a training ground for various early "wind engineers" Denmark and one notable pupil that deserves further mention.

Figure 5.8   Poul la Cour (left); la Cour-type windmill, Denmark.

Johannes Juul, who may be seen as the "godfather" of Danish wind industry was one of la Cour's pupils before he would go on to eventually design the famous 200 kW Gedser turbine. This design, a third in a series that Juul had developed, featured an upwind 3-blade configuration, a gearbox connected to an asynchronous generator, stall regulation, an active yaw system, and both mechanical brakes and tip brakes. This very design, with enhancements in the blade design and other features, is what we now commonly associate with the "Danish Concept" of wind energy which has become a dominant design for wind energy that persists to the present day. Though Juul's prototype design constructed in 1957 was never turned into an effort of mass

[8]See chapter *The Aerodynamic Research on Windmill Sails of Poul la Cour, 1896–1900* by Povl-Otto Nissen.

production, the machine would run for more than 10 years without a major failure and thus would serve as an example of wind energy technology that produced a significant amount of energy both reliably and safely. In general, the line from la Cour to Juul laid a foundation of knowledge that would form the basis for the post oil-crisis movement to establish Danish energy independence through wind energy.

Figure 5.9   Johannes Juul (1887–1969) (left); The Gedser wind turbine (right).

5.3.1   The Founding of NIVE and Its Early Activities[9]

Just as with the rest of the industrialised world, Denmark experienced a crisis in response to the OPEC oil embargo of 1973 and again in 1979. However, Denmark, which relied on oil for 92% of its energy consumption and possessed few indigenous fuel resources, perhaps suffered even more greatly than other nations. The crisis that began in 1973 led directly to the formation of a small network of actors in Western Jutland of Denmark who were motivated to find alternative energy sources for electricity and heating. Preben Maegaard, one of the central actors in the Danish wind movement, describes how the crisis catalysed the formation of that actor-network:

---

[9]This work again focuses on a few key actor-networks within the history of Danish wind energy development and the "Danish concept". Other key networks deserve mention and their histories are included in other chapters. See *History of Danish Wind Power* by Benny Christensen and From *Energy Crisis to Industrial Adventure: A Chronicle* by Preben Maegaard.

"That team you see—it was actually founded in January of 1974. This was just when the oil crisis was at its highest. It was winter and there was a real concern in this country on how we could get through the winter because people were relying on heating their houses by using oil. They had all thrown out their old stoves, and we lived in modern times where oil was available, and it was a complete shock that this supply of oil was suddenly interrupted. And when you live in a cold climate here where we have these cold winters, we really feared to freeze (...) The Minister of Trade appeared Saturday evenings in primetime and reported on the supplies of oil and how much was in storage (...) he would tell people to go to the forest and collect some wood, and he would say you should close some of your rooms (...) and only use one room to save energy." (Maegaard, 2010)

The team that formed in 1974 referred to above was "NIVE", the North-Western Jutland Institute for Renewable Energy.[10] It consisted of a few engineers, blacksmiths, and teachers from a local technical school. The focus of NIVE was not exclusive to wind energy by any means and in fact, their initial projects focused on other technologies with particular emphasis on biogas and a few projects in solar energy. Their aim was "optimal use of local human and technological resources" (Maegaard, 2010). The group had received a grant of DKK 50 000 from UNESCO for which it was trying to establish a use in a meeting held at a local Ecumenical College on 12 January 1974. Despite the energy crisis, the college leadership was hoping to focus on spiritual development, however, Maegaard and others were more concerned about the need to search for immediate, pragmatic, and concrete solutions to address the crisis. There was a farmer, Poul Overgaart, at the meeting, who wanted support in developing a biogas plant and he had gone to the technical university and consulted with other experts in search for information. He could not find anyone to provide him with the needed practical information for designing and constructing such a facility, and he recommended the group to use part of the grant to bring experts to Denmark who might be able to provide the needed information. The outcome was that an organisation was needed that could coordinate the access and development of practical information on alternative

---

[10]Various interviews with Preben Maegaard informed the history of the formation of NIVE provided in this section.

energy solutions. During the next few years, the NIVE actor-network became involved with a series of projects to develop and provide information on design for small-scale distributed energy technology including biogas, solar thermal and very soon after, wind turbines.

As part of the biogas development program, Preben Maegaard had gone to the libraries in Copenhagen to search for any useful information on biogas technology as well as to develop a contact list of experts to invite to meetings at the Ecumenical College. In that process, he came across the proceedings from the UN 1961 conference on new energy sources.

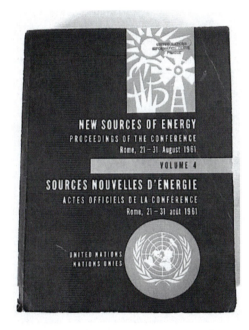

Figure 5.10   Cover of the *Proceedings of the United Nations Conference on New Sources of Energy: Solar Energy, Wind Power and Geothermal Energy*, Rome, 21–31 August 1961.

There were several volumes in the set, including one on biogas but also one on a workshop dedicated to wind energy which featured articles and presentations both by Johannes Juul and Ulrich Hütter.

Maegaard bought the whole set at the local bookstore. The workshop proceedings on wind energy would eventually become

NIVE's "bible" on wind energy development (Maegaard, 2010). Very little was published at the time on wind energy technology and system design, but in his search Maegaard managed to find a book *Vindkraftboken on wind turbine design,* by Swedish engineer/scientist Bengt Södergaard.

Back in Thisted, Maegaard made two early attempts during 1975–1976 to construct a turbine based largely on the Swedish book's design of a 2-bladed downwind machine with free yaw, but he found it to be a "paper tiger" that would become so powerful and unstable in high winds that he needed a tractor to yaw it out of the wind so that the machine would stop. After one scary experience with the machine, he decided to not have it run again. Instead, a new turbine was made with three blades that had sail wings upon the suggestion of another NIVE actor, Henrik Stiesdal,[11] who would eventually become an important wind technology engineer at Bonus and later at Siemens Corporation. The importance of safety and reliability in the design was impressed upon the NIVE team from these early experiences.

Experiments with wind energy technology at NIVE continued and these early activities would provide the actor-network at NIVE with the foundation for future work on wind energy development. In late 1976, NIVE actor Bjørn Rossing, a relatively wealthy man from Copenhagen who had decided to retire in Western Jutland, had the idea of purchasing one of the new Riisager turbines.[12] The carpenter Christian Riisager had recently become successful selling a 22 kW wind turbine with wooden blades. Unfortunately, his success meant that the demand for his turbines quickly outpaced supply and the price tripled from its initial costs. As a result, Rossing approached the NIVE group with the idea that they could make such a turbine themselves and even improve upon it. Two of the NIVE members, Ian Jordan and Preben Maegaard, set forth on the task of designing the blades for the turbine. This activity set in motion important developments for the NIVE group that would play a significant role in the expansion of the overall wind energy actor-network in Denmark. In addition, the group also would rely on recent developments of a related actor-network near Ulfborg, Denmark at the Tvind Schools.

---

[11]See chapter *From Herborg Blacksmith to Vestas* by Henrik Stiesdal.

[12]See chapter *From Energy Crisis to Industrial Adventure: A Chronicle* by Preben Maegaard.

5.3.2 Anti-Nuclear Movement, OVE and the Tvind Turbine

The unique developments at the Tvind folk high school stand out as inspiration to community activists across the world and have resulted in numerous awards for their grassroots design, construction and operation of a 2 MW wind turbine that still operates even today. The development of this turbine, however, was grounded in the cultural and political context of the period and this is where its story begins.

Since the 1950s, Denmark had an on-going discussion over the potential use of nuclear power as a source of electricity for the country. The government founded the laboratory Risø in Roskilde with a focus on research for nuclear power, but the initiative quickly met with community resistance led primarily by the Organisationen til Oplysning om Atomkraft, OOA (The Organisation for Information on Nuclear Power). The organisation of activists had local chapters across the country that would distribute information and lead opposition to nuclear development. The OOA's slogans included things like "Atomkraft? Nej tak!" ("Nuclear power? No thanks!"), and "Hvad skal vaek? Barsebäck! Hvad skal ind? Sol og vind" ("What shall be gone? Barsebäck![13] What shall come in? Sun and wind!") and these were chanted time and again by thousands of activists (Maegaard, 2010).

Figure 5.11   OOA's logo, Smiling Sun, and slogan "Nuclear Power? No thanks!" (© OOA Fonden/smilingsun.org).

---

[13]Barsebäck—a Swedish nuclear facility near the border of Denmark.

However, it became clear to the leaders of OOA that saying "no" to nuclear was not the same as saying "yes" to the alternative (Maegaard, 2010). They decided on 2 February 1975 at a nation-wide meeting in Bryrup, Jutland, to form a new organisation. Lars Albertsen, who would be a key figure in the policy-space for wind energy, would set forth a new term "vedvarende energi", or sustainable energy, to describe the goal of the new organisation.[14] This new organisation, OVE (the Organisation for Renewable Energy) was initially organised similarly to the OOA—disparate local networks of activists led meetings, called "VindTræf", or sit-ins, to discuss and share information about renewable energy—primarily wind energy (Maegaard, 2010). Members of both the NIVE and Tvind networks attended and contributed to the various OVE meetings. Maegaard recalls that he went to almost all OVE meetings across the country which included two to four a year from 1975 to 1980 (Maegaard, 2010). Beyond activism, the OVE openly promoted technology development and information sharing. Thus, the earlier network that had brought different actors together around the theme of anti-nuclear power was an important catalyst in the development of the OVE network which would become a critical component of the wind energy development effort in Denmark.

One prominent group of soon-to-be OVE members came from the Tvind folk high school of Western Jutland. Rising out of the folk high school tradition in Denmark, the Tvind School was actually a hub for distributed activity of the so-called "travelling high schools" that began in 1970. The schools' principals involved a few key tenants such as the world as a classroom, the integration of practice and academics, and students as the drivers for teaching and learning.[15] The teachers founded the school from their common savings and everything became shared resources both from an economy and a time standpoint. The school also emphasised solidarity with those in less-developed communities and part of the studies included bus caravans to Turkey, Iran, Afghanistan and finally Pakistan and India where students and teachers could experience first-hand working with community

---

[14]See the "Winds of Change" Web site for more detail on the organisation's history (OOA, 2009; OVE, 2009).

[15]Much of the history of the Tvind experience is based on interviews with Britta Jensen of Tvind, December 2010, as well as the Maegaard's (2009b) article and the Tvind Web site.

development. Everything at Tvind was done as a community: teachers took responsibility of different project areas but all were linked together in a common effort and involved in all aspects of school life. The site of the Tvind wind turbine today is on the school grounds near where a campus was first developed in 1972.

Figure 5.12   Tvind turbine today (left); View from the turbine on Tvind School (right) (Photos: Folkecenter).

Shortly thereafter in 1974, the oil crisis would affect the young school and its limited financial resources. The collective teacher community decided that they would need to develop some in-house energy sources and considered both solar and wind as potential options; in the end, the strong wind resource in the area would persuade the group to develop a wind turbine. Indeed, today you can see the trees all over the local area which have been permanently swept to side by the consistently strong winds in the area.

The group at Tvind included some teachers like Amdi Petersen who were strongly against nuclear power (he had even been arrested in Germany during protests against nuclear energy in the 1950s). Nuclear power enthusiasts of the day claimed that wind turbines that were currently under development were so small that they would never have a substantial impact on mitigating Danish dependence on oil, and so the Tvind high school decided to go big, 2 MW, in order to show Denmark the potential of wind energy. Led by the Tvind high school teachers, a group of students, interested community members, engineers and affiliates from universities, institutes and industry came together to build the Tvind 2 MW wind turbine using a 3-blade downwind configuration, variable speed operation, an asynchronous generator, full pitch control, an active yaw system, and a fully-rated converter for grid interconnection—all on top of a concrete cylindrical tower and nacelle (Tvind, 2010).

The group was inspired by the earlier Gedser turbine of Johannes Juul, but more so, by the design and work of Professor Ulrich Hütter of the University of Stuttgart. The group reached out to Hütter, as a consultant, and even made a trip to Germany at one point to gather his input on their design. Hütter's unique blade design allowed for the fibreglass strands of the blade root to be wound around the bolt holes so that the entire piece could be connected easily to the hub (a task that had been accomplished on the Gedser turbine by a set of supporting cables).

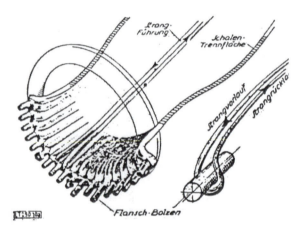

Figure 5.13   Ulrich Hütter's blade design with fibreglass strands twisted around blade root bolts, drawing from 1960.

Figure 5.14   Making blades at Tvind (Photos: Tvind).

Aside from Hütter, the various consultants on the project included a number of future wind industry actors and the project played a key role in influencing the Danish popular imagination about the potential of wind power. The overall design process was truly collective and the Tvind design is truly unique. There were a few people that persisted in their involvement throughout the whole project: Amdi Petersen who provided much vision and initiative for the project, Jens Gjerding from Tvind's Vestjysk Energikontor collected information and pulled the system design together, Hans Jørgen Lundgaard Laursen was an advisory engineer on the structural dynamics, Lars Svanborg served as supervising engineer concerning the machinery, and the chief welder and teacher Henning Jønsson.

However, many individuals contributed to the overall project design. A team of up to 21 students were actively working on the turbine development at any given time and they held regular meetings where the two or three engineers hired by the school for the project would provide different potential solutions for design aspects of the turbine and these would then be discussed at length until a unanimous decision was reached on a final turbine design (Jensen, 2010).

Figure 5.15 Meeting at Tvind to discuss the design and construction process (Photo: Tvind).

The school also relied on minimal funds for the overall project and made the components themselves whenever possible. For the tower, nacelle and foundation, they poured their own concrete after acquiring an elevator and concrete mixer at heavily discounted prices—this involved both developing the correct concrete mixture as well as performing the complete assembly task. They also made their own blades by first learning from a local carpenter/boat builder how fibreglass structures were manufactured and then applying the techniques they learned to create their own set of blades.

Figure 5.16   Collective work on foundations for the Tvind wind turbine (Photo: Tvind).

They designed their own active yaw system which included a unique design of two reciprocating 5-toothed arms that would work the turbine around its vertical axis. The pitch system was equally unique since as a full-bladed pitch system for variable speed operation, the technology was ahead of its time, and also the pitch system provided the entire braking mechanism for the turbine and thus included three redundant systems (two collective and one independently enabled actuator set). Finally, the decision to use variable speed technology was due in part to the large size of the turbine and the concern over connecting such a large system to the grid without fully rated power conversion electronics. Soon into operation, the Tvind group found that operating the turbine at higher speeds would tend to induce dangerous harmonics in the tower and a lower rating was found suitable for safe operation and long-term reliability.

Figure 5.17   Tvind wind turbine, 1970s (Photo: Tvind).

Countless people were involved in the design and manufacture of the successful Tvind turbine and the loosely connected actor-network surrounding the turbine's development would soon begin to develop their own technology and set in motion the early development of the Danish wind industry. Thirty years later, the Tvind turbine, having been designed with relatively conservative design principles and limited to operation well below the original rating, still operates and generates about 1 MW of power to provide electricity to the facilities at the school.

Excerpts of original texts from posters describing the process of building of the Tvind turbine, 1978. Posters are located in Rotunde, beneath the tower of the Tvind windmill.

- At Tvind in Western Jutland a number of workers, teachers and students are building their own power station. With their own money. They build it together, for the sake of natural energy, for the sake of a human society—and against slavery, against monopolisation and against nuclear power.
- The hole was dug at first. The iron was bound afterwards. The concrete mixed and poured around the iron. The base of the tower completed slowly. And then one day we glided upwards leaving the tower under us. Metre by metre, hour by hour, and after 22 days at the very top. So this was the position of the blades. At this height, well, well. From here they were to go round and round.
- We had heard about experiences like these from a number of countries all over the world, such as China, but now we learnt from our own experience that it was true: People can do anything when they unite about their common future. No problem, no obstacles are too big for the united people.
- Difficulties have turned up in scores. From the very beginning we could count on them. Ever since the cutting of the first turf, where we 400 people were digging together, we have met problems. And the joy of solving them has all the time been greater than the day before. By now we are really wild about them. Because we have discovered that it is through the solution of such difficulties we move forward.
- Ourselves from the lack of confidence that common people like us could do things like building a big wind power station, from the fear of the materials, from the ignorance of the physical laws and the actual size of the natural forces, from the ignorance of even elementary concepts of the technological world, this disintegrated, fragmented expert world, which no one today can overlook, from the traditional view on girls' ability to build anything.
- Therefore we were sometimes scared in the beginning of our work building the power station. As when we first found out that we had made wrong calculations. Twice as much concrete as we had thought necessary at first was needed. So we could not afford to buy ready-mixed concrete from the nearest mixing plant. We solved it together.
- At a meeting we found out that we could mix the concrete ourselves. At half the price. But then it would also take twice the time. Unless we

ourselves had a large concrete mixer. So we found one at a very cheap price, because the economy of our society is in a crisis—therefore the means of production are for sale cheap.

- While F. L. Smidth's concrete experts advised us not to try to pour the tower without their expensive help, their so-called "wind power expert" abused us on the phone. He called us "You who are destroying the whole wind power cause by trying to build a windmill". We answered back by adding seven people to the windmill group.

- While boards of one foundation after the other, governmental or private, pried into our applications for financial support for our work to find a tiny reason for saying no, no and no again, common people from the whole country flocked to the building site to look, to talk, to tell us their thoughts about energy.

- We continue building until the end, until the beginning. The blades have started to go round, the production of electricity increases with every month while neighbours gather and ask questions and get answers. And tea.

- And while the Danish Television Company time and again gave broadcasting time to our Minister of Commerce revealing the blessings of nuclear power, at least 500 people a week confided to the mill builders and the students and pupils at the schools that all people are for wind power—and that it is only "the high-ups" who want nuclear power.

- "It's the Government and the whole industry who want all this nuclear business—because they will profit from it. That is why". And so we continued building, supported by the common opinion. By the opinion of the many Danes.

- Nature is represented by the wind. The forces of society are planning nuclear power plants, motorways, military bases and a lot of other crap around us. While we were binding the first iron, the Industrial Board were discussing where to place Denmark's first nuclear power plant. We entered the debate, in writing, in speech and continued iron binding.

- The government is preparing a proposal for nuclear power plants. It is to be put before the "Folketinget", the Danish parliament soon. Maybe we cannot stop them doing it. But we and many people are learning that together we can change anything we want to. Even if the next tower has to be built into the skies.

Figure 5.18   For their achievements Tvind team received European Solar Prize in 2009; to the right—Hermann Scheer, EUROSOLAR, Berlin.

5.3.3   From Tvind, NIVE and Others to the Development of the "Danish Concept"

When researching wind energy technology innovation and design, the term the "Danish Concept" is used frequently. For some, this may simply mean an upwind turbine—facing into the wind—with three blades turning around a horizontal-axis. This definition can be expanded to encompass a number of other characteristics of early Danish wind turbines such as stall regulation, a gearbox connected to an induction generator, and even the fibreglass material for the blades and independent braking systems. Finally, one may add to the "Danish Concept" not just subsystem characteristics but overall design principles for the technology such over-designing for cases of uncertainty and the use of certain historical precedent from the Gedser wind turbine.[16] For many, the idea of the "Danish Concept" would invoke images from single or grouped wind turbines powering rural farms and

---

[16]While these "design principles" are not explicitly mentioned as part of the "Danish Concept", they certainly embody the concept in the abstract.

communities in Denmark to vast wind farms at Altamont, Tehachapi, and San Gorgonio passes in California, in the United States, where many so-called "Danish clones" (from various companies but all looking nearly identical to the naked eye) would be seen almost littered across the hilly landscapes. This "Danish Concept" was born out of the recently formed wind energy actor-network in western Denmark which combined the smaller actor-networks of Tvind, NIVE and the OVE as well as other groups not mentioned here, including the Riisager wind turbine community and others.

Returning to Rossing's suggestion that NIVE build a wind turbine to alleviate the shortage in supply of Riisager turbines, Maegaard and his team revisited the UN report and their experience with the Tvind turbine development. This report, with its articles by Morrison, Juul and Hütter, would be used by NIVE as their main source of guidance for designing their Gedser-inspired turbine rated at 22 kW, similar to the popular Riisager turbines. Just like the Tvind turbine, the design featured 3-blades made of fibreglass material winding around the bolt holes at the blade roots for a good connection to the hub. Maegaard, in particular, had been somewhat involved with the Tvind wind turbine—he had consulted with them on different aspects of the design and accompanied Amdi Petersen to Sweden to acquire what would be the Tvind drive train: An ASEA generator and gearbox that was a spare from one of their industrial facilities. The Tvind team decided to use a downwind turbine configuration because of the advantage for the use of a more stable yaw system and the coning of the blades away from the tower under thrust loads—both features that the Tvind team believed would make the turbine more reliable (Jensen, 2010). Maegaard, who had read in detail the 1961 UN report from the conference on new energy sources, had noted in an article on wind testing that the author Morrison had found that tower shadow induced cyclic loading on the blades which could affect the reliability of downwind turbines. He shared this with the Tvind team though they had already committed to a downwind design (Maegaard, 2010). For the NIVE design, however, an upwind configuration was deemed preferable. Thus, the design featured the basic characteristics of the Gedser and Riisager turbines with the idea the incorporation of the new blade design based on the Tvind turbine.

Figure 5.19   Smedemester/NIVE 22 kW wind turbine.

Many attributes of the Gedser, aside from the basic configuration, were also used in the 22 kW design including the asynchronous generator, electrical yawing and automatic control system. To build the turbine locally, the NIVE team saw that the critical component would be the blades. Each component was readily available, could be ordered, or could be easily constructed by local companies and blacksmiths except for the blade and the control system. Maegaard worked with a local controls engineer from Thisted, Finn Bendix, to design the system of relay controls for start-up, shut-down, grid connection and so on, and at the same time, his team began to construct a mould for the 5 m blade which was based on the Gedser blades more so than the more aerodynamically optimised Tvind blade design.

Before the group at NIVE had a chance to complete the model, the team came into contact with Erik Grove-Nielsen of Økær in Viborg, also in Jutland. Erik had realised from early OVE and NOAH meetings that wind turbine self-builders had problems with creating the blades.[17] In 1977, Erik had attempted to make a smaller 11 kW version (4.5 m blade) of the Tvind windmill. He

---

[17]See chapter *Økær Vind Energi—Standard Blades for the Early Wind Industry* by Erik Grove-Nielsen.

then offered to make the blades for NIVE himself given his prior experience using the Tvind based design. From his own experience with Tvind, however, Maegaard knew that those blades were designed to run at a fast tip-speed ratio (when sized down from the Tvind) and thus would likely be too noisy. Erik Grove-Nielsen then agreed to produce a modified design for NIVE which featured blade dimensions similar to that of the Gedser turbine but which incorporated Hütter's modern blade root design compatible with fibreglass materials.

The rest of the design was thus inspired by Gedser from the basic configuration, to the sizing of the various components for expected loads. From the 1961 UN report, various rules of thumbs were incorporated such as a degree of loading on the rotor that the structure would have to withstand, 30 kg/m^2 swept area, or the size of the main shaft, 1% of the diameter of the rotor (Maegaard, 2010). Selection of the rest of the components and construction of the first NIVE wind turbine in 1978 involved a highly collaborative process between the group at NIVE and various local companies, blacksmiths and engineers. Maegaard himself became very interested in the component selection process and his team would call up companies across Denmark to order catalogues for selecting the turbines components:

> "We took the bearings from the catalogues here"—there are a few large cabinets in the Folkecenter which hold the catalogues with notes used in the early design work—"... and then you needed a yaw ring... this was more difficult to find... you had to think about where you used such a ring, and they were used on cranes and on some vehicles... and then you started calling [the yaw ring manufacturers]. These companies, they had never heard of wind power. You called them and they said wind power—what is that? And then you got the catalogues... and it was also a problem, who you talked to about wind power.... The company would say wind power? We have heard about that, does it mean that you are some that are against nuclear power? Ah, then they would say, well, we have some policy... we will not deal with that because we don't believe in this wind power. Now this doesn't matter for selling some bearings; you sell to many other sectors also. We just want your product, whether it's for wind power it makes no difference... but often you could not get them to send the catalogue once they found it was for wind power... and years later the same company, I would talk to them, and they would say we made a big BIG mistake..." (Maegaard, 2010)

And through this process of design work and construction of a set of component suppliers, various industrial actors began to engage with the wind energy technology development. While many parts were supplied by the local blacksmith community and masons, various parts were outsourced including the yaw ring, gearbox, generator, brake callipers, bearings, and most importantly, the turbine blades. This in turn led to the creation of a backbone supply chain for what would become the Danish wind industry—some of whom were already involved in the burgeoning wind industry through supplying parts for Riisager-style turbines, and some of whom were brand new to the industry. The idea of component supply was one of the fundamental aspects of the wind turbine. The ethos of the local small- and medium-sized enterprises (SMEs) in Denmark is important to the history. Unlike the United States and much of the rest of Europe which had fully embraced the American system and mass production, Denmark in the mid-20th century still had a strong local production base including a network of local masons, carpenters, painters and blacksmiths. The idea in the development of the NIVE turbine was to engage with the local production network to the extent possible for the development of the wind turbine.

5.3.4 Growing the Danish Wind Networks and Industry

Thus, the early windmill development involved local production and engagement from blacksmiths. This eventually led to the engagement of the Danish Blacksmiths Association (DS). As mentioned before, local wind turbine manufacturers met regularly at the OVE "VindTræf" and other local meetings to discuss best practice for wind turbine design from their field experience.

It was at the fifth of these OVE "VindTræf" meetings in Brandbjerg that two gentlemen from DS approached Preben Maegaard and were interested in how they could turn wind energy into an economic boon for their members. This led to the establishment of a formal relationship where NIVE would provide a series of standard design manuals to DS that covered every aspect of the design of a modern wind turbine. Over the years, the design team at NIVE and later Folkecenter created a design handbook accessible to anyone and based on the "Danish Concept". It began with the early 22 kW and 55 kW designs and extended to a full design series including DANmark 8 (13 kW), DANmark 11 (22 kW), DANmark 17 (75 kW), DANmark 19 (95 kW), DANmark 22 (150 kW), and the DANmark 25 (200–270 kW) (FC, 1990).

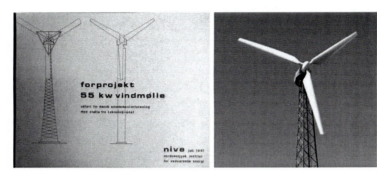

Figure 5.20 Project cover for the 55 kW DANmark wind turbine, NIVE, 1981 (left); DANmark 17 at the Folkecenter's Wind Test Field (right) (Photo: Folkecenter).

From this beginning, all of the major Danish wind turbine manufacturers of the 1980s through today can trace their origins to this "Danish Concept" established at NIVE along with the Riisager turbine development and based on the combination of Tvind and Gedser turbine design concepts. The early companies, all working with Økær, took advantage of the developing supply chain and leveraged the design basis that had been established. Each new company, however, brought to the table its own area of expertise to improve component design and contribute to the overall Danish wind turbine "supply chain" (Maegaard, 2009a; Maegaard, 2010). For instance, one oversight of the first turbine designs had to do with the omission of the blade tip brakes, which had been present on the Gedser turbine in the 1950s. As Maegaard noted, for the Tvind tower, they also omitted tip brakes and used parachutes brakes in addition to blade pitch for an emergency stop mechanism—which also made an upwind configuration prohibitive because for an upwind system, the parachutes would hit the tower. The braking system was one of the more challenging aspects of the first system and in Maegaard's own words, the tip brakes "in the first [turbines] were not considered. They were neglected. I can tell you they were really neglected... we did not read Juul carefully enough!" (Maegaard, 2010).

However, two over speeding situations where the mechanical braking system failed brought the issue back to the fore. The Herborg (later Vestas) turbine introduced by Karl Erik Jørgensen and Henrik Stiesdal[18] in 1978 would also use the same basic 22 kW blade design from Økær, but only a few months after installation, the prototype ran into an over speed situation and the blade and hub system collapsed to the ground.[19] A second failure occurring a few weeks later caused Økær to stop production and focus on a braking solution that would be stable—eventually this resulted in the creation of a tip braking system similar to that of the Gedser turbine. The system on the Gedser turbine was based on hydraulics but Økær used a spring loaded system which would activate whenever the turbine lost power connection from the grid and the mechanical brake failed.

Figure 5.21 Hütter-type flange of 5 m Økær blade made interfacing to the hub uncomplicated (left); tip brake (right).

Another innovation of the Herborg turbine included the use of two differently sized asynchronous generators (at a lower and higher power rating) in order to improve overall turbine performance. This as well became part of the *de facto* Danish wind turbine standard and the "Danish Concept" of the 1980s. In 1979, the Herborg design was licensed to Vestas which was at that time a medium-sized enterprise that made former equipment truck cranes and heat exchangers—the company that would later become the largest wind turbine manufacturer in the world.

The stories of other Danish manufacturers are similar.[20] All of the manufactures were originally focused on agricultural or

---

[18]See chapter *From Herborg Blacksmith to Vestas* by Henrik Stiesdal.

[19]See chapter *Økær Vind Energi—Standard Blades for the Early Wind Industry* by Erik Grove-Nielsen.

[20]NIVE and the extended wind industry network provided the basis on which initial designs were selected by several companies (Maegaard, 2010).

transportation machinery. Nordtank, originally a tank producer, started as well with the 22 kW turbine and the same blade profile but used its expertise in rolling and welding large tank sections in order to produce a "welded tube" tower which eventually replaced the lattice towers of NIVE and Herborg designs to become part of the standard design configuration. The last major company, Danregn Vindkraft A/S or Bonus Energy A/S,[21] originally tried to license the Nordtank technology but then developed their own design with Maegaard serving as consultant to the firm as well as a board member for a time in the early years. From these origins, all linked to NIVE, Tvind, OVE and Riisager, the "Danish Concept" became pervasive in early Danish wind manufacturing and eventually obtained its status as the dominant design within the industry. Even Tacke, a German wind turbine manufacturer, has its origins in a jointly funded program by the EU for the creating a new 400 kW generator. In 1988, both Nordtank and Folkecenter submitted proposals but Nordtank won the contract due to their manufacturing base. However, when Nordtank went bankrupt, the project was turned over to NIVE's descendant, the Nordic Folkecenter, provided that they work with a manufacturing firm, the German company Tacke, with the project cost share of 2/3 and 1/3, respectively. From this project, Tacke moved into the wind space and was eventually acquired by Enron Wind and then by GE Wind, when Enron went bankrupt.

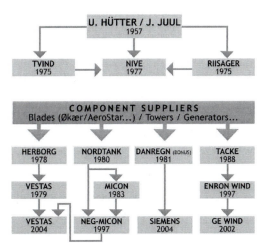

Figure 5.22   Danish wind power evolution.

---

[21]See chapter *From Danregn to Bonus* by Egon Kristensen.

5.3.5 The Influence of Uncertainty on the "Danish Concept"

Development of wind turbines touches on almost every area of natural science and engineering. From the advanced composite materials used in rotor blades, and the characterisation of a turbulent and even chaotic wind resource, to the aerodynamics associated with the rotor, to the design of the electric generator and interface with the electric grid, to the dynamic loads associated with the overall structure, wind turbines have been and still are the subject of a broad set of research and development efforts. Bringing all of these technical innovations together is the overarching system design for a wind turbine which is inherently complex and involves a large amount of uncertainty in terms of the physics of the wind as well as the machine itself. This uncertainty has had a significant impact on the history of wind energy technology and the "Danish Concept" embodies design principles meant to address this uncertainty.

Designing the "Danish Concept" turbine involved many structural considerations—designing towers that would not buckle and collapse, drive shafts that would not crack and fail, gearboxes and other components that could handle the applied torque, and loads translated from the rotor through the rest of the system. Aeroelasticity involves the study or science of aerodynamic (dynamics of air flow) forces that induce load on a structure causing a range of response behaviours from the structure. The term and the science are inherited from aeronautical engineering and in particular parallels are often drawn between wind turbine and helicopter technology. Aeroelastic codes for wind energy design are so called because they take a technically holistic perspective including a meteorological model of the wind field, aerodynamic models characterising the interaction of the wind field with the blade and the resulting forces, and structural dynamic models which capture the effects of these forces on the entire turbine structure as well as individual components. Today, there exist several commercial and publically available software packages that encompass the codes needed to do full aeroelastic analysis in particular for horizontal-axis wind turbine configurations of various types. However, only a few pieces of simplified models from the current analytic suite were available when the Gedser, Tvind and NIVE prototype turbines were designed and built. In particular, the coupling of meteorological models with the aerodynamics and structural dynamics had not been achieved for wind specific applications and thus the structural behaviour of a turbine under extreme conditions and over its lifetime was highly uncertain. This translated to conservative design principles for the NIVE turbine and other contemporary Danish wind technology development efforts. As Maegaard highlights, "There were many such rules of thumb. They were more on the safe side". Without well-developed physical models, the NIVE team relied on information in particular from the Gedser project which influenced the design of the NIVE blades then adopted and subsequently developed by Økær. The UN conference on new energy again became a reliable resource in guiding these "rules of thumb":

> "We know from the Gedser windmill that [the turbine] should be induction type—asynchronous type—but you know the size of the rotor, and we had some standards... [From] New Sources of Energy we had all the information from the Gedser windmill. There were some rules of thumb—30 kg/m^2—this was the axial pressure on the rotor, and from that one could calculate well the tower—what kind of steel, how the lattice system should be. You could calculate the size of the foundation, and once you have these basic figures, the rest of it is conventional. There will always be some sellers you can go to once you have the loads and you could go to an engineer saying, can you help us to know the size of the steel."

The measurements taken during the initial operation of Gedser windmill did in fact provide a basis of information over expected loads that may occur both during operation and with the wind turbine stopped. Measurements from the Gedser windmill indicated that axial pressures, the aerodynamic pressure on the rotor, during operation ranged from around 20 kg/m^2 at 25 m/s wind speeds to nearly 35 kg/m^2 for wind speeds in excess of 30 m/s (Juul, 1961). As mentioned before, it was a factor of 30 kg/m^2 pulled from the report that NIVE used in its dimensioning for the rest of the system design. Thus, combining the dimensions of the NIVE turbines with the rules of thumb obtained from the Gedser experiments served as a basis for design when understanding *via* physical models were not yet available. Of course, these rules of thumb were known even at the time of the UN conference to be quite conservative. Experiments from the Gedser turbine indicated that "the dimensions of blades, tower and cabin [were] rather on a liberal scale" (Juul, 1961). This "over dimensioning" was then translated to the NIVE turbine which itself, at 22 kW, was a much smaller turbine which would result in even more conservative design dimensions. However, without a better understanding of the physics and dynamic interaction of the structure with the wind field, the over dimensioning was seen as a good design principle.

> "[Wind manufacturers] wanted to make a reliable blade. It was the company that wanted to have a reputation for good quality. The uncertainty over how to dimension such a turbine was quite big, so let's make it strong enough. Later it was calculated that the blades had a design life of 70 years when normal design life of such equipment is 20. But, not knowing how to calculate it better, it got a design life that was later proved to be 70 years—which was too much you can say... half would be sufficient. Now [the turbine at the Folkecenter facility] has been here for 25 years, so it's good it was not 20 years... it seems to be of a durability that it can be there for several more years. The gearbox is also over dimensioned, the generator is over dimensioned, etc... the procedure we followed, was if we were uncertain about the loads, then we took the highest value, so it will never go down, not break down and so on. Where we could see something critical we took the most cautious solution." (Maegaard, 2010)

This in part, Maegaard describes, was done because their client base for the designs—the blacksmiths and small and medium enterprises—were selling directly to the customers:

> "These blacksmiths here were selling to their local clients, and if they lose confidence, it could be a disaster of course. So we would rather use too much material, make it too strong, than to lose confidence in the use of it. I think this was an important principle to have, because these wind turbines set the standard of high reliability... they were part of setting the standards for what is wind power in Denmark." (Maegaard, 2010)

Indeed, during the years, the number of wind turbines that would be manufactured by Danish companies would climb until the Danish manufacturers all but dominated the entire international market for wind turbines. While the parallels to the aviation sector were important for wind turbine development, especially for blade aerodynamics, when it came to the overall structure and the loads, the analogies began to break down. Many researchers highlight that the aviation-inspired predominantly US technology had a design focus that was less well suited to the wind turbine industry due to the different operational characteristics of the sectors. Maegaard notes that while a helicopter undergoes overhaul every 200 hours of operation, "a wind turbine has to run for thousands of hours before having an overhaul" and the principles for over dimensioning and emphasis on reliability were more suited to the wind turbine industry (Maegaard, 2010).

Figure 5.23 Testing of un-twisted modern blades at the Folkecenter 1984. Civil engineer Jacob Bugge constructed the hub, and the nacelle was designed by Preben Maegaard. Long blades were gradually shortened to verify optimal proportions.

One of the earliest employees of NIVE, engineer Jacob Bugge published *Bogen om Vindmoeller* (*Book of Windmills*)[22] targeted at wind turbine designers with a number of formulas and equations to calculate, among other things, the design loads for wind turbines of different types. The book's chapter on "dimensioning" of wind turbine components highlights the design principles utilised by NIVE and similarly by other adopters of the "Danish Concept" in its chapter introduction:

> "A fairly safe fixing of the dimensioning loads, for example as seen in the different loading standards, requires fairly extensive practical experience. Such experience with wind turbines has not yet been fulfilled. It is therefore important to consider such uncertainty in developing a proposal for design at present. The main three sources for this uncertainty are:
>
> • There is no certain knowledge of the wind loads acting on a windmill.

---

[22]Bugge, J. (1978) *Bogen om vindmøller*, Clausen Bøger, Copenhagen.

- In many cases there is a new use of materials and components. The way they behave, and the strength they may be conferred, can be more different than expected.
- It can be difficult to predict which strains or strain combinations are really serious. A weak effect which enables harmonics to develop in the mill can thus be devastating.
- The design, described below, should be taken with the above reservations. It should only be considered as an interim proposal for the design, until there is a larger body of experience." (Bugge, 1978, translated from page 151)

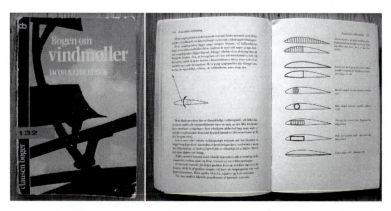

Figure 5.24 Cover and pages from Jacob Bugge's *Bogen om Vindmoeller* (1978, Clausen Bøger, Denmark).

The emphasis from small- and medium-sized Danish wind turbine manufacturers was to use a margin of safety commensurate with the understanding of system loads and behaviour at the time—that is, a very large margin of safety. The small wind turbines of the early 1980s in Denmark and the 55 kW turbines that were the main Danish export to the California market all were based on that prior incremental experience dating back to the Gedser turbine and the consistent use of conservative design principles in the face of large uncertainty.

5.3.6 The Risø Test Station: A "Mega-Network" for Danish Wind Energy

The wind industry actor-network that encompassed NIVE, Økær, Riisager, Tvind, the OVE and the new entrants of Vestas, Nordtank and other companies continued to develop the technology through shared best-practices and trial-and-error development. At the same time, attempts to advance the science and understanding of the physics relevant to wind energy design were underway in several countries. Of particular importance was ensuring the safety of turbines as they were deployed for use near or in residential areas. The assurance of safety led to the notion of a certification process for particular turbine designs. Risø National Laboratory, which had managed the testing and evaluation of the Gedser turbine put back into operation in the mid-1970s and had also consulted on the Tvind project along with DTU (the Danish Technical University), was well positioned to take on this new role as the wind turbine certification centre for Denmark. An addition to the developing Danish wind energy actor-network, thus, was the development of a test facility for small wind turbines at Risø.[23]

Figure 5.25   Risø National Laboratory. The research centre is located north of Roskilde with direct access to Roskilde Fjord. The two circular barns contained small nuclear reactors for research use (Photo: Risø National Laboratory).

During the early years of the wind industry development, no formal design or safety standards existed in Denmark. The OVE established a wind turbine safety group at its fourth "Vindtræf",

[23]See the reports by Pedersen (1981, 1983, 1984) and Petersen (1980) for specific information about the early wind turbine testing at Risø.

but this group hardly provided regulation to the industry. However, after 1979, Denmark required certification of wind turbines in order to have access to a subsidy that was essentially an investment grant of 30%. The system was set up to additionally encourage participation by making the certification process free for Danish manufacturers and providing payment to the companies for the time the turbine was at the test station. The test centre for small windmills was established in 1978 and was given the licensing task by the government in 1979. The first certified turbines were the Gedser-type machines, or Riisager-inspired designs, that produced 22 kW or 55 kW of power which were considered to be "second-generation types".

Most of these machines included the NIVE-inspired design types using Økær blades with variations for the different manufacturers. The test station noted that while "design principles exist that ensure mechanical integrity and proper functioning on at least a short-term basis", longer-term performance was still under question which limited the ability of the test station in terms of establishing design requirements (Petersen, 1980; Pedersen 1984). Initially, requirements from the test station allowed for both air-brakes as well as braking by yaw, rotating the turbine out of the wind, but this was amended in subsequent documentation which required a "fail-safe" braking system that is disengaged when the system is active and is kept in series with each element of critical safety importance for both the electronics as well as vibrations. In addition, two independent braking systems were required to ensure that over speeding, as experienced with the early Herborg and other turbines, was not a problem. Other early guidelines address manually operated breakers for grid disconnection, relay systems for automatic grid disconnection, a designated "operator", as well as metering and payment rules for electricity generated by small wind turbines owned independently and connected to the grid. These fundamental requirements regarding braking and grid interconnection were then extended under the 1981 subsidy extension act to include a link from the fundamental requirements to the operational status of stopped, normal operation and emergency shutdown, as well as specific requirements regarding system security, static tower and foundation loads, and load tests as performed by the test centre.

The "Smedemestermoellen"—Blacksmith's turbine—designed by NIVE, was one of the first turbines to be certified by the new test station, having been installed there in September of 1979.

Figure 5.26 Testing of 22 kW Smedemester (Blacksmith's) wind turbine at Risø.

The same turbine was also used to publish the first set of "standard measurements on windmills at the test station" that would become the evolving Danish test standard throughout the 1980s. Early operation of the test station identified some design flaws such as varying stall behaviour of the NIVE blades or overloading of the drive train by the Riisager windmills. In 1982, the test station was then extended to accommodate a common set of foundations for the new 55 kW series of turbines which were fast coming to market. It was this very series of 55 kW Danish turbines that would take the California wind market by storm in the mid-1980s and thus establish Denmark as the top international supplier of wind turbines, all based on the Danish Concept. Up until the end of the decade, Risø served as the single certifying authority for wind turbine manufacturers in Denmark, which allowed them to become established as "a centre of knowledge directing itself to the wind turbine industry, public authorities, and to users of the technology" (Petersen, 1989). This knowledge fed back into the industry reinforcing the development of the technology and the overall Danish mega-actor-network for wind energy technology. The era of the "Danish Concept" involving a country-wide set of actor-networks from small companies of 1 to 2 people to large-scale industrial–academic complexes had come to fruition.

Figure 5.27   Meeting in a café to discuss another Blacksmith's wind turbine design. From left, Peter Hjuler Jensen (Risø) and Knud Buhl Nielsen, Preben Maegaard and Jørgen Krogsgaard.

5.4   Conclusions

Regarding the history of the wind energy industry and associated technology, the development of the two is neither independent nor unidirectional. As many historians and theorists of technology and society have noted, typically there are sources of influence moving in both directions from the society to the technology and the technology to the society. A step further, and technology and society are no longer isolated worlds where interactions occur across fixed boundaries. Instead, various material and human agents create actor-networks that may be maintained over considerable lengths of time and morph and grow and decline and split and interact with each other and other actor-networks. This flexibility of analytical perspective is particularly relevant when breaking down certain entrenched narratives regarding the wind industry. Dominant narratives where a social force influenced an innovation process or an innovation process was an intrinsic foregone conclusion may both overlook the complexity of the actual history of interest.

There are various actor-networks that arose during the development of the wind energy technology from the 1970s up to 2010. Only a few are treated here. Nonetheless, these few actor-networks can provide an interesting basis for analysis. In particular, there is the actor-network clustered around the "Danish Concept". Catalysed by the events of the oil crisis and the fuel shortages in Denmark, this group began to form around the notion of renewable or domestic energy for Denmark (including both wind and biogas). The group began to grow with association of Tvind, Riisager, NIVE and OVE as its nexus and links were tied to various domestic manufacturers and local masons. Related networks, including the Organisation for Windmill Owners, allowed for an ever-expanding Danish wind energy actor-network. The interaction with Erik Grove-Nielsen and Økær led to the establishment of a loosely tied but related network centred on the all-important blade component of the "Danish Concept" wind turbine. The Økær-centric wind turbine industry, NIVE, Riisager and OVE groups mutually reinforced the growing network and the "Danish Concept". This interaction then led to the use of the NIVE-designed Blacksmith's wind turbine and similar Danish concept turbines to be the basis for standard testing at the newly established Risø test laboratory. The use of the NIVE design in the very first set of formal standards for wind energy, as well as the Bonus, Vestas and other wind turbines, further established the "Danish Concept's" centrality in the Danish wind turbine industry. The establishment of Risø as a test centre was creating a new source of accessible information on the design of wind energy technology and the proliferation of the Økær blades was complemented by innovation at a number of other companies. Thus, though NIVE (later the Nordic Folkecenter), Økær, Riisager and the OVE continued to serve as the centre for Danish wind energy design, the development of the theory and science of wind energy began to take on increased importance with Risø as a new nexus for a related and intersecting actor-network. The intention then of Risø was to continue to grow throughout the 1980s and combine with a larger network of international players to promote the development of standard test procedures and the science and engineering behind wind energy design (Risø, 1988).

In contrast to the various organisations mentioned above, the material agents within these actor-networks could also be seen as frames of reference for analysis of the developing actor-networks. Beginning with the Gedser turbine of the 1950s, the "Danish Concept" for wind energy technology began to take shape. The well-documented success of this turbine was a prime motivator for the designs of the late 1970s when more detailed information was not available. It influenced directly the development of the Riisager, Tvind and NIVE turbine designs which then inspired the development of a relatively standard design across the entire Danish wind industry. The "Danish Concept" performance in terms of reliability even in the harsher environments imposed by the California wind plants ensured the continued success and growth of the overall Danish wind energy industry. The performance of the turbine in the face of uncertainty is in essence its "agency" exerted on the network. While Juul himself designed the turbine with principles of reliability in mind, the uncertainty of the physics influenced the overall process and were beyond the control, indeed the comprehension, of the actors involved in the development of the "Danish Concept". Even today, the uncertainty surrounding wind turbine and plant performance is substantial and surfaces in unexpected failures, lower realised levels of energy production, and more frequent need for turbine repairs.

Thus the perceived success of the Danish wind energy actor-network is dependent on many smaller actor-networks containing the technology itself and many organisations from OVE to NIVE and even to Risø. This history provided an introduction into many of the complex and interwoven relationships of the Danish wind energy actor-network which itself is part of a global actor-network for wind energy. Even so, there is much more detail involved in just the Danish story than this work can highlight. In deconstructing a history of technology, the complexity quickly compounds. This work hopefully brought forth, if nothing else, an appreciation for that complexity and the notion that to tell a relatively complete history of one single technology, such as a wind turbine, will involve many actors, networks and networks of networks in order to provide an adequate depiction of that history.

References

Bugge, J. (1978) *Bogen om vindmoeller*. Clausen Boeger, Denmark.

Christensen B. 2009. "Danish Windmill Production before and after Poul La Cour—and the first golden age of wind power 1906-1920" in Wind Power—the Danish Way, Poul La Cour Foundation, Askov, Denmark.

FC (1990) *Nordic Folkecenter Brochure*.

Garud, R. and Karnoe, P. 2003. "Bricolage versus breakthrough: Distributed and Embedded Agency in Technology Entrepreneurship." Research Policy 32 (2): 277–300.

Gipe, P. (1995) *Wind Energy Comes of Age*. Chelsea Green Pub. Co., Vermont, USA

Heymann, M. 1995. "Die Geschichte der Windenergienutzung: 1890–1990".

Heymann, M. 1998. "Signs of Hubris: The Shaping of Wind Technology Styles in Germany, Denmark and the United States", Technology and Culture, 39(4), 641–670.

Jensen B. E. (2010) Personal interviews (December 2010).

Juul J. (1961) "Design of Wind Power Plants in Denmark." In *Proceedings of the United Nations on New Sources of Energy*, Rome, August 21–31, 1961.

Lundsager, P. and Jensen P. H. 1982. "Licensing of Windmills in Denmark." In Proceedings of the Fourth International Symposium on Wind Energy Systems, Sept. 21–24, 1982.

Maegaard P. (2009a) "The New Wind Power Pioneers and the Emergence of the Modern Wind Industry 1975-1979." In *Wind Power the Danish Way*, Poul La Cour Foundation, Askov, Denmark.

Maegaard P. (2009b) "The Tvind Windmill Showed the Way." Nordic Folkecenter publications.

Maegaard P. (2010) Personal interviews (July 2010 and December 2010).

Mumford, L. (1934) *Technics and Civilization*. Harcourt Brace Jovanovich, USA.

Nissen P.-O. (2009) "The Scientist, Inventor and Teacher: Poul La Cour." In *Wind Power—the Danish Way*, Poul La Cour Foundation, Askov, Denmark.

OOA (2009) "OOA: Organisation for Information about Nuclear Power" (windsofchange.dk).

OVE (2009) "OVE: Organisation for Sustainable Energy" (windsofchange. dk).

Pedersen, T. F. (1981) "Standardmaalinger paa vindmoeller opsat paa proevestationen for mindre vindmoeller" (Risø-M-2325).

Pedersen, T. F. (1983) "Nordisk Katalog Over Vindmoeller." Proevestationen for Mindre Vindmoeller, Risø.

Pedersen, T. F. (1984) "Experience with Testing of Windmills at the Test Station for Windmills at Risø." In *Proceedings for the AWEA Wind Expo*, September 23–26, 1984, California, USA.

Petersen, H. (1980). "The Test Plant for and a Survey of Small Danish Windmills" (Risø-M-2193).

Petersen, E. L. and Skrumsager, B. 1989. "Meterorology and Wind Energy Department Annual Progress Report 1 Jan–31 Dec 1988", Risoe-R-576.

Putnam, P. (1948) *Power from the Wind*. Van Nostrand Reinhold Company, New York, USA.

Righter, R. (1996). *Wind Energy in America: A History*. University of Oklahoma Press, Oklahoma, USA.

Risø (1988) "The Test Station for Windmills: 1978–1988."

Serchuk, A. 1996. Federal Giants and Wind-Energy Entrepreneurs: Utility-Scale Wind Power in America (Dissertation), Virginia Polytechnic Institute and State University, Virginia.

Tvind (2010) "Tvindkraft" informational website (www.tvindkraft.dk).

Van Est, R. (1999) *Winds of Change: A Comparative Study of Politics of Wind Energy in California and Denmark*. International Books, Utrecht, Netherlands.

About the author

 **Katherine Dykes** is project manager for the development of a wind energy systems engineering (WESE) software research tool at the US National Renewable Energy Laboratory (NREL) of the National Wind Technology Center (NWTC). She has also worked as adjunct faculty in a Power Systems Master's Degree Program at Worcester Polytechnic Institute (WPI).

Katherine's academic passions lie at the cross-section of engineering and social science. She has a BS Cum Laude in Economics and Statistics from the Wharton School, University of Pennsylvania and BS Cum Laude in Electrical and Materials Science, University of Pennsylvania. She did her MS in Electrical Engineering and in Agricultural, Environmental and Developmental Economics at the Ohio State University and received a PhD from the Engineering Systems Division of Massachusetts Institute of Technology in 2013. The dual background in engineering and economics allows her to employ methods from both fields for an integrated approach to the research of technology innovation, diffusion and design.

Katherine's research interests and projects include system dynamics simulation–based modelling of socio-technical systems, with emphasis on technology innovation and diffusion; power systems operation and planning, with emphasis on new technology integration (utility-scale wind energy, distribution-side technologies); integration of technical and financial models for system design; and design of complex technical systems and their social influences, technology standards, and co-evolution of science and technology.

Chapter 6

Danish Pioneering of Modern Wind Power

Niels I. Meyer

Department of Civil Engineering, Technical University of Denmark, Lyngby, Denmark

nim@byg.dtu.dk

It is not obvious why a small European country like Denmark, with about 5.5 million inhabitants, managed to place itself in a global role as pioneer of modern wind power in the 1970s. Why did this role not go to large industrial countries like the United States, the United Kingdom and Germany with their manifested results as technological innovators including the field of aerodynamics applied to helicopters and airplanes?

This chapter attempts to clarify part of this question where the author has been actively engaged in promotion of the modern Danish wind power project since the early 1970s. My text will focus on the period from 1974 to 1985 where large parts of the official Denmark were promoting the introduction of nuclear power and natural gas, while wind power and other forms of renewable energy sources (RES) had low priority. Wind power had to rely on enthusiastic NGOs and small private innovators in an alliance with a few independent university researchers.

Wind Power for the World: The Rise of Modern Wind Energy
Edited by Preben Maegaard, Anna Krenz and Wolfgang Palz
Copyright © 2013 Pan Stanford Publishing Pte. Ltd.
ISBN 978-981-4364-93-5 (Hardcover), 978-981-4364-94-2 (eBook)
www.panstanford.com

The surprising outcome of this uneven battle over a decade was a victory to wind power and a defeat to nuclear power. This illustrates the importance of broad public participation in democratic decisions. It is my personal theory that the combination of long democratic traditions and the modest size of our country have been significant factors in the final success of the Danish wind power adventure. This has influenced my choice of concrete examples in this chapter. I have left out a number of technological, legal and commercial details, as they are treated in the following chapters.

6.1　It Began Already in the 1890s

History illustrates that new technological development is often initiated by talented and original individuals with visions of alternative solutions for society. In the case of wind power it is fair to give much credit to the Danish physicist Poul la Cour for his pioneering work at Askov Folk High School in Jutland in the 1890s.

La Cour developed and built a wind turbine for electricity production with a rotor diameter of 22 m, including mechanical speed control. He even tested a number of rotor profiles in wind tunnels and provided energy storage based on hydrogen produced by electrolysis of water. The hydrogen was subsequently used for lighting purposes. He deserves the credit of laying the ground for modern wind power development—and for introducing production of hydrogen as an energy carrier based on RES. More details on the story of la Cour and his contribution to the development of wind power are given in other chapters in this book by Povl-Otto Nissen and Benny Christensen.[1]

The concepts and technologies developed by la Cour provided a basis for wind electrification in Denmark during the first two decades of the 20th century. In 1918, a total of 120 rural wind power stations were established with rated turbine powers between 20 kW and 35 kW, yielding a total installed wind capacity of about 3 MW, compared to a total Danish electricity capacity of about 80 MW. With the typical capacity factors of that time, this corresponds to around 3% coverage by wind of the Danish

---

[1]See chapter *History of Danish Wind Power* by Benny Christensen and *The Aerodynamic Research on Windmill Sails of Poul la Cour, 1896–1900* by Povl-Otto Nissen.

electricity demand in 1918. Even today, only few nations have exceeded this coverage by wind.

During the following four decades, wind turbines were further developed and tested in Denmark and elsewhere, especially in Germany, the United Kingdom and the United States. In Denmark this period culminated with the 20 kW Gedser windmill, in operation from 1959 to 1967 (more details in the chapter by Benny Christensen). The operation was successful, but in the 1960s oil and coal was becoming so cheap that wind power could not compete and most electric utilities in industrial countries forgot about wind power. Luckily, Danish wind power enthusiasts did not forget about the technological turbine solution promoted by the Gedser windmill and characterised by three blades on a horizontal axis in an upwind position (Fig. 6.1). The Gedser windmill became the mother of modern Danish wind turbines in the 1970s after the global oil crisis.

Figure 6.1   The 200 kW Gedser windmill.

The early innovators combined the courage of economic risk-taking (of their own private money) with a characteristic Danish feature of common sense based on a principle of precaution, which may be expressed as follows: *When you have found a reliable and efficient turbine technology, stick to it and start with modest-sized structures that you can manage. Go on from there and increase gradually the turbine capacity.*

This was exactly the strategy of the Danish wind power pioneers after the global oil crisis as illustrated in Table 6.1.

Table 6.1 Development of rated power for commercial Danish wind turbines

Year	1978	1983	1985	1988	1991	1996	2001	2008
Typical rated power of new turbines (kW)	20–40	55–75	100–150	200–250	300–450	500–700	1000–1500	2000–3000

Note: The number ranges refer to the typical capacity range installed in a given year.

In contrast to this, most wind projects sponsored by government in large industrial countries like the United States and Germany focused on high capacity turbines in the MW range, often based on experiences from especially production of helicopters. This was not a success and illustrated that wind power technology cannot be copied directly from other fields. In Denmark the electric utilities were copying this focus on relatively large turbines in their demonstration project with two 630 kW turbines (1979) near the city of Nibe in northern Jutland. The Nibe turbines were providing useful technological knowledge, but no commercial interest was found in Denmark for following up on the concept of the Nibe turbines (Fig. 6.2).

Figure 6.2 The Nibe wind turbines.

Ironically, the only successful large capacity turbine in the 1970s was the Danish Tvind windmill that has been producing at a capacity of about 800 kW since that time (Fig. 6.3). This turbine was built by a group of technological amateurs at the Tvind schools in Jutland over three years from 1975 to 1978 with some help from a few professional engineers with previous wind power experience.

Figure 6.3   Tvind wind turbine.

The main goal of the Tvind schools was to change Danish society. It was not the main goal to build large wind turbines, but the Tvind windmill is a surprising example of the potential of idealistic people—also in the field of technology. However, this potential should not be romanticised based on the Tvind experience. The working environment at the Tvind schools both psychologically and in relation to health had its dark sides (Østergaard, 2000), but the fact remains that the Tvind windmill was an impressive technological achievement at that time and an inspiring example for Danish wind power enthusiasts.

6.2   Global Oil Crisis

Most industrial countries did not have an official energy policy before the so-called oil crisis in 1973/74. It was left to the oil companies to provide the necessary energy supply for most sectors, including a significant part of the electricity production. In the case of Denmark, cheap oil was fuelling about 90% of the electricity production in the years before the oil crisis, which made Denmark especially vulnerable to the jump in oil price and to the insecure supply situation in the Middle East. This triggered a new phase of official energy planning in Denmark where energy supply security became of central importance. The electric utilities in Denmark focused on three main solutions: shift from oil to coal, introduction of a natural gas system and introduction of nuclear power.

The shift from oil to coal and the introduction of natural gas, was not politically controversial (global warming was not an issue) and the utilities implemented the shift from 90% oil and 10% coal (rounded numbers) to 10% oil and 90% coal over a few years. Most of the remaining electricity production on oil was supposed to be replaced by nuclear power over a number of years. This strategy was supported by the utilities, large industry and the majority in the Danish parliament. However, in the Danish tradition of open democratic discussion this impressive support did not suffice. The proposal of nuclear power gave rise to a broad and heated public discussion over a decade which brought RES and especially wind power high up on the political agenda. The history of the Danish nuclear debate is important for an understanding of the situation of Danish wind power during the decade up to 1985.

6.3   Nuclear Controversy in Denmark

The introduction of nuclear power was an essential element in the first official energy plan from the spring of 1976 (Danish Ministry of Industry and Commerce, 1976). An alternative energy plan without nuclear power and with a higher contribution from RES was published in the fall of the same year by a group of energy experts from Danish universities (Blegaa *et al.*, 1976). A summary in English of the alternative energy plan was published in 1977 (Blegaa *et al.*, 1977).

Soon after the oil crisis a couple of new NGOs [The Organisation for Information on Nuclear Power (OOA) and the Danish Organisation for Renewable Energy (OVE)] initiated campaigns describing the problems with nuclear power and promoting RES as an alternative, but they were lacking support from "official Denmark".

As president of the Academy of Technical Sciences (ATV) from 1971 to 1977, I decided, however, to establish a committee that analysed the potential for wind power in Denmark. The committee published its first report in 1975 proposing a broad wind energy program in Denmark (Lykkegaard *et al.*, 1975). This was followed by a second report from the Academy in 1976 outlining a five-year program in the field of wind energy (Hvelplund *et al.*, 1976). Due to the general prestige of the Academy these reports were important as a supplement to the efforts of grassroot movements and small private wind energy innovators.

Some of the Danish utilities apparently regarded the two Academy reports as a serious threat to their planned introduction of nuclear power. To my surprise they responded with a direct attack on me as president of ATV. They argued that the Academy could not have a president who was trying to prevent technological development in Denmark. The utilities were apparently so focused on nuclear power that they completely overlooked the significant technological challenge in developing modern wind power. This was the first time in the history of the Academy that a group of members had launched an attempt to get rid of the president before the end of his period. However, the large majority of the Academy members did not agree with the lack of confidence in the president so I stayed in office to the end of my term. But it was an early illustration of a new style in the technological discussion concerning potential energy systems.

In retrospect it is surprising that this technological question could give rise to so strong feelings and to the introduction of a new style with personal attacks on people of a different opinion. This was mainly performed by the pro-nuclear technical representatives from utilities and universities while the NGOs focused on the problems of nuclear power and the advantages of RES. Here are a few examples of statements in the media by supporters of nuclear power:

- The primary Danish engineering journal in a leading article in 1974 described the opponents to nuclear power as "political swindlers, doomsday prophets, and pseudo-Marxist loud-mouths."

- A professor from the University of Copenhagen who was acting chairman of the Danish government's Council for Energy was cited (1976) for the following description of the opponents to nuclear power: "the opposition is part of a typical 'religious' wave created by 'barefoot walkers' and others related to bio-dynamics and home confinement. All of them are governed by feelings only."
- In 1979 the chairperson of an organisation campaigning for nuclear power publicly accused the opponents for being connected to communistic groups that might be preparing both sabotage and terrorist actions.
- Also in 1979 the managing director of the largest electric utility claimed in a Danish newspaper that it was the intention of the opponents to nuclear power to use the energy controversy as basis for a revolutionary change of the Danish society.

In several cases my name and the names of some of my colleagues were included in these accusations in spite of the fact that we never had any connections to revolutionary communistic groups nor to terrorists. I have not found a credible psychological explanation for this change of style.

The Danish parliament granted money in the fall of 1974 for a balanced information campaign on pros and contras in relation to nuclear power. The project was placed in a new independent organisation called the Committee for Information on Energy Systems. This organisation quickly developed pedagogical information materials that were well received by local study groups all over the country. A special publication was written in collaboration between a supporter and opponent to nuclear power with a joint text when they agreed and parallel texts on the same page when they did not (Linderstrøm-Lang and Meyer, 1975).

The information campaign was a success, but the outcome was not what was desired by the government: more and more people became convinced that Denmark should not have nuclear power. As a consequence, the government decided in April 1975 to discontinue the information campaign after about one year of work—at a time when the debate was really taking off.

The end result was that the Danish parliament in 1985 (one year before the Chernobyl accident) decided that nuclear power should not be included in the Danish energy supply system. This opened up new political possibilities for Danish wind power and

RES in general. However, before 1985 a number of other important policy changes influencing wind power had taken place in Denmark.

6.4 Test Station for Wind Turbines

In spite of the general scepticism towards wind power from official Denmark, some important projects and support schemes were initiated by the government in relation to wind power from the late 1970s.

In 1978, it was decided to establish a test station for wind turbines at Risø National Laboratory and about a year later the test station was appointed the official authority for certification of wind turbines. This secured a high reliability of Danish turbines exported the booming Californian market in the beginning of the 1980s—in contrast to many competitors from other countries. I visited the Californian wind parks on a blowing day in 1983 and noticed that half of the turbines were not spinning in spite of the strong wind. The other half was of Danish origin with certification from Risø (Meyer, 1983).

In 1979, the new Danish Ministry of Energy introduced an economic subsidy covering 30% of the capital cost of new wind turbines with certification from the test centre at Risø. This subsidy was gradually phased out during the 1980s after a total investment subsidy of about EUR 38 million, but it did promote Danish wind power during a difficult economic period, partly related to the collapse of the US market in the second half of the 1980s. It is also worth mentioning that the concept of a *wind atlas* was developed at Risø (Petersen *et al.*, 1981). This is now internationally accepted as an important tool for the penetration of wind power.

6.5 Official Danish Committees for Promotion of Renewable Energy

The Danish Ministry of Industry had established a committee for promotion of new technologies in the late 1970s. As a member of this Technological Council, I proposed in 1980 to establish a subgroup with focus on the Danish development of RES, and in 1982 it was finally decided to establish the Danish Steering Committee for RES with me as the chairman.

In the Danish tradition the members of such committees are usually selected according to a politically balanced combination of representatives from labour unions, industrial organisations, ministries, universities and professional organisations involved in the subject in question. This is a guarantee that no new and original schemes are proposed—disturbing the political decision makers.

As a consequence, I insisted that the Steering Committee should be different and that all members should be experts in the field of RES and be enthusiastic about a sustainable future with RES as an important factor. To my surprise this was accepted.

The annual budget of the Steering Committee started at EUR 2.7 million in 1983 and was gradually increased to EUR 4 million in 1990. In this period there was a "green majority" in the parliament and this majority was a guaranty for our budget. The total governmental subsidy to the Steering Committee over the nine years of operation from 1982 to 1991 amounted to about EUR 30 million.

The Steering Committee became an important factor for the development of Danish RES. Its goal was to support the development of RES production in small- and medium-sized enterprises (SMEs), and to promote technological development, demonstration projects and information projects concerning RES. The Steering Committee was different from the usual political correct committees due to its open and loud-speaking discussions of strategies and technological questions. The fact is, however, that we nearly always reached a constructive consensus that could accelerate the penetration of RES in Denmark. Our support covered a broad spectrum of RES activities including biogas, solar heating, wind power and local and regional demonstrations of systems based on RES (Beuse *et al.*, 2000), for example, a project describing how the Danish island Bornholm could be transformed to a "green island" including some wind parks (Jørgensen *et al.*, 1986). This project was partly implemented.

In the same period from 1983 on, the Folkecenter for Renewable Energy (FCRE) in northern Jutland (Thy) in a constructive way supplemented the work of the Steering Committee by practical demonstration of wind turbine technologies combined with construction manuals for SMEs that wanted to produce wind turbines. FCRE was initiated by Preben Maegaard in 1983, and has now promoted wind power and other forms of RES (including wave power) for nearly 30 years. The concept of a development centre supporting SMEs in the field of RES has been copied internationally in a number of countries as described in more detail in the chapters by Maegaard and Dykes.

In 1988 the Steering Committee funded five draft projects for offshore wind farms in Danish waters. The results were presented at a seminar in 1989, and one of the projects was considered so promising that funding was provided by the committee for a detailed project design. However, the project was never realised due to opposition from the Ministry of Energy. I had several meetings with representatives from the ministry but they turned out to be exceptionally creative concerning obstacles to the whole idea of offshore wind farms. Their worries included concern of possible damage to sunken (unknown) old ship wrecks and problems for the Danish military that wanted free access to shooting exercises over the sea (Meyer, 2000).

Luckily, Danish wind power industry had more influence on the ministry and the first offshore wind farm in the world was made operational in September 1991 at a site northwest of the island of Lolland in the Baltic Sea (Fig. 6.4). For the next decade Denmark was leading the global development of offshore wind parks of increasing sizes as illustrated in Table 6.2. Today, the waters around the United Kingdom have taken over this role.

Table 6.2   Offshore wind farms on Danish waters

Year	Place	Number of turbines	Total capacity (MW)
1991	Vindeby	11	5
1995	Tunø Knob	10	5
2001	Middelgrunden	20	40
2002	Horns Reef 1	80	160
2003	Nysted	72	166
2003	Rønland	8	17
2003	Samsø	10	23
2009	Sprogø	7	21
2009	Horns Reef 2	91	209
2010	Avedøre Holme	3	11
2010	Rødsand	90	207
2012	Anholt	Under construction	400

In the fall of 1990, the Ministry of Energy decided to change the organisation for promotion of RES. This was used as an excuse to close down the unconventional Steering Committee and to replace it by a traditional type of advisory council without project money. The operation of this council during the following years confirmed my concern that a council without project money and populated by members with balancing vested interests is not able to promote new policies (Meyer, 2000).

Figure 6.4   The Sprogø Wind Farm, north of the Great Belt Bridge, 2009 (Photo: Fxp42).

6.6   The Golden Nineties for Danish Wind Power

With the change to a centre-left government in January 1992 the attitude towards wind power took a drastic turn in a positive direction (see Fig. 6.5). The credit for this change should to large extent be given to the new Minister of Environment and Energy (Svend Auken).

After a number of disagreements between utilities and wind power producers over conditions for grid connection and tariffs, regulations were introduced by the Danish government in

1992. This included a variation of a feed-in tariff which was fixed at 85% of the utility production and distribution costs. On top of this tariff from the utilities, the private wind power producers would receive a tax refund ("environmental premium") of DKK 0.27 per kWh (3.6 eurocents per kWh). After a drop in the yearly growth of wind capacity in the early 1990s, this initiated a strong growth in land-based wind capacity in the last part of the 1990s.

A competing tariff scheme in Europe is called "trading of green certificates" (TGC) with certificates issued to individual wind power producers in accordance with a specified government target for the total national wind power production and with sanctions to utilities if the target is not fulfilled. In the late 1990s the civil servants in the Danish Energy Agency convinced the minister (Svend Auken) that the TGC scheme was in better agreement with the dominating preference for market competition in the EU. This was a misunderstanding (Hvelplund, 2001, and Meyer, 2007), but the politicians decided to switch to a TGC scheme from the existing tariff system with its several features similar to the feed-in scheme. To add to the general confusion, this switch was never implemented and the Danish tariff system continued to be based on a strange mixture of features from the feed-in scheme supplemented by environmental premiums.

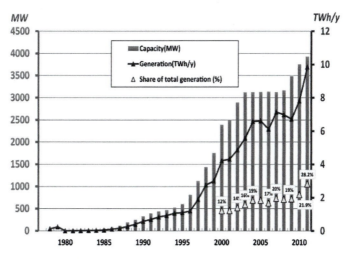

Figure 6.5   Installed wind capacity, annual generation and capacity factors in Denmark from 1977 to 2011.

The stagnation in installed wind power from 2002 is mainly due to a shift to a conservative/liberalistic government in the fall of 2001. This government discarded most of the previous government's support for RES and changed to a policy relying on market forces. The stagnation has remained during the whole period of this government until a new centre-left government was elected in September 2011. The new government has published an ambitious plan for wind power and other forms for RES. If these plans are realised, Denmark may return to its pioneering role.

6.7 Conclusions

The strategies for promotion of wind power have been different for different European countries (Meyer, 2007). From the early 1990s three countries relying on some variation of the feed-in scheme have been dominating: Denmark, Germany and Spain. However, in a historic perspective a number of other factors than the tariff scheme have also been important. Thus, the Danish strategy for promotion of wind power in the period from the mid-seventies to the end of the nineties has combined a number of different elements:

- common sense precautionary strategy with gradual increase of turbine capacity
- local co-operative ownership of wind turbines and careful selection of sites
- investment subsidies in the early period of development
- long-term government support for research, development and demonstration
- national tests and certification of wind turbines
- government sponsored wind energy resource surveys (wind atlases)
- feed-in tariffs and other regulations of electric utilities
- priority to wind electricity in the grid
- government energy planning and official targets for wind power

The combination of these constructive wind policies may offer an answer to the introductory question in this article about the background for Denmark's pioneering role in the history of modern wind power.

References

Beuse, E., Boldt, J., Maegaard, P., Meyer, N. I., Windeleff, J. and Østergaard, I. (2000) *Renewable Energy in Denmark 1975–2000* (in Danish), OVE publishers, Aalborg, Denmark.

Blegaa, S., Hvelplund, F., Jensen, J., Josephson, L., Linderoth, H., Meyer, N. I., Sørensen, B. and Balling, N. P. (1976) *Alternative Energy Plan for Denmark* (in Danish), report with support from OOA and OVE, Copenhagen, Denmark.

Blegaa, S., Josephsen, L., Meyer, N. I. and Sørensen, B. (1977) Alternative Danish Energy Planning, *Energy Policy,* June 1977, pp. 87–94.

Danish Ministry of Commerce (1976) *Danish Energy Policy 1976* (in Danish), report from the Danish Ministry of Commerce, Copenhagen, Denmark.

Linderstrøm-Lang, C. U. and Meyer, N. I. (1975) *Facts and Evaluation of Nuclear Power* (in Danish), report published by the Council for Energy Information, Danish Ministry of Commerce, Copenhagen, Denmark.

Lykkegaard, K., Meyer, N. I., Myhre, T., Petersen, B. M. and Sørensen, B. (1975) *Wind Power* (in Danish), report published by the Danish Academy of Technical Sciences, Copenhagen, Denmark.

Hvelplund, F., Jels, H. A., Johansson, M., Kofoed, S. S., Meyer, N. I., Petersen, B. M., Petersen, H. and Sørensen, B. (1976) *Wind Power 2* (in Danish), report published by the Danish Academy of Technical Sciences, Copenhagen, Denmark.

Hvelplund, F. (2001) "Political Prices of Political Quantities?" *New Energy* 5, pp. 18–23.

Jørgensen, K., Meyer, N. I. and Pilgaard, H. (1986) *Energy Plan for a Green Island* (in Danish), Borgen publishers, Copenhagen, Denmark.

Meyer, N. I. (1983) *Wind Parks and Solar Power in California* (in Danish), report from Physics Laboratory III, Technical University of Denmark, Copenhagen, Denmark.

Meyer, N. I. (2000) *Development of Renewable Energy in Denmark* (in Danish), Chapter 3, pp. 75–110 in Beuse *et al.*

Meyer, N. I. (2004) "Renewable Energy Policy in Denmark", *Energy for Sustainable Development",* Vol. VIII, No. 1.

Meyer, N. I. (2007) "Learning from Wind Energy Policy in the EU: Lessons from Denmark, Sweden and Spain", *European Environment,* **17,** pp. 347–362.

Petersen, E. L., Troen, I., Frandsen, S. and Hedegaard, K. (1981) *Wind Atlas for Denmark*, Risø National Laboratory, Roskilde, Denmark.

Østergaard, I. (2000) "The Tvind Mill—A Joint NGO project" (in Danish), Chapter 4, pp. 132–139 in Beuse *et al.*

About the author

Niels I. Meyer (1930) is professor emeritus of physics at the Technical University of Denmark (DTU). Prof Meyer has been member of the Club of Rome (1972), president of the Danish Academy of Technical Sciences (1971–1977), vice president of DTU (1969–1974), member of the Energy Supervisory Board (Regulatory Committee) for the Danish energy sector (1997–2004), chairman of the Danish Steering Committee for RES (1982–1991). He has also been member of a number of Danish and international Science Councils. Since 1972, Prof Meyer's research focuses on sustainable energy development and promotion of renewable energy sources. He has initiated new curricula at DTU on Energy and Resources.

Chapter 7

From Energy Crisis to Industrial Adventure: A Chronicle

Preben Maegaard

Nordic Folkecenter for Renewable Energy, Kammersgaardsvej 16, Sdr. Ydby, DK-7760 Hurup, Thy, Denmark

pm@folkecenter.dk

PART I
1974: After the Oil Crisis

7.1   Politicians' Relations with Wind Power

We are in the winter of 1974. The Danes have been told to economise on oil, as oil covers almost 100% of the country's energy consumption. And all of it comes from remote Middle East.

The first plans for wind energy emerge already during this winter, that is, long before anyone has heard of OVE, Chr. Riisager, the Risø testing station, a Danish natural gas net and all the rest of the terms that are constantly present in the newspaper headlines during the following years. But already in 1974 two camps had been formed: it was either windmills or nuclear power.

Wind Power for the World: The Rise of Modern Wind Energy
Edited by Preben Maegaard, Anna Krenz and Wolfgang Palz
Copyright © 2013 Pan Stanford Publishing Pte. Ltd.
ISBN 978-981-4364-93-5 (Hardcover), 978-981-4364-94-2 (eBook)
www.panstanford.com

As early as during the war years 1942–1945, a leading industrial group F. L. Smidth & Co., had built advanced aeromotors, and now they were going to start afresh with analyses based on the most recent technology, encouraged by many interested in wind power.

On 28 February 1974 leading Danish daily *Berlingske Tidende* reported on a large-scale windmill project, which would dominate the debate for the rest of the year. F. L. Smidth & Co. obtained a loan from the state in order to analyse the viability of developing commercial windmills. But on the same day two persons publish their critical arguments. Inventor Karl Krøjer bluffly announces that, "windmills are just crazy. It is unrealistic thinking. If just a few percent of the country's energy consumption should come from windmills we should need so many of them that we would not be able to see the country because of the steel skeletons with whirring blades," goes the vision of the inventor.

Neither does Steen Danø, managing director of Thrige-Titan, then Denmark's heavyweight in electrical machinery, perceive any future for windmills. His expectations for the future are of a completely different sort: "We should not just produce more electricity for ever increasing consumption. That is not a solution. We shall have to think afresh and look for a way of using the raw materials in a better way." Already the notion of resource consciousness and low growth has crept in as an alternative to wind energy.

In September 1974 F. L. Smidth & Co. presents their first result. It originates with Jean Fischer, managing director, who had already started to work on wind power long before the energy crisis. He presents a model windmill with an elegant 130 m concrete column with six built-in Darrieus rotors, a type which has always been surrounded by much fascination. The size of the windmill is unprecedented, 1 200 kW, and is predicted to become a tourist attraction. Five hundred of these wind turbines, placed along the Western coast of Jutland, could replace one nuclear power station.

Once more Steen Danø expresses his scepticism: "The danger of the romantic attitude towards wind power is that in 25 years from now we shall live on windmills and other funny projects and have no electricity for the industry." Mr Danø warns against the windmill plans and suggests that an atomic power station should be built in Funen. "We must get cracking and produce cheap electricity. The waste heat can be utilised for heating up the many greenhouses of Funen," is his recommendation.

Jean Fischer continues his battle for large-scale wind power. Showing his model at a conference in Stockholm, he gains the support of Professor Niels I. Meyer, pro-rector of Denmark's Technical University and president of the Danish Academy of the Technical Sciences, who will, during the coming decades, become a leading figure in the field of renewable energy. Niels I. Meyer says, "We cannot afford to neglect any kind of energy source that we ourselves are able to control. Consequently it is evident that Jean Fischer's technology must be submitted to thorough practical tests along with other types of wind turbines. Wherever we find promising systems we must work on them. We should not let wind power to be wafted away with the wind."[1]

The sceptics, however, pushed themselves forward more and more. Immediately after Professor Ulrik Krabbe of Denmark's Technical University, had spoken enthusiastically about the windmill project at Tvind at a conference in Copenhagen, engineer Mogens Johansson from the research department of the central power utilities (DEFU), stated that wind power would only play an extremely limited role. In his opinion wind power might become a supplementary source of energy, but it would only provide a few percent of the total energy consumption, and only by means of investment of billions.

Thus the note was given out about the attitude of the central power utilities towards wind energy. Denmark was in the process of being divided between pros and cons: centralists against the soft energy paths.

E. L. Jacobsen, director-in-chief of the leading power utility, ELSAM, says at the end of 1975, "The power utilities are willing to shoulder the task of developing wind energy, as it is decidedly one of our concerns. We are also willing to spend money on it, but it should result in electricity at the right price." He does, however, consider nuclear power to be the most realistic alternative.

In this manner he hedges his bets, because at that time it was difficult to calculate a price of wind power that could compete with other kinds of electricity. The power plants showed a positive attitude because Erik Holst, spokesman on energy for the leading Social Democratic Party, wanted them to do something. "The

---

[1]See chapter *Danish Pioneering of Modern Wind Power* by Niels I. Meyer.

utilities must engage, not least in the interest of their own reputation. They have a moral duty to pursue other options than nuclear power," he said and was backed whole-heartedly by former Prime Minister Hilmar Baunsgaard. But they were up against powerful interests—the consumer-owned power utilities, the directors of which more and more become the ones to decide the direction. The power plants had become a state within the state.

Consequently, Ove Guldberg, spokesman on energy for the Liberals, is able to interfere. "When big industry cannot make windmills commercially viable, the power utilities should not be burdened with such tasks. The consumers will be the ones to foot the bill," says Ove Guldberg who was supported by many elected power utility representatives. And that is how it turned out. Not until 20 years later, when Minister of Energy Svend Auken definitively put an end to the building of coal-fired power stations, the utilities have finally realised that it would be necessary to take a greater part in the investment in wind power to avoid being left hopelessly behind as future producers of electricity.

7.2   Big Business

The report from the Wind Energy Commission, appointed by the Danish Academy for Technical Sciences (ATV), appears in July 1975. The report concludes that 5% of Denmark's electricity consumption could be covered by 250 big windmills placed on the Western coast of Jutland. The payback time would be 15 years. In addition, wind power could become important for local production of district heating. On farms and solitary houses it might be possible to set up perhaps 200 000 wind turbines! The report also mentions effects on the balance of payments, employment, foreign dependence, consumption of resources, pollution and inflation. The recommendation is that research and development in the wind energy area should be continued.

Despite the report's pointing to interesting new big business area for a potent group like F. L. Smidth & Co., it had its motives to definitively stop the engagement in wind energy shortly after. Danish big industry was not ready to enter the production of renewable energy plants, which was not particularly logical, as Denmark did not have capital interests of any significance in the areas of coal, oil and uranium to be considered. Consequently, it became quite different—virtually unknown enterprises in Jutland harvested the fruits of the new industrial sector, which in the course of some years would obtain a billion turnover while many of the old enterprises lost turnover or vanished into darkness.

When F. L. Smidth & Co. throw their hand in, many Danes felt that no one else had the capacity to carry through wind power. "But we cannot manage this task on our own," said CEO Benned Hansen from the industrial conglomerate, which had in 1942 successfully been able to develop a modern windmill from scratch in only nine months. F. L. Smidth were not interested in research grants from the state in order to continue work on the plans, but generously offered to place their knowledge at the state's disposal, if the latter would take an interest in wind power, said Benned Hansen.

The retreat of F. L. Smidth & Co. did not mean that the dream of wind power had failed. Wind power had a symbolic value far stronger than anyone could imagine at the time. Through hundreds of years, the Danes had become conversant with wind power as a source of energy, and particularly in times of crisis they turned to wind power. During the Second World War windmills provided light in many homes. With a windmill on one out of three farms, for the Danes wind energy was a symbol of the progress of the cooperative movements and farming.

7.3   Responsible Solutions

The energy crisis in the 1970s put a definitive end to the belief in progress. Whereas the interest in wind energy formerly had vanished, as soon as wars and periods of energy shortage were over, there was now a new realisation that the resource shortage and environmental problems had come to stay. It was a crisis without an end; the industrial era with its squandering of coal and oil could not last forever. Consequently, a change to renewable energy was the only responsible solution to the energy problems in the long run. So said the people, who had looked into the future. And popular logic agreed.

In local communities people struggled on with energy solutions. One of the well-known projects of 1974 was some grandiose plan to supply the small island of Tunø with heating from the wind. Being isolated from the mainland, an island is a manageable unit and the inhabitants have to cooperate on elementary things.

For engineer Hans Jørgen Lundgaard Laursen, local blacksmith Thomas Nørgaard and later researcher Frede Hvelplund, it was a challenge to persuade the people of Tunø to support the idea. The blacksmith was aware of local scepticism, which he could easily understand. It all must be handled in the right manner. "How comes that they try to make people believe that if we are going to have windmills, they are to be placed in a long row all the way down the Western coast of Jutland. In that case no one will want windmills," Thomas Nørgaard remarked.

Hvelplund and Lundgaard Laursen wrote a report, which attracted much attention. They described the first project for a community wind turbine in Denmark based on one big unit. For Hvelplund, it was the beginning of continuous research in the economy and politics of wind power, whereas Lundgaard Laursen was invited to Tvind, to coordinate the ambitious windmill project of their schools. There they had the money and the appetite for a big task, while the more cautious inhabitants of small islands were not ready yet.

The Tunø project was discussed in Folketinget (the Danish parliament). Several politicians had become aware of the great perspectives in supplying island communities with renewable energy. They just overlooked the fact that there were no wind turbines capable of functioning yet—at least none that fitted the needs of Tunø.

7.4 Prophets and Demagogues

One of the political friends of wind power at the time was social democrat Mogens Camre, member of the parliament. With great courage he recovered details from some old projects for sun and wind energy that had been smothered in red tape during their journey through the granting bodies.

One of the forgotten applications was a wind power project by a well-known industrialist Th. Myhre from Århus, who had applied for financing a highly efficient turbine blades and a new type of electric generators with variable speed. But this was refused. The courage of the granting bodies failed, according to Mogens Camre, "as it would not do to encourage that kind of projects, because that might create confusion among politicians and make them decide not to build nuclear power plants."

Mogens Camre felt the pulse in this matter: The multinational energy corporations had no interest in self-sufficiently and economically sustainable forms of energy. He found that it was an appalling instance for the decline of democracy. "Our energy policy is being determined by technocrats who are directly or indirectly, personally financially dependent on Denmark's implementing nuclear power plants. We politicians are becoming the slaves of a technical and economic development which is governed by multinational profit interests, and which even has the means to influence the opinion of the people by propaganda and advertising," was Mogens Camre's comment.

The demagogues of atomic power were on their way in. In autumn of 1975, the Danish daily newspaper *Børsen* wrote that now the people must realise that nuclear power equals welfare. According to Ove Guldberg, former cabinet minister and now spokesman on energy for the Liberals, "the wind energy prophets are hardly distinguishable from those who are trying to undermine and obstruct the necessary decisions in the labour market and the education sector." With this remark, Ove Guldberg had typecast the economists, professors and engineers who worked for wind power.

It is evident that the larger respectable enterprises would no longer come out in the open with plans for developing and marketing wind turbines. Bankers, business people, even employees would be of opinion that if their enterprise showed interest in wind energy it was gambling with its future which illustrates that business climate was already totally infected by nuclear power.

7.5   Getting Together

I harvested some personal experience during presentations in Danish assembly houses, inns, and folk high schools. My 1970s calendars show many months with four lectures a week. Off to evening and weekend meetings with slides, books and folders on wind energy.

The audiences were thirsting for information, for concrete knowledge on technology, yields and economy. They were windmill owners *in spe* and naturally wanted to know which wind turbine was the best, a question impossible to answer. So the task was to inform and enlighten, to make the coming owner of a windmill as knowledgeable as possible for choosing his first wind turbine.

Consequently, much time was spent on elucidating the advantages and disadvantages of upwind and downwind wind turbines; whether to have the mechanical brake on the fast or the slow shaft; whether glass fibre was ecologically acceptable as blade material; was the Darrieus turbine not the ideal; should the wind turbine be connected to the grid; did water brakes have a future; how great should the safety factor of the wind turbine gear be; could you build one yourself or would you need help from the outside; should you own the turbine all by yourself or together with others, etc.

Many of the questions are still completely relevant today. However, things developed so fast, that elementary experiences with the operation of one type of windmill had hardly been made, before it was out of production and been replaced by something new, bigger and more economical. When once in a lifetime, you had to enter the difficult and new market for wind turbine acquisition, you might have been an easy catch for a ruthless salesman.

7.6   The Pioneers' Pop-Up

Combating pollution, wastefulness, noise, private cars and atomic power were the themes of the NOAH summer camp near Grenaa in 1975. Here a wind turbine built in a week provided the power. Hot water came from a solar collector made from parts of a fridge. The cooking was done on a solar stove, and an oil barrel served as a biogas digester—all of it very primitive, but very inspiring.

So when it is feasible to build a windmill tower of thin spruce stems, and it works, it must be very straightforward to work with it all, was the conclusion of many visitors. On 15 July 1975, leading Danish daily newspaper *Politiken* wrote about the camp that, "it is possible to wrench all the energy that people need from sun and wind. (...) Nuclear power plants that Elsam and Risø want to force on us, are the wrong track."

Some people went home to show how it could be done in a technically and aesthetically professional manner. "It would really be impossible to make people support renewable energy with the low level of technology demonstrated on the camp," blade pioneer Erik Grove-Nielsen remarked much later to me. We had both been at the camp.

7.7   The Swedish Connection

When it comes to design, manufacturing or purchasing wind turbines, people often did not really know what to believe. In 1975, there was only one book specifically about windmills, and it was in Swedish. Its title was *Vindkraftboken* (*The Wind Power Book*), written by Bengt Södergaard, who sadly enticed many do-it-yourself people, amongst others myself, into building windmills. None of them ever came to work properly despite the many elaborate instructions and calculations. The author fatally lacked the basic experience. He sold dreams, and it is, of course, possible to learn from the mistakes one makes. Many of the later Danish windmill designers learned from Södergaard how not to do it.

By a funny coincidence, another Swedish dream was onstage at the same time. At Kolding Folk High School, architect Carl Herforth was engaged in building a zero-energy house, which was to get its power from a windmill. He had found a Swedish windmill, which, regarding price and efficiency, beat everything else. It was a sensation that made its way into the newspaper columns. Permission had been obtained from the town council to erect it, and from the power utility of Kolding to connect it to the grid, in order to make the electricity meter run backwards, when the windmill produced more power than the house consumed.[2]

An independent Danish daily newspaper *Information* reported that this was the first time such a permission had been given in Denmark, thus missing the fact that already in the 1950s, J. Juul had erected three grid-connected windmills on the islands of Zealand, Møn and Falster, with rated power from 11 kW to 200 kW, using the principle of the asynchronous generator. This illustrates clearly that, when not kept up, fundamental knowledge and experience

---

[2]By 2012, it is called net metering.

can be lost in less than twenty years. In 1975, everything about J. Juul's pioneering principles had apparently been forgotten, even despite the fact that the Gedser windmill (in Falster) had been made ready for experimental service again.

The Swedish windmill from Vindkraftbolaget in Sollentuna, that was to be installed in Kolding had been constructed by engineer Ivan Troeng. It had 8 kW and two blades with an 8 m span. The price, as well as the annual production, was sensational. The cost would be DKK 10 000 (EUR 1 400) when placed on a 10 m tower, and the annual production was calculated at 16000 kWh in Kolding! On the West coast at 10 000 kWh more! That would mean a price of 8 øre/kWh (EUR 0.011) or one-third of the electricity price of power utilities, which was then 27 øre (EUR 0.035).

Such figures startled all Danes interested in wind power. Other windmills cost 3 to 10 times as much, and now it was suddenly possible to predict a triumphal progress for wind power. But not everybody fell flat. "A check of the price of the Swedish company to make sure that a comma has not been placed wrong must be an absolute necessity," said manager Jean Fischer to the Danish Academy for Technical Sciences (ATV), according to *Information*. The price was found to be correct, but utopian, and after that, silence descended on the Swedish windmill. Danish manufacturers could breathe freely; there would be no Swedish entry on the Danish market.

The principle of the connection to the grid and the asynchronous generator had been launched in a manner that anybody could understand. About Ivan Troeng's windmill, architect Claus Nybroe wrote in *Information* that "the wind power system based on the asynchronous generator principle, has a great future. This is a technology that can be mastered by a great number of smaller enterprises all over the country." Claus Nybroe has been proved perfectly correct.

The entire later Danish windmill industry including Vestas, Nordtank, etc. became technically tied up with the asynchronous generator principle. And we must be grateful to Troeng from Sweden, because he reminded us of its advantages. Almost 20 years later, an alternative appeared in the shape of the synchronous ring generator, as a rival to the asynchronous principle with the advantages of the asynchronous generator's but without its evident drawbacks.

PART II

1975–1980: The Years of Breakthrough for Danish Wind Power

With the asynchronous generator rediscovered, this chapter will present various sides of the origin of the contemporary wind turbine design and who made the crucial conceptual and technological contributions. It is important to acknowledge the work of Johannes Juul and Christian Riisager, who in 1976 was the first to commercialise a wind turbine using Juul's design principles.

It was not the hybrid blade technology of Juul and Riisager, however, that lead the way to the development of the contemporary wind turbine in Denmark, but the 2 MW Tvind windmill, designed and built by a group of amateurs from the Tvind School.

In 1976, Tvind transferred Professor Ulrich Hütter's advanced blade technology from the DLR at the Technical University of Stuttgart to Denmark and made it available for the general public. Erik Grove-Nielsen's newly founded Økær Vind Energi brought to the market Tvind's downscaled 4.5 m blade, the basis of the emerging component wind turbine. Soon after that, NIVE defines the specifications for various wind turbine components (blades, controls) that enable small enterprises to manufacture 22 kW wind turbines.

This chapter describes how the combination of Juul's principles (heavy, upwind, 3 blades, asynchronous generator, stall regulation) with Ulrich Hütter's/Tvind's advanced blade design resulted in the winning wind turbine concept, called the Danish Concept. This process takes place 1975–1979 when the later successful manufacturers Vestas, Nordtank, Bonus, etc., commercialise the concept, which they, step-by-step, scaled-up and took to the world market that they dominated for the next 15 years. When other countries like Germany, Spain and later China, got into commercial wind turbine manufacturing, their designs were Juul-Hütter hybrids, that became industrial standard. Some manufacturers changed to synchronous generator, however, within the same basic concept.

This chapter highlights pioneering achievements in the development of the modern wind turbine and presents successful solutions that are in many ways related to the early efforts of developing MW-size wind turbines as well. The main issues of this development can be divided in following stages:

(1) **Juul's heritage:** Johannes Juul's design leads to Christian Riisager's wind turbine

(2) **The national wind power versus Tvind's bottom-up programme:** Tvind transfers Hütter's blade technology to Denmark

(3) **From NIVE to master blacksmiths: The follow-up on La Cour's tradition:** NIVE and the blacksmiths pave the way for the component wind turbine combining Juul's design and Tvind/Hütter's blade technology

(4) **Blades are the core of it all:** The importance and growth of the independent blade supply industry

(5) **The windmill industry goes international:** Wind turbine development of large corporations versus small enterprises

7.8 Juul's Heritage

The person, who paved the way for our modern 3-bladed windmills, was Johannes Juul (1887–1969), sectional engineer at the SEAS local power utility in Haslev. In 1951 he initiated a full-scale experiment with an 11 kW 2-bladed windmill. It was named after its location, Vester Egesborg. Juul harvested his early experiences from this windmill. One blade broke off, which made him conclude that it would be good to go from two to three blades, as it would thus give them the necessary strength.

As early as 1953, his first test set was working, and the 3-bladed windmill had come into the world, the type that became industrial standard and made wind power a realistic and economical form of energy. It is still striding victoriously all over the world: three blades placed in front of the tower, stall regulation, asynchronous generator linked directly to the grid and active yaw. They are the most important code words to characterise Juul's concept, on which innumerable variants have been embroidered, just as it has happened with other historical technological breakthroughs.

The F. L. Smidth & Co.'s Aeromotor from the early 1940s, installed on the island of Bogø, was later reconstructed by Juul. In its original version the wind turbine on Bogø had long, narrow, pitch-regulated blades. It was a fast runner, with a DC generator. During the Second World War, windmills of that kind were erected all over Denmark, in order to supplement power production from the many local power plants. Diesel oil was rationed, and by using wind–diesel systems, it was possible to maintain the necessary supply.

Figure 7.1   Remaining tower and nacelle of Johannes Juul's first windmill from 1951 in Vester Egesborg.

After the war, also the island of Bogø changed to alternating current. By reconstructing the FLS-windmill, according to the experience gathered in Vester Egesborg, it became possible to use a much simpler and more future-oriented wind power concept. Johannes Juul swam, so to speak, against the current. In theory a modern windmill ought to have narrow, fast blades that could be pitched in order to yield optimally. But already at the Vester Egesborg windmill, Juul had decided on fixed blades with a tip speed of only 38 m/s, whereas the fast runners might turn two to three times as fast. To ensure that the windmill would not run away it was equipped with a simple turning mechanism on the tip of the blades, called tip brakes.

Figure 7.2   At the redesigned Bogø windmill from 1953, the Danish windmill concept was successfully determined (left); The 200 kW Gedser windmill from 1957 operated for 10 years with high efficiency (right).

Another decisive innovation of the Bogø wind turbine was the use of the AC machine as a generator. It was just an ordinary 45 kW AC induction motor, running under-synchronously when functioning as a motor; but when connected to rotor and taken up into over-synchronous working area, it delivers power to the grid and maintains a constant speed. The more the wind blows, the more it produces. When the same electrical machine is used as motor, with increased load more power is drawn from the grid, without, however, changing the speed of rotations by more than a few percent. In this manner, Juul was able to keep the number of rotations constant.

By adapting the rotor size, the number of rotations and the capacity of the generator in an optimal way, Juul had solved a crucial aspect of windmill technology with very simple means. When the wind speed increases, more power is produced, until it reaches its maximum capacity at a wind speed of 12–14 m/s. At higher wind speeds the rotor will stall. The windmill is stopped only at wind speeds over 25 m/s.

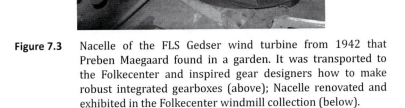

Figure 7.3 Nacelle of the FLS Gedser wind turbine from 1942 that Preben Maegaard found in a garden. It was transported to the Folkecenter and inspired gear designers how to make robust integrated gearboxes (above); Nacelle renovated and exhibited in the Folkecenter windmill collection (below).

7.8.1 Juul's Record Stood for Almost 30 Years

In 1961, Johannes Juul presented the production figures of his windmill at a United Nations conference in Rome, showing the results of the Bogø wind turbine before and after the reconstruction. The improvements were convincing. Whereas the FLS 1942 version had delivered 114 kWh/m² swept area per year, Juul multiplied the production five times to 610 kWh, on the same site with his version. Annually, the 45 kW AC windmill produced 80 000 kWh. And it was not as a single occurrence. Measurements exist from the years 1953–1960 with production figures so high, that we have to wait until the mid-1980s until the new generation of wind turbine manufacturers, using fibreglass blades and computer technology, are able to make windmills with the same output per annum, as Juul demonstrated in the 1950s.

With the Bogø wind turbine, the technical breakthrough was a fact. Johannes Juul had created a concept, which later became the new industrial standard for wind energy. It was logical and comparatively simple to scale up the Bogø windmill to the 200 kW of the Gedser wind turbine, erected in 1957. It operated for the following 10 years and was too a success. In other parts of the world various types of wind turbines were developed and tested, but none of these were based on a concept that would later have any commercial relevance.

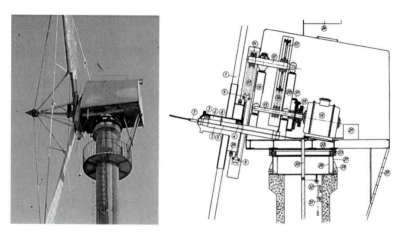

Figure 7.4   Nacelle of the 1957 Gedser 200 kW windmill, upscaled version of the Bogø windmill (left); Section (right).

The Bogø windmill was, technically speaking, of a more elegant design than the 1957 Gedser wind turbine, which for economical reasons got a chain drive transmission, whereas the Bogø windmill, although built at a time of war and scarcity, had an advanced FLS gearbox of the so-called integrated type: a compact unit, low noise and of the same high mechanical quality as the gearboxes that FLS manufactured in big number for their cement factory machinery.

In fact, the rather primitive chain drive was the weak point of the Gedser windmill. The chain enclosure was not 100% tight, which caused lubrication oil film on the surrounding crops. That did not make wind energy popular among the neighbouring farmers.

Figure 7.5   Two generations of gearboxes. To the left, FLS-type integrated gearbox from 1942 as used on the Bogø and the first Gedser windmill. The white gear to the right came from Tacke Getriebe in 1992. It is installed in Hanstholm, on the 525 kW windmill designed by Folkecenter and similar Tacke wind turbines.

Figure 7.6   Inside of the nacelle of the Gedser wind turbine: chain gearbox, anti-shaking system (red ball—when falls, the windmill stops). The nacelle is exhibited at the Danish Energy Museum, Tange.

The Bogø windmill was overlooked and forgotten in the history of technology of the post-war years, because the newer and bigger Gedser windmill was still standing there, when the interest in wind energy was revived in 1973. In the meantime, Johannes Juul died, and the engineers from the FLS epoch, who might have transferred his experience, either did not recognise the scope of Juul's achievements or they were engaged in quite different activities in various institutes or the industry.

Among energy researchers, the belief in nuclear energy was far stronger than their belief in a future for wind energy, at least in the technology that carried J. Juul's name. He did not in any way belong to the establishment of research or industry; he was a practitioner, a kind of James Watt or Rudolf Diesel, who was able to make things work. We do not find serious interest either in J. Juul's wind turbine concept in the many official projects that materialised in the United States, England, Germany, Sweden, the Netherlands, and Denmark after the 1974 oil crisis.

7.8.2   Windmill Pioneer Christian Riisager

The honour of rediscovering J. Juul's unique technical results and turning them into practice belongs to Christian Riisager (1930–2008). J. Juul's extremely elaborate and clear-cut descriptions of his wind turbines in an UN conference report were a kind of construction manuals which demonstrated and illustrated his design in a most educational manner.

In this way it was plain sailing for practical people like blacksmith Erik Nielsen and Christian Riisager, whose son was, it has been told, in the air force and could translate English texts, at that time when the wind turbine was conceived. Riisager and his son are likely to have studied the J. Juul's papers together. But as everybody else at the time, they failed to note that Juul had ensured his windmills against damage and runaway by using tip brakes on the blades.

Riisager deserves particular mention, as without him, perhaps the hurdles of the first difficult years might not have been passed. In fact, he made a very good start. In 1976, together with master blacksmith Erik Nielsen from Herning, he installed two 22 kW windmills for journalist Torgny Møller at Vrinners Hoved, Mols, and teacher Karsten Fritzner in Boddum, Thy.

Both sites had good wind conditions. As both windmills turned out reliable, and with a high production figures, Riisager became famous. People came from near and far, to see these two wind turbines. They were perceived as technological miracles.

This was still only a few years after the oil crisis in 1974, the greatest national Danish disaster since the Second World War.

Things were so dramatic, Karsten Fritzner told, that one night he had caught some persons, climbing the 12 m high tower. They had opened the nacelle to learn about the technical secrets that so efficiently delivered power from the wind.

Figure 7.7   The first two Riisager 22 kW windmills in Vrinners and Boddum (on the photo) paved the way for the later Danish wind energy adventure, 1976.

Figure 7.8   Riisager's very first design of a grid-connected wind turbine, installed in the garden of his private residence.

Figure 7.9   Riisager wind turbines appeared in many different configurations and sizes.

It is unknown to history, whether those uninvited guests were able to make use of whatever they found during their nightly espionage raid in Thy. But as I am now in possession of the wind turbine in question, it can be revealed that in the nacelle made of plywood there was a second-hand heavy-duty rear-shaft from a British military vehicle integrating main shaft, brake and gearbox.

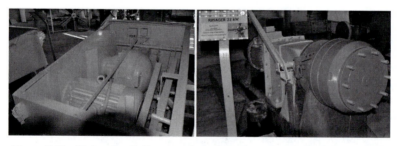

Figure 7.10   From the Folkecenter's Collection: the first Riisager 22 kW nacelle using military vehicle rear shaft (left); Riisager-clone made by Sonebjerg Maskinfabrik (right).

A simplex chain drive made the connection to a generator of the asynchronous type. Furthermore, there were some mechanisms consisting of a weight placed on the blades and bars which were able to connect the rotor to the yaw gear, and thus turn the windmill 90 degrees out of the wind and, consequently, safeguard it from over-speeding. The aerodynamic part of the blades had an outer shell of waterproof plywood. Evidently, the maker was a master carpenter. Later on the shell was made from fibreglass.

Riisager also stood to his principles when he made the tower. He preferred lattice towers and did not follow the fashion when in the early 1980s the fast growing industry introduced elegant tubular type towers, which made it possible to ascend safely on the inside.

It is hard to say how things would have turned out for Riisager, if the very first windmill had not been sold to a journalist of *Information*. Torgny Møller wrote enthusiastically about his fantastic windmill, which provided power and heating for the family and the Jutland office of the paper. Every month provided irrefutable production records, which were really tangible evidence. It produced more than 30 MW a year, seven times the power consumption of an ordinary family.

As this miraculous windmill cost less than DKK 50 000 (EUR 6 700), it became evident to the readers of *Information*, who were not used to reading much of this kind, that wind power could liberate the Danish people from the control of the power supply wielded by the Arab oil Sheiks, something that was ruining the country's finances and self-esteem.

That was not the only reason why Riisager came into focus, as one of the good sons of the country. The state had launched a wind power programme in conjunction with the power utilities. They made flashy brochures for the Folketing (parliament) committee, presenting the Nibe A and B wind turbines, developed by the leading industrial companies of the country. And this was in contrast to the alliance of Riisager, Møller and Fritzner, which in accordance to the good old Jutlandish custom of including the blacksmiths, the local entrepreneur and the electrician, had actually found the solution to the country's future energy supply by paying out from their own pockets. That was the kind of things that Danes liked to hear. It fitted perfectly with their ingrained scepticism towards experts and big capital.

Orders for Riisager's windmills poured in. He made an endless number of design changes throughout the years, with changing cooperation partners in the industrial environment. However, blades, tower and controls remained the same; but in regular intervals, Riisager created quite new technical solutions with new drive trains and nacelles for his windmills. At one time Christian Riisager preferred Italian planetary gears from big earth-moving machines that proved to be extremely noisy, at another a proper machine construction from Sonebjerg Maskinfabrik near Kolding.

Figure 7.11   Riisager 30 kW made by Sonebjerg Maskinfabrik.

An industrial adventure was expected when windmill maker Riisager turned up in Vordingborg, where in the mid-1980s, a wood industry was looking forward to become an important supplier of laminated wood blades for Riisager.

The dreams, however, never became true. At the time, several new competitors had turned up, and they worked by a quite different concept. They were able to provide an up-to-date industrial product with well-constructed mechanisms, wrapped in a modern design which harmonised far better with the spirit of the times.

The new people also had an enormous advantage—they did not have to design and make everything themselves; a new industry of specialised sub-suppliers with compatible components had emerged, and they did not fit in with Riisager's concept. The drawings and designs belonged to Riisager but he did not manufacture, he depended on his own sub-contractors, whereas the emerging independent sub-suppliers delivered components to the entire industry.

Although there were clear differences in the production methods between Riisager and those of the new industry, they had something very important in common compared to the Danish government and many foreign projects within wind energy: They were based on Johannes Juul's wind turbine principles.

The Riisager epoch ended around 1985, when the production stopped. By 2013, there are still some Riisager windmills standing here and there in the Danish landscape and abroad, providing their owners with wind-powered electricity.

Figure 7.12 Riisager first-generation wind turbine when it arrived at the Folkecenter's collection of windmills.

7.9 The National Wind Power versus Tvind's Bottom-Up Program

In 1975, the Danish Academy for Technical Sciences (ATV) recommended that work on wind energy should be continued; the country had strong traditions to build upon. It was no great matter to grant DKK 50 million (EUR 6.5 million) to a state wind power programme. But how should such a project be implemented? Whereas abroad there was an aviation industry, with some of the knowledge needed for wind turbine design; the Danish officials responsible for the realisation of the state wind power program had no obvious candidates from research and industry with relevant knowledge and experience. The result mirrored this fact.

An obvious comparison falls between the state Nibe A and B wind turbine program and the Tvind Schools' 2 MW windmill. Both were conceived in the wake of the energy crisis. Of course, Tvind did not possess any kind of expertise on wind power. They knew how to drive a bus, build houses or run a school. And they even planned to build a wind turbine three times the size of that of the state project; they had not got much money; they lived in a remote place in Western Jutland, far away from institutes and experts; they had poor relations to big industry. So one would think that all odds were against Tvind.

However, the Tvind people had one unique advantage, compared to the researchers in the state program: they recognised that they did not know anything about wind energy. An awareness of that lent them strength to go out into the world, search out and involve people with relevant expertise and experience and who wanted to join the task of lifting the Tvind wind power project.

It is not politically correct to give Tvind credits of having created important basis for the Danish windmill industry. Written stories about the Tvind windmill tend to point out the fact that Tvind received the same professional assistance from Denmark's Technical University and Risø (Denmark's former atomic energy test station). Also, often in the history of wind power, links are skipped and touching stories are told about individuals, who worked hard in the Tvind windmill team.

It is a fact that the industrial establishment did not believe in any future for wind power. Therefore, the big 2 MW Tvind windmill and its downscaled 18 kW version were of great technical and symbolic importance for the development of wind power. The Tvind people were innovative and courageous; they inspired many others to work with wind power, which in a number of years grew into a strong bottom-up movement.

Figure 7.13   Tvind windmill, created by amateurs from scratch. In operation since 1978 (left); In 2012 in red and white livery by Utzon Architects (right).

The resulting windmill was a feat, for which posterity has not shown proper appreciation. International publications on wind power tended to overlook it, and it was not presented at international conferences as the impressive example it actually was. In fact, the Tvind windmill would have been a magnificent success anyway. And it will undoubtedly go on, towering over the flat Western Jutland landscape for many more years, and provide the schools with clean energy even it was not designed by the most famous international institutes and industries.

7.9.1   Money Alone Makes No Difference

Neither the Nibe wind turbines, nor the Tvind wind turbine, were constructed with a purpose of industrial production. Tvind remained a single example, while the group of enterprises behind the Nibe A and B, disintegrated. Whatever was left over, was made into a new consortium, Danish Wind Technology A/S (DWT), initiated by Poul Nielson, the Minister of Energy. Now was the moment for the big companies, ASEA, Vølund, SEAS and the state to demonstrate how professionals could develop and manufacture wind turbines. Still that did not get them anywhere.

Helge Petersen had been nominated as a consultant with Danish Wind Technology, DWT (Dansk Vind Teknik, DVT), in Viborg, a new industrial enterprise, where he was responsible for designing and developing a range of various sizes of innovative wind turbines for DWT. The defeated competitor from the Thisted Airport was Petersen's favourite.

In 1981, the then energy minister Poul Nielson presented the new company DWT in Nairobi in Kenya during a big international energy conference together with powerful participants from big industry, such as ASEA and Vølund. Together with the Danish state each company owned one third of the shares. "Now the rest of you in the wind energy sector can just go back to bed," prophesied Mr Nielson, powerfully seconded by his chief of section, Ove Dietrich. They were convinced that they could design and manufacture windmills that were better and cheaper.

The 15 kW DWT windmill shown in Nairobi got a blade damaged after just a quarter of a rotation. René Karottki and I told a British journalist covering daily news of the conference about the accident and were told that we had done harm to vital national industrial interests. Later on the windmill was donated to Kenya as an official gift and erected near Lake Victoria. It never came to work as a windmill, however, the tower made a good use as a waste water pipe in a local fishing industry.

"All in all the new windmill giant, DWT, performed better in public relations than in the making of windmills," reported Ejvind Beuse, engineer and folk high school teacher, in a critical portrait in the journal *Vedvarende Energi* about another DWT windmill also created by Helge Petersen.

The next version of the unusual prototype—the giant 265 kWh windmill from Koldby—was erected next to the Nibe A and B 630 KW windmills, the joint Danish government and central utility project. The DWT wind turbine was almost a "freak" with its thick tower at the bottom, and knitting needle thin upper part of the tower carrying the nacelle and blades on top. With lack of proper proportions its appearance was simply ugly.

New versions of that controversial style windmill would not sell either, in spite of the strong industrial background. In the article

Beuse asked whether this was a case of manipulation with state support, and he saw this partly state-owned company as a threat to the other twenty windmill factories. But his worries turned out to be superfluous. The DWT products performed too badly and looked wrong, so no one wanted to buy them. At the technical school in Tønder 10 different windmills had been erected for teaching purposes. "We'll say, 'no, thanks,' to the DWT 15 kW, even as a gift," said *Jens Jensen* from Rejsby, then president of the school's board.

Figure 7.14   One of the peculiar concepts within the DWT program was the 265 kW downwind windmill. The prototype was installed in Koldby, close to the North Sea.

The windmill had been subject to comprehensive testing at Risø who gave this state-sponsored windmill a good rating of performance; however, its possible virtues did not convince potential customers and the interest soon vanished completely.

During the following years DWT had a turbulent existence. ASEA pulled out, and SEAS, a local utility company joined in, and everybody lost money on this affair. Not until the company was made part of Vestas was any progress made. Vestas made use of the name, became Vestas Danish Wind Technology and took over the big manufacturing halls in Viborg, which they used for a number of years.

Figure 7.15   700 kW DWT wind turbines at Masnedø Island.

Figure 7.16   Two generations of windmills are shown in the left image: Esbjerg 2 MW turbine from 1988 (left); Vestas 1.5 MW from 1996 (right) in Tjæreborg Enge. Grounding the Esbjerg wind turbine is shown in the right image.

This era of the wind power development came to an end, when the 2 MW Esbjerg windmill in Tjæreborg Enge was grounded by dynamite. The power utilities, the state and the EU had spent DKK 70 million (EUR 9.5 million) on it, 10 times of the cost of the Tvind wind turbine. It was a monstrous amount, compared to the funding at the disposal for developing the smaller wind turbines that crept up into the MW-class during the 1990s, but were based on a quite different design and philosophy of production.

7.9.2 Giants and Their Strategies

The only consolation and excuse for the lack of success of the Danish state wind energy projects, was that the British Wind Energy Group (WEG) had similar experiences. They were backed too, by the big corporations and had been pampered with almost all the Britons' funding for wind energy R&D. Just like the DWT, the WEG was unable, despite its impressive range of products, to obtain just a modest bite of the lucrative Californian market in the mid-1980s, in which otherwise anything that looked like a wind turbine, could have been sold at soaring prices.

Figure 7.17   WEG wind turbine (Photo: Arne Jaeger).

It requires a more profound historical analysis why the big funding, laboratories, well-paid researchers and goodwill from the state, failed against barefoot research and development which carried wind power through to a commercial success.[3]

Towards the end of the 1980s, I was invited to lecture on the Danish wind power adventure at the research centre of Espoo in Finland. David Lindley, director of WEG, was the other lecturer from abroad.

The Finns had some confidence in me, as I had been an advisor at the gear manufacturer Valmet. I sensed they were interested in having a windmill industry of their own. However, Lindley spent all his energy in trying to convince the Finns that the Danish concept was atypical and technologically too primitive, with many rather small suppliers, whereas his company aimed at research and professionalism.

Apparently, Lindley must have been a more convincing speaker. The Finns bought a WEG wind turbine and gathered the same bad experience as other WEG customers. The customers never returned; and without customers, no industry. This is part of the story why Denmark has a wind power industry, whereas England got none and has to buy almost all its equipment from abroad.

A story like the one told here about DWT can also be found in other countries, which invested enormous amounts of money in the development of giant wind turbines. The people in the state administrations systematically avoided giving financial support to anything just a little similar to the Danish J. Juul concept. The technology had no scientific potential and was not leading to technological innovation—were the answers to the applicants, who then changed their minds to satisfy the program evaluators and wrote new applications with concepts just as unfortunate as those of DWT and WEG.

In the United States, Boeing, Westinghouse and others, took part in the development of new wind turbines. In Germany the big corporation MAN took leadership, while having DM 100 million at their disposal for a single wind turbine—GROWIAN, as the big wind power plant was called. After 200 hours of testing, it was scrapped. Similar stories could be told about the Swedish giants Maglearp and Näsudden.

---

[3]See chapter *Networks of Wind Energy Enthusiasts and the Development of the "Danish Concept"* by Katherine Dykes.

Figure 7.18 WEG wind turbines (Photo: Arne Jaeger).

In the case of all these big and very ambitious projects, it was professionals from universities and big corporations, who designed and developed the constructions, and led the work, followed by great public attention, glossy publications and presentations at international conferences.

In the initial phase, the experience gained from the big test wind turbine was generally so depressing, the technology so complicated and costly, that wind power would hardly have been a relevant option today, if not for the Danish parallel development from below, starting with windmills of 20 kW, not 2 000 kW, made by independent enterprises and available without exclusivity to all potential manufacturers.

In the 1980s, German pioneer enterprises such as Enercon and Tacke or Spanish Ecotecnica got the idea of using components from the Danish supply chain. Without it, perhaps these countries would hardly have had their sturdy wind power industry, which in the 1990s became the only real competitor to Denmark.

In the end of the 1980s, it was difficult to estimate the technological influence of these costly public development projects on the contemporary commercial windmill industry with its small machines of 30 to 200 kW. The research projects, however, later proved to be all-important for the development of the later MW-size wind turbines.

7.10 From NIVE to Master Blacksmiths: The Follow-Up on La Cour's Tradition

In December 1976, Bjørn Rossing who lived in Vestervig in Thy, wanted to buy a 22 kW windmill from Christian Riisager. Rossing already knew from Karsten Fritzner, who lived in the same municipality, that Fritzner's price was lower than DKK 50 000. For Rossing, however, the price had gone up to DKK 75 000 (EUR 10 000). Rossing found that it was too costly.

This caused the recently established NIVE group in Thy, composed of local engineers, farmers, blacksmiths, teachers, and the author of this narrative, to join forces on a 22 kW windmill project. We were convinced it could be manufactured at a lower price than those DKK 75 000. Five sets of prototypes blades, with low tip speeds similar to Juul's models, were ordered from Økær Vind Energi. This was the starting signal for the Økær 5 m blade, which was later used by many producers for their windmill projects.

We had blades, but still a control was lacking. This part of the project was placed with a member of the group—engineer Finn Bendix of HM Automatic in Thisted, where he was working with controls for industries. When the parameters for the controls had been set, it was a minor task for Bendix to design windmill controls that would work.

Figure 7.19 Hydraulic brake testing for the NIVE wind turbine (left); Testing the Kumera gear for the 90 kW NIVE wind turbine, gear capacity 235 kW, main shaft 220 mm, August 1982 (right).

Probably the prototype NIVE wind turbine was the very first in Denmark to have a wind vane and electrically powered yaw motor, another of Juul's principles, whereas most other people preferred free yawing—the yaw motor consumed power, the argument went.

The NIVE contribution to the general wind power development was a low noise and efficient blade and the fact that HM Automatic could deliver windmill controls to the various wind turbine developers. Blades and controls now were available and supplied among others to the Herborg blacksmith Karl Erik Jørgensen and Henrik Stiesdal for their joint design work that was going to be the first Vestas windmill.[4]

7.10.1 Lessons Learned

High school teacher Kaj Fjendbo Jørgensen wanted to buy the first wind turbine of NIVE design. HP Maskinfabrik in Thisted assembled it, thus becoming the producer of the windmill; the same task could just as well have been undertaken by one of the many other blacksmiths in the local area. They would very well have been able to screw it together.

The wind turbine design knowledge, the selection of components and coordination was within the NIVE group members, who were neither teachers, like in Tvind, nor do-it-yourself people. It was a working group, a model of which was to be found in the beginning of the century in Askov, where Poul la Cour put his constructions at the disposal of the SMEs of the Danish people.

In the NIVE team we saw our efforts as a follow-up of the la Cour tradition. Our research and design work was available in the form of detailed construction manuals and drawings. There were no patents and all-important components were commercially available at competitive prices. With modest requirements for engineering and tooling SMEs, without previous experience in wind turbine manufacturing, within a few months could bring a state-of-the-art wind turbine to the market and as their own brand.

After having been exhibited in the local Sjørring Hall, Fjendbo's 22 kW windmill was erected and plugged into the grid. From the beginning it worked in a satisfactory manner with astonishingly

---

[4]Karl Erik Jørgensen's windmill development is recorded in chapter *From Herborg Blacksmith to Vestas* by Henrik Stiesdal.

few problems, considering that it was a prototype. The windmill was sturdy, with high mechanical safety factors. But unfortunately it made too much noise. The otherwise robust British Fenner gear, mounted on the main shaft, was not well suited for a windmill, and for that reason it was not used for the next version.

Like other windmills at that time it did not have air brakes to prevent runaways in case the mechanical brake failed. In that area we sinned against the dogma of Juul, which caused the Fjendbo family to feel uncomfortable about having the windmill operating during the nights, when the gales swept over Thy.

Even before the prototype was sufficiently tested, the HP Maskinfabrik began to sell the wind turbine, bypassing the NIVE group, which caused some complications, but it made no real difference. The foundation had been laid for something which aimed further than at one single small enterprise in Thisted, which had, by the way, done excellent work and had also been paid for it.

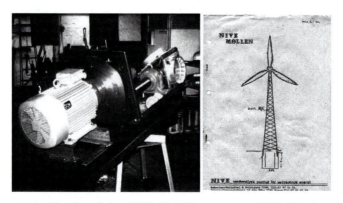

Figure 7.20   Nacelle of the NIVE 22 kW prototype, 1976. Wind turbine characteristics: Asynchronous generator, upwind, fibreglass, fixed hub, stall regulated, electric yaw (left); Cover of the NIVE brochure about windmills from 1977 (right).

The vision was that there should be free access to professional know-how about construction with SMEs as key players. The wind belongs to all of us. The knowledge of how to utilise it should also be available to the people. And during the following 15 years this came true through a wide range of public domain constructions from NIVE and since 1983 from the Folkecenter for Renewable Energy, for which public domain design became a core activity.

7.10.2   Knowledge for All

On 30 September 1978, the Danish Organisation for Renewable Energy (OVE), of which I was president, held its fifth wind get-together/workshop (VindTræf) since 1976. The big auditorium of Brandbjerg Folk High School was almost full. The most important topics were the report from the committee for windmill security and the presentation of Risø's test station for smaller windmills, introduced by its newly elected leader, Helge Petersen.

However, the most important challenge of the workshop was to discuss the large-scale implementation of the new generation of small windmills and provide them with a stamp of approval. As chairman of the meeting, it became my task to unite the widely diverging ideas. This was rather decisive because at this workshop we had stepped far outside the inner circle of the energy activists.

Professional associations, trade promotion officers, manufacturers, consumers' organisations, politicians and researchers as well as energy secretariats, consumers' foundations and associations and people who were building their own windmills, were among the participants. It is beyond doubt that this get-together was epoch-making for the wind power. For me it was especially relevant. I got in contact with blacksmiths Frits Sørensen and Jens Jensen, president and board member of the Danish Blacksmiths' Association (Dansk Smedemesterforening), later DS Trade & Industry, representing 2 200 SMEs, among those, by the way, also Vestas.

Two blacksmith leaders participated in the meeting to get inspiration and establish contacts to enable their members to manufacture wind turbines in the many small enterprises. The investment boom in the agricultural sector after the EC membership had brought good jobs for numerous blacksmiths, but later it died down and they now wanted a role in the new renewable energy industry.

During the industrialisation, the blacksmiths had already built up a fund of experience in making successful design manuals with standard constructions at the disposal for their member enterprises. Since around 1960, blacksmiths all over the country could manufacture state-of-the-art agricultural transport vehicles using the manual, which for instance Vestas did, grew by it and eventually designed their own agricultural machinery. Since 1978, drawings of steel beams for farm buildings became an important service for the members. So why should not windmills, biogas plants and solar panels have a future as something for what a construction manual could be made?

This sort of entrepreneurship fitted ideally into the NIVE visions of renewable energy, local production and openness. It has, in fact, been carried on ever since within the framework of the Folkecenter for Renewable Energy by embracing various renewable energy technologies offering them as technological and organisational support to many companies until they were able to go on their own.

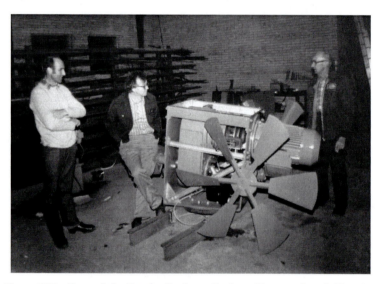

Figure 7.21   From left: Bendy Poulsen, Preben Maegaard and Henning Poulsen with the prototype 22 kW made 1978 at H. Poulsen & Søn's workshop in Lyngs, using the technical counselling of NIVE.

It was quite unusual to find support for renewable energy in the business establishment in the 1970s. The Federation of Danish Industries represented the big industries staked in 100% on atomic power with all its influence. But the Danish Blacksmiths' Association's boldness and foresight also won the Danish Federation of small and medium-sized enterprises for renewable energy. One of its leaders, Jesper Bøge, put forward these views with zeal in places where the young and small renewable energy organisations would never have gained an ear. The importance of that kind of alliances in a society goes far beyond what can be measured immediately and is part of the process, which in 1985 caused Denmark ultimately to give up plans of introducing atomic power.

It was for the sake of their commercial interest, rather than for the sake of energy policy, that the blacksmiths, *via* their organisation favoured wind power. For this reason it was vital to motivate the members into manufacturing wind turbines. A prototype 22 kW was made at H. Poulsen & Søn in Lyngs, using technical counselling of NIVE. In his youth Henning Poulsen had worked with la Cour-type windmills. For his son, Bendy, wind power soon became an engagement, which took up his working hours as well.

The design was to belong to the Danish Blacksmiths' Association, but in principle it was at the disposal of all interested parties.

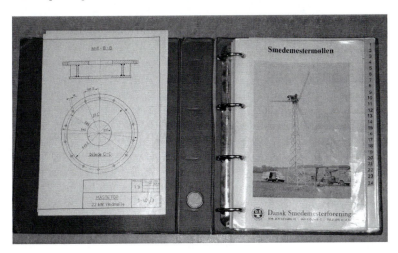

Figure 7.22   Blacksmith's handbook for windmill design from the Danish Blacksmiths' Association.

There were some changes in comparison to Fjendbo's first NIVE windmill: The windmill was supplied with yaw rotors (they did a good job with Riisager) and it was to be blacksmiths, not electricians, who were going to produce the windmill. However, the choice of yaw rotor was a mistake, it was not a technology that pointed forward, it represented something *altmodisch*. On the other hand, the windmill got a sturdy and in particular a low-noise gear from German Bauer, which became the standard for low acoustic noise of the time.

Figure 7.23 Anders Petersen's 22 kW NIVE-designed Blacksmith's wind turbine, Rangstrup, 1979.

Figure 7.24 Heavy-duty low-noise Bauer gear.

The prototype was erected at the Risø test station in order to get its approval. That happened in September 1979, a year after the first contact to the Danish Blacksmiths' Association was made.

The windmill was installed without controls. The plan was to experiment with new controls, developed by Professor Ulrik Krabbe, DTU. The university did not, however, succeed in making the new controls to work properly, and the wind turbine got conventional controls from Hårby El. They worked impeccably.

Figure 7.25   Blacksmith's 22 kW windmill being tested at Risø.

The testing at Risø took an unreasonably long time. Different opinions had arisen about what such a test station was to be used for. Risø considered the wind turbine to be fully developed, whereas many smaller producers expected that the test station with all its equipment, experience and professional staff, would take an active part in getting the adjustment of the blades to optimise performance, etc. The fact was that the windmill had never been in operation before. It was erected for testing and optimisation, and thus the designers and producers had had no occasion for fine-tuning. That part of the process of development was definitely not in harmony with the la Cour tradition.

During the process, the test windmill was equipped with a second generator, which in theory would lead to enhanced production, according to the ideas of the time. But in practice it only augmented annual production by 2 MW on top of the measured 36 MW on one generator. This would not justify the use of two generators. It was just a fashion.

Figure 7.26 Risø test station in the late 1970s with a mixture of types. To the left a NIVE design for the Blacksmith's . The 2-bladed wind turbine developed by Ulrik Poulsen was called the "Giraffe" or the "Stork". It had long extended aluminum blades. This example illustrates the wide variety of types and concepts in Denmark by that time: they all got a chance.

7.10.3 Local Manufacturers

Along with the visionary president of the Danish Blacksmiths' Association, Frits Sørensen, I organised presentations all over Denmark, in my capacity as NIVE coordinator. The message to the smaller enterprises was that now they could all become wind turbine manufacturers for a rapidly growing market with great opportunities for export as well. No technological impediments existed; we could furnish a complete design manual and support. The windmill had the approval of the authorities.

Various enterprises all over the country were turned on by the engagement in renewable energy of the Danish Blacksmiths' Association and began to produce the 22 kW windmill. However, the technological development was running fast forward. There were already 55 kW windmills being produced, and now for the blacksmiths the next initiative was a 90 kW wind turbine. It was a big jump, too big, as it turned out.

The 10 m Alternegy fibreglass blades which should be tried out for the first time, delayed in materialising. Therefore a new project featuring wooden blades, supported by the Ministry of Trade & Industry, was initiated in 1982. These blades were produced in a new workshop hall which in 1983 became part of the Folkecenter.

Figure 7.27 Production in 1982 at the Folkecenter (above) and transport (below) of wooden blades for the 90 kW Folkecenter-designed Blacksmith's windmill for Rudbjerg, Lolland.

The blades were made from solid, laminated spruce, based on experience from a NIVE experiment in Alaska in 1980, and followed by a number of experimental blades, that is, some 6.5 m for the experimental windmill of Danebod Folk High School, under the leadership of René Karottki. Many expected that solid wooden blades would be far too heavy, but in actual fact they were able to compete with the rather heavy fibreglass blades of that period.

Figure 7.28 The 90 kW Blacksmith's windmill (Rudbjerg) with wooden blades.

The 90 kW wind turbine was produced at master blacksmith Anders Davidsen's in Lolland, according to drawings by a design team under the auspices of NIVE, who here learned to design and dimension a wind turbine bigger than any other in the young wind power industry (apart from the projects of the power utilities).

Figure 7.29 Manufacturing the lattice tower for the 90 kW windmill in the workshop of Anders Davidsen, Naesby, Lolland, 1982 (left); President of the Danish Blacksmiths' Association, Frits Sørensen speaks at the inauguration party of the 90 kW Blacksmith's wind turbine in Rudbjerg. In 1982 it was the biggest commercial wind turbine in Denmark. Anders Davidsen was the manufacturer, and wooden blades came from Folkecenter (later AeroStar blades were used) (right).

At Vestas, Nordtank, Bonus, etc., the biggest type was 55 kW. The cooperation with Davidsen was highly enriching. He involved 100% in the windmills, personally and financially. Davidsen was a truly idealistic activist in the most positive sense.

Figure 7.30 Anders Davidsen arriving at his office, Lolland (right); Davidsen's company logo (right).

The design team were civil engineer Jacob Bugge (author of a popular wind energy book[5] and with experience from the Nibe windmills); engineer Henrik Kirchheiner (later of Hanstholm-møllen ApS), Kim Andersen and Knud Buhl Nielsen. During the process, Knud and Kim were employed by Anders Davidsen to design wind turbines of 22 kW and 60 kW, intended to be of his own design.

Figure 7.31 Lolland 65 kW prototype (left); Lolland wind turbines in the United States (right).

---

[5]Bugge, J (1978) *Bogen om vindmøller*, Clausen Bøger, Copenhagen.

In 1982, Davidsen had been far-sighted enough to order some smaller gears when the Kumera gear for the 90 kW was ordered in Finland. Soon afterwards he reaped the fruits of this decision, as he was able to export about 75 machines of his own design to California under the name of Lolland.

For Kim Andersen and Knud Buhl Nielsen, the cooperation with the blacksmiths became the jumping board to big industry. They designed Nordex's very first wind turbines (150 and 200 kW). This Danish company used to make oil tanks and boilers, but had never done anything within wind power and nevertheless got on the top 10 list of wind power at the end of the century.

According to its purpose, the State Steering Committee for Renewable Energy (Styregruppen for Vedvarende Energi), with Professor Niels I. Meyer as its dedicated and visionary president, master blacksmiths were an obvious target group. The next interesting project was the local production of 10 × 90 kW wind turbines for a local wind farm. The island of Mors was the first chosen location, and a group of SMEs to produce the 10 wind turbines was formed. For a start this was a fine order. As the entire windmill industry was booming particularly on the American market, the prospects of export were promising, too.

But no windmills were built in Mors. The local enterprises were too greedy, they thought that they could get easy state money and made unrealistically high calculations. Consequently, a group of seven SMEs in the island of Bornholm, formed the Baltic Power, and got the opportunity to carry out the visions of local production. The first windmill made in Bornholm was a 99 kW prototype with wooden blades of the type developed by NIVE in 1982.

As these blades could have been produced locally, the Bornholm people had a strong wish that the other nine wind turbines should also have wooden blades. But in the meantime enhanced fibreglass blades had become available, and that meant the end of wooden blades.

The commercial risk and coordination of the project was in the care of Dansk Andelsenergi which employed engineer Per Nielsen as the local anchor man. He did an excellent job. He had to get the local companies together and start a local production by means of the Folkecenter drawings.

Figure 7.32 Prototype of the Folkecenter-designed Baltic Power 99 kW wind turbine with wooden blades, Bornholm. The following nine windmills had fibreglass blades.

Ebbe Münster and Olaf Erichsen of Dansk Andelsenergi were the driving forces in the cooperative company, which did not, however, get the influence that had been expected on local production. Commercialisation and competition had begun to break through in a serious manner. The windmill project was part of the Green Island project for Bornholm, which gradually lost momentum, and no wind turbines were built beyond the original nine for local use.

Figure 7.33 Wind turbines of Folkecenter design, produced by local manufacturer's consortium, Baltic Power, Bornholm.

It was disappointing that they did not succeed in making use of the evident opportunities for export to the United States. But the Bornholm project offered particular challenges. New tower was constructed specially for this windmill. Cor-Ten steel, which shows a natural, rusty patina when untreated, was chosen as the material. Apart from the point of liking this or not, the Bornholm people would never send a tower all the way to Western Jutland to have it galvanised. They were true island people who wanted to strengthen their local economy and make their own things. It did not matter to them whether a wind turbine came from the Vestas domicile, or from Japan, as long as they had to spend their own money. It was also an issue of cooperation, something inherent in the culture.

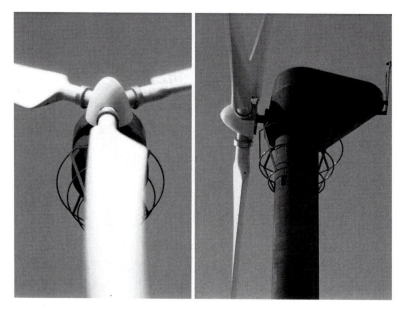

Figure 7.34 Baltic Power used 9 m blades from AeroStar and locally produced Cor-Ten steel towers.

Master blacksmith Kai Hermansen was the representative of the Bornholm consortium. He had a strong engagement in the project, so strong that sometimes we had to meet at the office of the trade association in Odense in order to sort things out. Anyhow, the project led to cooperation and investments in the consortium, so the values for the island community cannot be unilaterally measured just by the number of manufactured wind turbines.

At Mors the local production of windmills was replaced by ten 75 kW Vestas windmills, which were at that time the biggest in the market. This was the first wind farm in Denmark to be organised as a cooperative. The project was activated and supported financially by the Steering Committee for Renewable Energy and the Ministry of Industry. It was important to demonstrate that profit making, which more and more was occupying the souls, could be rivalled by something different.

Figure 7.35   Vestas 17/75 kW wind turbines on the South-West Mors Wind Farm.

7.10.4   The Blacksmiths' Heritage

When the Folkecenter for Renewable Energy was established in 1983, the direct engagement of the Blacksmiths' Association in the design and development of wind turbines was passed on to the Folkecenter. This was how the president, Frits Sørensen, wanted it to be. From this time on, the design of new types of wind turbines was intensified and professionalised. It was no longer work to be done in people's living rooms.

As this was a period when the entire wind power sector was expanding vigorously, many enterprises joined the bandwagon and utilised the Folkecenter designs. It was now possible to buy important components such as blades, towers and controls from independent, specialised sub-suppliers. In this field we went one step further; for small enterprises it must be easy and with low investment levels, to get started. This was the reason why we had to find an alternative to the costly welded base frames, on which the established windmill producers installed the bearings, gears, generators, etc.

Figure 7.36   Only one unit was built in the island of Møn of this 150 kW wind turbine of Folkecenter-design. The idea was local production and local ownership. Car mechanics Kaj Hansen and teacher Svend Olaf Højlund mobilised the island's community. Both have passed away but in 2013 the wind turbine still produces clean power for the islanders that celebrated its inauguration.

The Folkecenter wanted to integrate all of this in a compact unit and in the technical quality, which was now required by the market and the approval authorities. We were in close contact with a number of big European gear factories, which had many years of experience of designing and producing custom-made gears for a multitude of purposes (like locomotives, ships, propulsion drive trains, elevators for mining, drives for rolling mills and much more).

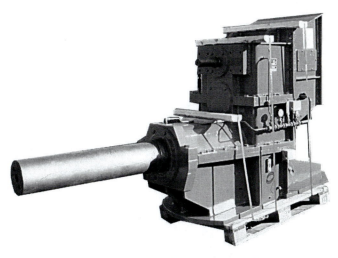

Figure 7.37 Integrated gear for 150 kW windmill. Folkecenter introduced this gear concept allowing the SMEs to manufacture windmills without the more complex nacelle frames with separate main bearing, gear, generator and other basic components. The gears were custom-made by the gear manufacturers Valmet, Kumera, Tacke and Penig.

Figure 7.38 Finnish Kumera gear installed on the DANmark 11 windmill (13 kW) with fibreglass blades and electric yaw.

We defined the capacity and the exterior design, and the gear factories would then shape the gears according to the specific needs of our wind turbines, as if they were cutting butter. In that way we gained access to gear design engineers in Finland and Germany, who possessed a degree of experience with construction, which did not exist in Denmark.

Figure 7.39 Folkecenter-designed Hanstholm 525 kW wind turbine, in operation since 1992, placed on the beach (left) and the wind turbine's integrated Tacke gear (right).

The manufacturers delivered gears to the entire windmill industry. We met several gear designers that shook their heads discretely at the way the Danish manufacturers pieced wind turbines together on a long, costly base frame. In particular, they did not understand the argument that it must be possible to replace the gear easily.

"Everywhere where we provide gears, for ships, for locomotives, we make them in a manner that they will last until all the rest is worn out," they said. "Can you really not afford to make such sturdy gears that they will last for as long as the wind turbine?" So, once again, it was the ever-ongoing issue of the under-dimensioned, vulnerable wind turbine gears that continued to ruin manufacturers and owners of windmills.

Figure 7.40 Good quality has a long life: Drive train and other parts of the 150 kW Folkecenter-designed Vind-Syssel from 1986 got a new term in Sri Lanka in 2012. Russel de Zilva, director of Vallibel, decided to have local production windmills using updated Folkecenter design of wind turbines. First, three second-hand 150 kW windmills were installed, followed by four windmills in 2012 which were made in Sri Lanka. Thirty additional will be manufactured for the Sri Lanka project. Later, 500 kW and 1500 kW types are planned for local production in Sri Lanka.

7.11   The Blades Are the Core of It All

7.11.1   No Blades, No Windmills

The idea of the independent blade producer first appears in an article in the Danish daily newspaper *Information* on 22 March 1976. Here Amdi Petersen, leader of the Vestjysk Energikontor in Tvind, says that, "(...) our next step will be the development of some quite concrete models. For example, we can make moulds in which people can make their own blades. They are often the biggest problem. So we will start doing that very soon... That is quite in the spirit of the wind. It cannot be monopolised. Nobody needs to hold back. Just come... all our experiences are there for the benefit of everyone."

Figure 7.41 Production and assembly of blades on the Tvind windmill, 1977.

Figure 7.42 Delegation from Chinese XEMC visiting Tvind.

But Tvind looked beyond and with their openness worked on the same ideological basis, as did the Danish co-operative movement at the end of the 1800s. At the passing of the first Patent Bill in 1895, the forerunners of the time made sure that agricultural processes and technology could not be patented. Inventions were not to enrich individuals, but be at the disposal of everybody. Especially Poul La Cour turned this production philosophy into reality. He was no Bill Gates.

7.11.2 The "Winning" Blade Concept

The most decisive for the Danish wind power industry was the concept used for the 27 m long blades of the Tvind windmill. Amdi Petersen insisted that exactly that solution to be used, and so it was. The concept was German, as mentioned earlier in this chapter, developed and tested in practice on wind turbines and helicopters by Professor Ulrich Hütter from Deutsche Luftund Raumfahrt Institut at the Technical University in Stuttgart.

Figure 7.43   The original Tvind blade from 1978–1992, with Hütter designed root system. This last original Tvind blade (replaced in 1991 by modern Vestas blades) is exhibited at the Folkecenter.

From wind turbine R&D projects in various countries it was known that connecting the blade to the hub was a challenge where successful solutions were the exception. This can be illustrated by two early Danish blade development projects. A single detail in this comparison between the state's two 630 kW Nibe wind turbines A and B, and the Tvind windmill, shows how the two projects solved this challenge.

Nibe A had a rotor of fibreglass combined with some sophisticated welded steel constructions, which could not be verified in relation to the alternating loads of the rotor. Nibe B was less clumsy; it had an inner beam, but that did not last long either, despite a generous and costly effort of the development. Some years later, both Nibe A and B were provided with wooden blades, which after being damaged by lightening, were later replaced by fibreglass blades.

Figure 7.44 One of the Nibe windmills (A) got a rotor of fibreglass in combination with some very ingenious welded steel structures, which in fact could not be verified.

At one OVE wind energy workshop, VindTræf, in November 1976 the idea of the independent blade production for the later component windmill was presented. A number of wind groups were formed, of which one focused on fibreglass blades. Here I suggested that "Twenty windmill builders should join to invest and build a mould, for example, for an 8 m blade." In the group were among others Rio Ordell, Erik Grove-Nielsen, Jens Gjerding, Claus Nybroe and Peter Andersen, all to become important pioneers of wind energy.

Jens Gjerding from Tvind did not hesitate. They did not make an 8 m blade but a downsized 4.5 m version of the big windmill's blades to be tested on their small 18 kW windmill, the so-called PTG windmill that was developed for this purpose. Tvind did not aim for industrial production but allowed private builders to have a blade for copying.

Figure 7.45 Tvind PTG wind turbine with first fibreglass blades (down scaled Hütter-type blades), around 1977. Children playing are the author's sons, born 1971 and 1974.

The self-built blades were installed on the windmill of electrician Leif Nielsen in Gredstedbro, but already in the summer of 1977 the windmill lost a blade and the interest in self-made fibreglass blades vanished. Erik Grove-Nielsen of Økær, Viborg, took over the mould and he commercialised the Tvind blade, the first one from an independent blade supplier. Thus, the component windmill had become a reality. From now on interested windmill producers could buy blades where the manufacturer had the copyright. At a stroke this made it much simpler to introduce a windmill to the market.

The first set of 4.5 m Økær blades were delivered at the end of 1977 to mechanic Svend Adolphsen in Knudstrup at Viborg. However, the 4.5 m blade had two important shortcomings. It had no air brakes, it ran too fast and was noisy which was not satisfactory for the users. Soon the newly established blade supplier had no customers.

This might have been the end of the independent sector of blade production, if Erik Grove-Nielsen had given up at this point, where his financial basis was rather weak. But various forces interacted. In Thy, the NIVE group was starting up local production of 22 kW wind turbines. However, we were not going to use the noisy Tvind 4.5 m blades from Økær and therefore, at the later Folkecenter, Ian Jordan and myself embarked upon building the mould for a 5 m blade with specifications similar to those from the Gedser wind turbine from 1957, but scaled-down.

An experienced fibreglass company in Vinderup wanted to produce the blade. The development work had gone quite far when we were approached by Erik Grove-Nielsen who politely and quite correctly pointed out that there was no reason to launch new blade projects. He could not sell his products and had a workshop in Økær equipped precisely for blade production.

We wanted a 5 m blade, but he hesitated to invest in a new product running the risk not to be able to sell the blades afterwards. In a week I made orders of five sets of blades! NIVE got its blade—using Tvind blade structural design and Johannes Juul's blade operational principles with respect to the optimal blade. The first wind turbine with the new blades was erected east of Thisted in 1978.

Figure 7.46 The bolt flange of the 5 m Økær blade for the 22 kW NIVE windmill.

Thus this part of the history is complete: From now on no one had to hold back. Amdi Petersen's vision turned out to give Denmark world leadership in the most victorious form of the clean energy types of the future. It was the first step on a long march. Denmark got independent suppliers of blades, others could supply control systems. Now, future wind turbine producers could purchase the basic components.

Figure 7.47 Blades of Økær, KJ, Olsen and Svaneborg from Folkecenter's collection. The Økær blades are unique because they belong to the period before tip brakes became mandatory.

They soon seized the opportunity. The 5 m blade was installed on the successful 22 kW Herborg wind turbine that was licensed to Vestas under the name Vestas HVK, the beginning of the world leading company. Soon also Nordtank, Danregn/Bonus and others entered the market with their first generation wind turbines using the 5 m Økær blade.

7.12 The Windmill Industry Goes International

The Danish blade concept was now established. Improving details and building production capacity was the next step. Erik Grove-Nielsen showed an astonishing talent for innovation; in the Aeroform regime he developed a complete repertoire of blades from 2.5 m upwards. Especially the 7.4 m blade was a hit. It was used on the popular 55 kW windmills from 1980 and onwards, paving the way for the modern wind industry. Later followed the development of even longer blades of 9 m, 10 m, 11 m and 12 m. Development and testing stayed in its native area in Sparkær, where Erik Grove-Nielsen had built a testing centre.

Altenergy's 12 m blade from about 1986 was the end of it. Only a few 12 m blades were made. With the shutting down of the American market the conditions in the windmill industry were extremely difficult; all wind turbine manufacturing, apart from Bonus, went into financial problems. However, in the meantime other blade producers had in due time entered the strongly expanding market. While AeroStar during the Californian boom suffered from quality management of their products the newcomers organised effective quality control and took over.

7.12.1 LM: The Blade Giant

LM in Lunderskov made a very discrete entry as blade supplier around 1982. Their starting point was many years' experience within fibreglass as an industrial product—a background which was to prove just as important as the knowledge of wind energy. They started cautiously with 7.5 m and 8 m blades, which were largely compatible with the then dominating blade product. The first LM blades for 55 kW and 75 kW wind turbines appear on Wind Matic that changed from Riisager type blades. At the time, a 3.5 m LM blade could also be found on the Wind Matic 7.5 kW. LM adapted to the needs of their customers and also supplied different sizes of blade shells for their Riisager clones to Sonebjerg Maskinfabrik.

LM made steady progress. The blades grew in size and in 1986 they delivered 11 m blades for 150 kW wind turbines. One can find them on wind turbines from Bonus, Vind-Syssel and others. LM's blades at the time were designed to turn anti-clockwise. They had air brakes of the spoiler type at the front edge of the blade, whereas other fibreglass blades turned clockwise and had tip brakes like the Gedser windmill.

LM finally succumbed. The spoilers disappeared and in the long run the public could not tolerate that wind turbines rotated in different directions. So from 1988, LM blades turned clockwise, which from a historical perspective, really was the wrong way. The wooden blades of the past turned to the left because the natural twist of the wood was determined by the sun passing over the sky from east to west, in the Northern Hemisphere at least.

Figure 7.48 Tip brake from Juul's Gedser windmill (left); Tip brakes on the 525 kW windmill in Hanstholm (right). In the early Riisager windmills and windmills with early 5 m Økær blades the tip brakes were "forgotten". After serious accidents due to runaways, tip brakes became mandatory. Økær blades became tip brakes inspired by the Gedser windmill.

Learning from other's experiences with fibreglass blades LM started producing extremely durable blades. Therefore by 2012, 8 m LM blades are still turning on old 75 kW wind turbines. Later calculations showed that they had a calculated life expectancy of 70 years and the dimensioning was adjusted to a more realistic life expectancy.

At the end of the 1980s Danish wind turbine manufacturers were seriously weakened after the collapse of the American market and hesitated to develop new, bigger windmills. But at LM they were ready for the next step, where the size jumped from 200/300 kW to 500 kW. This led to the cooperation with the Folkecenter about specifications and testing of the 17.2 m blade for a 500 kW joint development project, led by the Folkecenter (66%) and the German wind turbine manufacturer Tacke (33%), as partners. Tests that started in 1992 showed that the new LM blade was successful.

Generally speaking, a blade supplier such as LM has set the tempo of the growth of wind turbine sizes far more than the wind turbine manufacturers themselves. If a manufacturer hesitated to introduce a wind turbine using a new blade type, it might well be suddenly losing market share; relentless conditions of competition, over which the individual wind turbine manufacturers had no say.

Figure 7.49   LM 61.5 m blade installed 2009 on REpower 6 MW wind turbine in Elhöft on the German/Danish border (above); In 2012, a 73.5 m LM blade was installed on 6 MW Alsthom Haliade wind turbine (below).

Each new generation of blades used to increase by one or a few metres, but from the middle of the 1990s when MW-sized wind turbines entered the market, several metres were added to each new blade type. In 2004, LM made world record by supplying a blade of 61.5 m for a 5 MW German REpower wind turbine erected at the Elbe estuary and for REpower's 6 MW version.

For many years LM has been the world's biggest producer of wind turbine blades with factories on all continents. The blades have become so big that they are expensive and difficult to transport. That is one of the reasons why blades are produced in the country of installation when the market is big enough for a local production to be profitable. Often there is also a national condition that a specific share of the value of the wind turbine to be locally produced.

Figure 7.50   LM Wind's modern manufacturing facilities can be found in many countries of the world.

LM has been a blade supplier for a large number of wind turbine manufacturers all over the world. Manufacturers are dependent on a supplier, just as LM has an interest in having many purchasers for their blades, which is often unnoticed but an important part of the Danish wind power success. Many emerging economies are on the threshold of using wind power. For them creating work places and building new industries is often of primary importance, and there it would be of mutual interest asking a company such as LM to build a blade factory. With blades from an experienced supplier, newcomers within wind turbine manufacturing can also better overcome potential scepticism towards the quality of their product.

7.12.2 Vestas also Becomes Manufacturer of Blades

It is part of the history that Vestas, being one of the first and biggest users of Økær blades, around 1981 found themselves in a difficult situation after some blade accidents. The company acted accordingly and started their own blade production with own design, root system, etc. Since then Vestas has been almost self-sufficient in blades.

Vestas started out with a 7.5 m fibreglass blade for the 55 kW wind turbine, which was later extended in two stages. Blades were needed for the 75 kW and 90 kW wind turbines, and therefore the 7.5 m blade was given rather primitive "stilts". Since then Vestas has always been in the front regarding blade design and advanced production technology.

Figure 7.51   Vestas repowering in Thy; dismantling a 225 kW turbine with 29 m diameter rotor (left) and installing a new 2 MW unit with 90 m diameter rotor (right).

With the globalisation of the wind power industry, Vestas established its own blade factories in Denmark, Germany, China, the United States, etc. In 2011, Vestas announced the introduction of a 7 MW wind turbine with a rotor of 164 m diameter—the biggest in the world at that time—to be launched in 2014. Each blade is as long as nine London city buses.

Figure 7.52   Vestas 3 MW wind turbines with 112 m rotor diameter on the North Sea beach, Hvide Sande, Denmark. Each wind turbine produces 13.5 million kWh per year.

7.12.3   A Zealandic Blade Factory

Micon, the Danish wind turbine manufacturer had grown big during the Californian wind rush, and joined up with Ole Larsen in Ringsted and Helge Petersen, to make a complete series of blades. The alliance, called Micon Airfoil Technology (MAT), was logical.

Micon needed the blades, Ole Larsen had a reputation for delivering fibreglass work of high quality, and Helge Petersen had gone the whole way from blade design with F. L. Smidth & Co.'s Aeromotors during the Second World War, Risø to Danish Wind Technology (DWT).

Figure 7.53   Three 11 m MAT blades (left) and three 19 m LM blades (right) in a row at the Folkecenter Blade Expo.

The MAT blades were standard type, efficient and of sufficient strength. The finish was perfect, but unfortunately an untraditional type of air brake was chosen. Instead of the well-proven tip brakes, which were industrial standard already at the time, they got the idea of using parachutes which were hurled from a lid in the blade for emergency braking. Naturally you run into trouble refitting such a bundle of cloth in the blade ready for the next release. That is what a wind turbine buyer would think. And so MAT disappeared from the market again, being an example that due to such a minor error of judgement on a detail like aerodynamic brakes, an otherwise excellent product sank into oblivion. When around 1990 Egypt planned domestic wind turbine manufacturing, locally made blades used MAT technology. Today MAT's production moulds and blades can be seen as museum pieces in the Folkecenter.

Figure 7.54 Siemens 6 MW prototype with ring generator, installed at the Østerild test centre, Thy, Denmark. The rotor diameter is 154 m, world's biggest by 2012 (left). The B75 blade is the world's largest fibreglass component cast in one piece. The manufacturing process posed several challenges for the project team. In particular, the mould had to consist of two parts so that it could be transported (Photo: Siemens AG, Munich/Berlin) (right).

Figure 7.55 Blades continue to grow.

In August 2012, the first prototype of SWT-6.0 with a rotor diameter of 154 m has been erected on the test field in Østerild, Denmark.

About the author

 **Preben Maegaard** is a Danish renewable energy pioneer, author and expert. Since the oil crisis in 1974 he has worked locally, nationally and internationally at the organisational, political and technological level for the transition from fossil fuels to renewable energy. From 1979 to 1984 he was chairman of the Danish Renewable Energy Association (OVE), and since 1991 vice president of EUROSOLAR (www.eurosolar.org). In 2006, he was promoted as the senior vice president. Since 1994 he is member of the jury of The European Solar Prize. When the World Wind Energy Association (WWEA, www.wwindea.org) was founded in 2001 he became its first president, a position that he held till 2005. In 2001 he became the chairperson of the Committee of the World Council for Renewable Energy (WCRE, www.wcre. org). Since 1999 he has been a board member of the European Renewable Energies Federation (EREF, www.eref-europe.org). In May 2006 the World Wind Energy Institute (WWEI, www.wwei.nu) was initiated in Kingston, Canada, involving seven institutes from China, Brazil, Cuba, Canada, Russia, Egypt and Denmark. Preben Maegaard was appointed the first president of the WWEI. Since 1984 he has been director of the Nordic Folkecenter for Renewable Energy (www.folkecenter.net). In this capacity he has been responsible for the technological innovation of windmills, including design, construction and implementation of sizes from 20 to 525 kW, farm biogas digesters of volumes from 50 to 1.000 m^3 as well as integrated energy systems, including hydrogen and biofuels for transport. He has served on several Danish national governmental committees and councils. For over three decades, Preben Maegaard has been a conference director, organiser, speaker and/or participant of numerous national and international seminars, workshops and conferences and chairman of the World Wind Energy Conferences, WWEC 2003 in Cape Town, WWEC2004 in Beijing and WWEC2005 in Melbourne. In 2009 he became chairman of "The 1st World Non-Grid-Connected Wind Power and Energy Conference 2009" in

Nanjing, China. Since 2005 he has been an advisor for Chinese wind turbine manufactures, universities and authorities. He has authored and co-authored numerous reports, books and articles in Danish, English, German, Chinese and Japanese periodicals within the field of renewable energy and sustainable development and has received a number of awards.

Chapter 8

Økær Vind Energi: Standard Blades for the Early Wind Industry

Erik Grove-Nielsen

Windsofchange.dk, Følvigvej 8, 7870 Roslev, Denmark

grove@windsofchange.dk

In 1895, Danish meteorologist, inventor and teacher Poul la Cour worked hard to bring renewable energy sources in the form of wind energy to practice in the dark rural areas of Denmark. His principal aim was to bring prosperity to the people living in the countryside. A few decades prior, thousands of wind-powered ships took care of the transportation of goods between continents.

However, just a few years later the first aeroplane took off from the ground and Henry Ford started serial production of fossil energy motor cars, driven by the explosion motor. Oil together with coal became the main source of energy, enabling mankind to transport people and goods in growing numbers by cars, boats, trains and planes. Fossil fuels became the main driver of American and European societies, and only a few people questioned this development then. During World War II, when Denmark was occupied, it became difficult to import fossil fuels. As electrical utilities relied on domestic sources, engineer Johannes Juul, employed

Wind Power for the World: The Rise of Modern Wind Energy
Edited by Preben Maegaard, Anna Krenz and Wolfgang Palz
Copyright © 2013 Pan Stanford Publishing Pte. Ltd.
ISBN 978-981-4364-93-5 (Hardcover), 978-981-4364-94-2 (eBook)
www.panstanford.com

by electric utility company SEAS, was occupied for some time by prospecting local peat sources for securing the electricity supply south of Copenhagen. Being an early student at the wind courses of Paul la Cour, Johannes Juul began wondering, whether it would be possible to improve the situation by building wind turbines for electricity production. SEAS allowed him to build several prototypes. The last prototype was inaugurated in 1957 at Gedser. This 200 kW turbine was operating successfully until its decommissioning in 1967, only 6 years before the first "oil crisis" in 1973. Just 10 years after the last turn of the Gedser turbine's rotor, its design became the blueprint for the first commercially produced wind turbines in Denmark.

8.1   Nuclear Power—No Thanks!

I was born in 1949 and raised on a farm with cows, pigs, hens and ducks. My grandfather was a Christian missionary in Nigeria. Thus fighting for an idea was nothing new to me. In the wake of the oil crisis, electric utilities as well as the Danish political establishment in Copenhagen in 1974 insisted on the rapid development of nuclear power reactors in Denmark, in order to secure a stable delivery of electricity to the Danish industry. We did not agree on that idea, and we wanted to work for alternatives to nuclear power.

As a teenager, I was a nerd designing and building rubber powered model airplanes. While studying to become an engineer in Copenhagen, I was a member of the Polytechnic Glider Club, flying gliders. In 1973 I dropped out of engineering studies and moved to western Denmark with my would-be wife. We bought a small farm and I started developing a solar powered water heating system, which I intended to manufacture. We wanted to create new products for a greener planet, and we never asked if our goals were realistic.

In 1976 I made my first wind turbine blades for a 600 W turbine for a "Friends of the Earth" summer camp. The blades were 1.7 m long, made of fibreglass, and very primitive.

In the years after the 1968-youth revolution, the general anti-authoritarian spirit between young people helped the new anti-nuclear power organisation, OOA, to attract supporters and activists. OOA was known for the "smiling sun" badge. In 1975 OOA gave birth to a new organisation for renewable energy called OVE.

This organisation aimed at promoting renewable energy all over the country as an alternative to fossil fuels, and to prevent nuclear power from becoming part of the power supply mix. OVE arranged bi-annual meetings for self builders, politicians and other interested stakeholders. These meetings and the general network made by OVE became an important catalyst to the development of a renewable energy market place and technological development in Denmark.

Figure 8.1   600 W turbine at NOAH summer camp 1976.

Hundreds of enthusiasts building solar ovens, biogas facilities and wind turbines met at the gatherings each year. At Tvind, a group of schools managed to build their own megawatt wind turbine, and the faith in wind power gradually became stronger.

Figure 8.2   Skourup's self builder turbine at Vorbasse.

8.2   Learning by Doing

It was clear to me that many self builders establishing wind turbines did not know much about aerodynamics. I possessed some knowledge of aerodynamics from my years designing and building model airplanes and my later glider plane training. So I thought why not start producing wind turbine blades? I had never worked with fibreglass design or production, but at the beginning I received great help from the "Vestjysk Energikontor" at the Tvind Schools. Prior to building the blades for the big turbine, Tvind learned working with fibreglass by building blades for a small 15 kW turbine, and some fishing boats. Already back in 1957 German wind pioneer Professor Ulrich Hütter of Stuttgart designed and built an elegant 100 kW wind turbine with fibreglass blades. In 1976, Tvind learned and incorporated this design in the Tvind turbine's blade design, for the small blades, as well as for the megawatt blades.

Figure 8.3   Svend Adolphsen 11 kW prototype (1977) carrying my 4.5 m blades.

In summer 1977, aged 28, I decided to begin a production of rotor blades. After some time experimenting with laminated wood, I learned that a group of five individuals at Fouslet, in southern Denmark, had made a mould set as a copy of the blades for the small 15 kW Tvind turbine. At first I borrowed the mould from them, but later I purchased it at a price of DKK 2500 (USD 500). Thus my first blades in 1977 were copies from these 4.5 m long Tvind blades with German "ancestors". My first customers were two members of the Fouslet group: Henry Jørgensen and Svend Adolphsen. However, these 4.5 m blades were noisy and had poor aerodynamics.

8.3 Against the Wind

In 1976 Christian Riisager began production of the Riisager turbine of 22 kW and 45 kW carrying wooden blades. The first competitor to the Riisager turbine was the Kuriant turbine, developed as the Adolphsen turbine, for which I delivered the first fibreglass blades in November 1977. The Adolphsen turbine later, as the first wind turbine, achieved the type approval numbered A-1 by the Danish Risø test station. However, after selling the blades to the Fouslet group I had no new customers. I tried to improve the situation by travelling around the country in our Volkswagen bus, with a blade pointing out of the rear hatch, visiting blacksmith shops, and trying to get some blades sold. I had no success doing so. But a new initiative saved my business: a group of individuals led by Preben Maegaard created the NIVE Group in Thy, on the Danish west coast, and gathered to build five private turbines for their own households. In January 1978 Preben Maegaard asked me if I would design and fabricate 5 m long blades of certain specifications to their common project. These orders arrived as sent from heaven.

I then began designing my first 5 m blade. Before I made the mould, I had to decide which way the rotor should turn. The old Danish tradition prescribed counter clockwise rotation. This tradition was adopted in 1975 by the wind turbine pioneer, Christian Riisager, and the same year by the Tvind Community for their initial 15 kW turbine named PTG. But my wife Tove, being an individualist, and not at all attracted by the Tvind community, suggested that our blade design should turn the opposite direction of the Tvind PTG blades. My younger brother Johannes was part of the Wind Turbine Crew at Tvind, and I often may have tried to compete with their results: At Tvind they worked together as a collective, whereas we were individualists working together with other individualists across the country. Tove convinced me to mount the plywood airfoil sections on the steel tube for the plug, in a way so that the Turbine would turn in a clockwise direction.

This discussion and decision of design—taken place around the kitchen table—accompanied by music of The Rolling Stones and Bob Dylan, became defining for the direction of rotation for most of the world's new wind turbines, as our later blade customers: Vestas, Bonus (now Siemens), Nordtank and Enercon became technology and market leaders in the international wind power industry at a crucial time. For some years descendants of the Riisager turbine: Wind Matic and Tellus—all turning counter clockwise—were seen in the Danish landscape alongside with turbines from Vestas, Bonus, Nordtank and other turbines all turning clockwise. Eventually the three major Danish turbine manufactures conquered not only the Danish market but to some entent also the international wind market. As the successful "Danish Concept" was delivered with a clockwise turning rotor, this direction eventually became the global standard.

Figure 8.4   Delivering the first set of 5 m blades in May 1978.

Having designed the blade and made the plug and a mould, I delivered the first set of 5 m blades in May 1978. New orders came in from self-builders around the country, and soon the first 10 rotors were up and running.

One of the very important early customers was Herborg Vindkraft. Owner Karl Erik Jørgensen was a "devil-may-care" inventor and entrepreneur, who devoted everything that he and his family had to the development of his wind turbines. He received good help from the young student Henrik Stiesdal from a nearby village. Henrik was very smart, innovative and able to quickly make the needed calculations and improve the design.

Figure 8.5   Gathering of the grassroots people in summer 1978. To the left, my new blade; to the right, Henrik Stiesdal carrying his sailwing blade.

On 1 July 1978 I delivered the first set of 5 m blades to Karl Erik. He quickly got the turbine up and running; everything looked prosperous. Only one month later, on 2 September Karl Erik's new turbine had a runaway in a storm. The hub disintegrated and the blades flew in three directions. Two weeks later, yet another of my customers: blacksmith Krogh of Stoholm suffered a fatal runaway in another storm. The turbine was totally destroyed.

After these accidents it was not easy to sleep during windy nights: what could happen if more blades broke? I stopped the production and began designing an airbrake system. My first intention was to copy the tip-parachute system that was used as an airbrake on the Tvind megawatt turbine. However, a wise self-builder of the Fouslet

Group, Henry Jørgensen, urged me not to do so. He said, "if people are troubled by the parachutes coming out too often, they will just cut the wires, and then you have no security at all built into your blades; why don't you use the system designed by Johannes Juul for the Gedser turbine, the tip vanes." And so I did, though omitting the hydraulic system of the Gedser turbine, and designing a spring loaded release system for the deployment of the tip brake. My main reason for using a simple spring system, embedded in each blade was that I would like the blades to be delivered as failsafe units. Security should not depend on the buyer's ability to make a sound failsafe hydraulic control system. Security should be delivered with every single blade. But as no blades were then sold for some time, no payments came in, and we were in dire straits and near bankruptcy. After some struggle for funding I asked a private Inventors Fund. After waiting only seven days, a cheque of DKK 50 000 (USD 10 000) came into our letterbox one morning. This donation saved my company and tears of joy came to my eyes. On 13 November 1978 the first set of blades with pivotable tips were finished and delivered to Karl Erik Jørgensen, and then quickly mounted for the prototype test. A few days later Karl Erik was ready to perform the test of the safety system on a windy day. For the testing day I had arranged with a credible public engineer from the Danish Technological Institute of Aarhus to attend and witness the crucial test of the emergency airbrake. I hired a flatbed plotter and connected it in order to monitor the rpm of the rotor as it speeded up.

Figure 8.6   Failed 22 kW turbine at Stoholm.

As the controller relay for the generator and the shaft brake was mercilessly pulled out of their sockets, the powerful rotor lost its connection to the electricity grid, and very quickly speeded up with a scary noise—but after a second of doubt, the tip brakes were deployed with a swishing and roaring sound. What a relief to hear this noise, and see the rotor reduce its speed into a more safe rotation speed: now I could go on producing blades, and my customers could go on building turbines. Our home would be saved from bankruptcy for some time again.

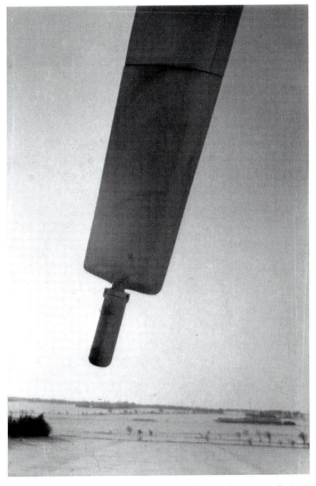

Figure 8.7   The prototype tip had outboard placed springs (winter 1978, in Herborg).

8.4   Big Players

One year later, in August 1979, a visionary "green majority" of the Danish parliament passed a Bill that entitled people who wanted to buy a wind turbine to be eligible for a grant that would cover 30% of the turbine price. To ensure, that only soundly engineered turbine designs were subsidised, the government established a *test station* for wind turbines at Risø. As one of Risø's first acts, they made it mandatory to employ an airbrake system on the turbine to be approved. For some time this gave my company a sort of monopoly, as my most serious competitor, the Riisager turbine was not equipped with airbrakes. In the following years my company, named "Økær Vind Energi", had a growth in turnover of more than 400% each year, and I frequently employed new staff. Most blades were delivered to self builders and small blacksmith companies.

In 1979 Vestas entered a license agreement with Karl Erik Jørgensen and Henrik Stiesdal, and thus their turbine design became the beginning of the Vestas wind turbines. As Økær Vind Energi delivered the blades for the Herborg turbines, we then became a supplier of blades for Vestas. Vestas was a well renowned family owned company, with some international sales in the agricultural machines business as well as the crane-producing department.

Figure 8.8   Økær delivering 7.5 m blades at Vestas in 1980.

Økær Vind Energi was a small company employing around 13 people at the most. In the beginning it was mostly long haired "hippie looking" folks. The people I hired in the early years were mainly left wing activists from a nearby town named Viborg. Ideologically we were on common grounds, and we deeply felt that we were struggling together to help build a greener future.

Figure 8.9   My foreman Erik Tøttrup doing finishing work on a 5 m blade.

For the first years for Vestas employees, especially white collars, they may not have felt at ease relying on the delivery of blades from such a company. For my company it was crucial, that we had demanding customers. In those early years I had two main drivers: my bank director, who spurred me to work harder, as my account was always overdrawn and first and foremost the demanding customers. Customers, such as self builders, small companies or bigger companies such as Vestas, Nordtank and Bonus constantly asked for new and better designs, and bigger blades. In 1979 Karl Erik Jørgensen of Herborg Vindkraft asked me to design and manufacture a 7.5 m blade for his new 55 kW turbine. At the same time we were applying for a loan from a Danish Regional Development Fund for building an improved and bigger production facility. The development fund made a feasibility study, which stated that I should focus on the production of my 5 m blades and not enter into new dangerous projects such as production of a "big" 7.5 m blade. I did not listen to their advice, but rather went on to design and build the plug and mould for the 7.5 m Økær blade in the winter and spring of 1980. On 13 March I delivered the first set of 7.5 m blades to Herborg Vindkraft for the HVK Vestas 55 kW turbine.

Figure 8.10   Økær 7.5 m blades on the Herborg/Vestas 55 kW prototype.

Two years later, the 7.5 m blades were turning on hundreds of Danish turbines from Vestas, Nordtank and Bonus, that soon conquered the Californian market during the "Californian Wind Rush". We had a "no" from the Regional Development Fund, so instead I managed to buy an industrial hall. In August 1980 we moved our production from Sparkær to this facility in a nearby city named Mønsted. As my customer portfolio shifted from self-builders to industrial companies, I started hiring more educated craftsmen for the production.

8.5 Standard Components Available to All, No Exclusive Sales

In 1979 the Wind Matic Company, which then produced the former Riisager turbine, came for a meeting at Økær Vind Energi in order to have us producing a special blade designed for Wind Matic solely. I wanted to produce standard blades for all, and did not want to use precious time for developing special designs and moulds for blades that would be sold exclusively to one company. I declined their offer, and never became a supplier to Wind Matic. But soon my order backlog was filled with orders from new companies that wanted to make turbines, without having to bother with their own blade development, and orders from Vestas, Nordtank and Bonus dramatically increased.

This standard component concept was later continued, when Coronet/Alternegy entered into a license agreement, and produced my blades now named AeroStar in Denmark and the United States. Besides Vestas, we soon delivered blades to Nordtank for their 22 kW, and from 1981 7.5 m blades for their 55 kW turbine. In December 1980 we delivered the first blades to Danregn Vind-kraft, later named as Bonus Energy, and today known as Siemens Wind Power.

Figure 8.11   Økær 7.5 m blade delivery in 1981.

In spring 1981 a young German, Aloys Wobben, then long haired, rolled into Sparkær in his old Mercedes with a trailer behind to pick up a set of 5 m blades and a spinner for his first 22 kW prototype turbine. Later he purchased 7.5 m blades for a 55 kW turbine and his company's name is now Enercon.

8.6   The Second Serious Blade Failure

After having delivered 7.5 m blades for 20 Vestas turbines, my company, and Vestas as well were suddenly shaken by a blade failure on a Vestas 55 kW turbine in Hinnerup in Denmark. The accident happened only nine days before Christmas in 1980. At that time a 7.5 m blade was considered big, and no test rig was ready for the test of the blade. Consequently, in less than a week we constructed and built a blade test rig. Another 7.5 m blade was bolted to the test rig and sandbags were stacked on the blade in an ultimate test. My blacksmith, the manager of Vestas and myself performed the test on a late evening. The test clearly showed that the blade did not have sufficient buckling strength, it collapsed two metres from the root, and I was responsible.

During that night the manager of Vestas tried, two times to persuade me to sell my company to Vestas. I knew that the price would be low, taking in consideration, my obvious design fault in the 7.5 m blade. At the same time my other good customers, such as especially Nordtank and Bonus, would be left out with no blade supplier. My lawyer helped me to avoid the takeover of my company by Vestas. Økær Vind Energi now was in a bad situation as we used all our resources in redesigning the blade in a common effort with the Risø Material Research Department and a Vestas engineer. The production of our main product, the 7.5 m blade was temporarily halted, sales and incomes were small, and my 13 employees still had to get their salary.

In early spring, around Pentecost time, we were no longer able to pay our electricity bill. On the Friday before Pentecost, the electricity company cut the power supply to our home and at the same time to the production site in Mønsted. I drove to Viborg and rented a 27 kW emergency power generator, which was installed at the production hall on Saturday, and the tank filled with diesel. On the next working day my employees were more than surprised, as they thought it was all over. However with the emergency generator we could go on with the blade production. At our home neither the heating system nor the private water pump was working, so my wife would lower a bucket into our water well to hoist the water needed for our household. Our boys were 2 and 4 years of age at that time.

A local business school manager and entrepreneur, Niels Aage Bjerre, saved the situation by proposing cooperation with a major boat producing company "Coronet Boats". The company, earlier owned by B&W Shipyard in Copenhagen, had been declining sales, and wanted to enter into a new market. On 27 April 1981, I signed a license agreement between Coronet and myself. I will never forget the ferry ride back home from Sealand that day. After several weeks we had the first payment from Coronet. We were able to pay the electricity company again, and the electricity supply could be restored.

In June the first moulds and production equipment were transported to Coronet in Slagelse, near Copenhagen. One of my craftsmen was staying for some time in Slagelse, at the Coronet factory, to make a fluent transfer of the production technology. Coronet made supplementary moulds, and after the summer holiday season in 1981, Bonus, Nordtank and Vestas had blades delivered from the new production site without delay. The new blade company was now named Alternegy, and the blade brand was AeroStar.

Following the Hinnerup blade accident, Vestas decided to develop its own blades, but still in 1985, it was purchasing AeroStar blades for its 55 kW workhorse. As Aloys Wobben founded the Enercon Company in 1984, he also chose AeroStar 7.5 m for his first E-15 55 kW turbine. Just one year after the Coronet took over the Økær blade design, Vestas, Bonus and Nordtank started exporting 55 kW turbines to California in great numbers. All these early turbines were equipped with AeroStar 7.5 m blades. These early blades were heavy and clumsy, but the simple and rugged design of the turbines made them survive in the Californian desert and mountain passes. The most critical and weakest part of these blades was the German Hütter Blade Root, which became an Achilles heel for the AeroStar blade. However the problems were mended with careful service, and in some cases root fixes. As a result of good technical servicing, out of 1 100 Bonus turbines erected in the Californian Wind Rush 1982–1986, some 1 045 turbines flying AeroStar blades were still producing electricity 25 years later, and many are still running.

The AeroStar blades were produced by three factories in Denmark and two new factories built in the United States between 1981 and 1986. AeroStar shared my vision of producing a standard blade that everyone could buy. No exclusive deals were done. That way they paved the way for new turbine designs and manufacturers worldwide. AeroStar blades were sold to 26 countries around the Globe. Mrs Anne-Marie Lundsberg was a tough frontrunner for this international sales work, as she tirelessly travelled the world andcreated an impressive international network for AeroStar.

8.7   The Danish Concept

The turbines from Vestas, Nordtank, Bonus, Micon, Danwin, DWT and Wind Matic exported to the United States in big numbers during the Californian Wind Rush, were all based on the same concept. Three-bladed stall controlled turbines with the rotor on the upwind side. This rugged and simple design was later named "The Danish Concept". The initial design was made by Johannes Juul in the years 1948 to 1957, and showed its qualities at the Gedser turbine, and now 15 years later in the Californian desert, where visionary American politicians paved the way for the world's first wind farms. We must be thankful to the Americans for welcoming wind turbine technology and companies from another continent. Many companies and designs matured in this period.

In 1986, the Californian Wind Rush came to an abrupt end when tax credits in California and nationwide in the United States expired. Vestas, Nordtank, Micon and AeroStar went bankrupt. Before 1982, LM Glasfiber did not market its own blade design, but manufactured blades that were designed by Riisager and later Wind Matic. From 1982, it developed its own blades, and in 1987 it expanded its business substantially with sales to the major players Bonus and the reconstructed Nordtank as well as German Enercon.

During the good years with the AeroStar cooperation, I built up a test centre for wind turbine blades at Sparkær. After the fall of the AeroStar Company Alternegy in 1986, I lost my primary income here, and had to lay off my employees. From 1991, Risø hired my test facility and made it a national blade test centre. In 1996, I sold the Sparkær Centre to Risø, but continued to work as a manager for the facility for some years.

About the author

Erik Grove-Nielsen was born in November 1949 on a family farm at Hejnsvig in Middle Jutland, Denmark. His grandparents were Christian missionaries to Nigeria. At the age of 8 years he saw a fighter jet plane crash land in front of the family farm. This inspired him to build model airplanes for the next decade, and study aerodynamics. In 1969, after high school, he began studies in chemical engineering at the Polytechnic High School of Copenhagen. After four years of studies his growing involvement in the emerging environmental movement forced him and his later wife to move back to Jutland to work for greener alternatives instead of ordinary engineering work. Living near Viborg he worked part-time as an environmental activist from 1973 in NOAH (Friends of the Earth), from 1974 in the anti-nuclear organisation OOA and from 1975 in the Renewable Energy Organisation OVE. From 1977 to 1981 he developed and manufactured wind turbine blades for self-builders and the emerging industry. From 1981 to 1986 Erik entered a license contract with Alternegy and designed the AeroStar blades, which were produced in Denmark and the United States and sold to 26 countries. In 1986 he built the independent blade test facility "The Sparkær Centre", which was later purchased by the government-owned institution Risø.

From the year 2000 to 2007 Erik developed a route for recycling glass fibre and carbon fibre from worn-out wind turbine blades. Since 2007 he has been with Siemens Wind Power, where he has been associated with the development of new wind turbine blades using advanced technology.

Chapter 9

From Herborg Blacksmith to Vestas

Henrik Stiesdal

Siemens Wind Power, Borupvej 16, DK-7330 Brande, Denmark

henrik.stiesdal@siemens.com

It was towards the end of 1976 that I began to take serious interest in wind power. I got my A-level certificate in the same year, and after a summer where I was doing various jobs I spent the autumn doing England on my bicycle. When I returned from my journey, a call up for the military was awaiting me. I was, however, not to be there until May 1977, and consequently I had the rest of the winter and a bit of the spring available for doing sundry jobs at my parents' farm near Vildbjerg between Herning and Holstebro.

While I was away the initial work on the Tvind wind turbine was reported in the newspapers, and in late December 1976 my father and I went to Tvind for the first time, and like everybody else we were fascinated by this group of evident amateurs who, from something that looked like scratch, were building the world's biggest wind turbine.

Wind Power for the World: The Rise of Modern Wind Energy
Edited by Preben Maegaard, Anna Krenz and Wolfgang Palz
Copyright © 2013 Pan Stanford Publishing Pte. Ltd.
ISBN 978-981-4364-93-5 (Hardcover), 978-981-4364-94-2 (eBook)
www.panstanford.com

9.1   Theory Is Not Enough

In early 1977 we visited Tvind several times. And there I found various literature on wind power. The most inspiring book was *Sol og vind* by Claus Nybroe and Carl Herforth,[1] a classic manifestation of grassroots energy and still worth reading because of the technology as well as the gusto.

The theoretical aspect of the wind turbine technology appealed to me from the very beginning. My mathematics-physics A level made subjects such as aerodynamics, electric technology and calculations of loads accessible at a simple level; these issues were challenging, at the same time by being about something to be used here and now, of a suitable complexity, but not sufficiently difficult to discourage me.

Of course, theory is not enough, and we soon carried out some practical exercises. The first experiments were about making a 2-bladed rotor out of a thick piece of laminated wood. The diameter was just over a metre, the width of the blades about 10 cm, and we used a bit of 3/8-inch water pipe as shaft. We ran the rotor simply by holding the shaft with our hands, using our palms as the *bearing*, and the hand in front served as a thrust bearing. After a bit of fine honing and balancing the little rotor ran quite well, and it turned out to have many of the characteristics that we came later to know in real wind turbines.

When the rotor was set moving by means of a brisk grasp around the shaft it began to accelerate slowly until it reached a certain speed—then it began to accelerate madly, running like blazes, making sounds like a helicopter in strong wind. It was a great experience to feel so directly how the rotor's behaviours changed radically as soon as it gained speed. The wind pressure rose dramatically, and we felt distinctly how the rotor reacted to all small turbulence eddies. To stop the rotor at low wind, we just pressed the shaft harder and braked it. But this did not work with strong winds, as the momentum was too high, and then we had to yaw by turning the rotor so that the shaft was 90 degrees out of the wind.

Even at moderate winds the small rotor ran so fast that it was impossible to follow the blades with your eyes, and at stronger

---

[1]Herforth, C., Nybroe, C. (1976) *Sol og Vind* (Sun and Wind), Informations Forlag, Copenhagen.

winds we measured the speed of rotation at more than 500 turns a minute. We really got quite "high" by running the rotor with a good wind, it was like a living creature reacting to invisible commands from the wind, our palms became hot with friction, and the blades whirred just past our noses at a speed of 100 km/h. Looking back one must wonder how come we did not have any accidents during one of the many times when we just had to feel the whirl.

9.2   My First Accident

The next addition was a 2-bladed rotor with sail wings, diameter 3 m. In *Sol og vind* I read about the experiments at Princeton where a simple frame construction with an aerodynamically shaped leading edge and a steel wire as trailing edge was covered with fabric, and where the aerodynamic forces contributed to shaping the fabric. A "sail wing" was reported to offer the same good aerodynamic qualities as a solid blade, but at much less effort. We had to try that.

Figure 9.1   Two-bladed experimental rotor, 1977.

Each blade was supplied with a frame made of steel pipes with a leading edge of wood and with a steel wire as trailing edge. The pipes were assembled with ordinary threaded fittings. The hub of the rotor was a T-piece with a pipe for the blade in each of the lateral branches of the T, and with the central branch screwed onto the rotor shaft. The sail wing was made from Dacron, the material that is also used for sails on boats.

For the rotor shaft support I made a simple shaft box arrangement with two ball bearings mounted on a lattice structure fitted to an ordinary farm wagon. We could then take the wagon out on a field, place it with the rotor in the correct position in relation to the wind and in various ways measure the yield of the rotor in order to see how the sail wing worked.

Figure 9.2   Testing of the rotor on the field, 1977.

With this experimental rotor I experienced my first (but not last) wreck. As in the case of so many other accidents I could have told myself that this *must* go wrong. The fact was that I, faithful to the tradition, had made the rotor running anti-clockwise. Traditionally wind roses would rotate in the clockwise direction, whereas Dutch windmills and windmills with adjustable narrow vanes would rotate anti-clockwise. In normal circumstances the direction would have been without practical importance, but that did not apply for the experimental rotor, as it was assembled with ordinary pipe fittings—with a clockwise thread like all other normal fittings. And it was just the combination of the anti-clockwise rotation and the right-handed screw thread that per definition resulted in the T-fitting "rotor hub" screwing itself loose from the rotor shaft. This risk did not occur to me until one day when I wanted to look into the efficiency of the rotor in hard wind. Suddenly, as I was measuring and applying load to the rotor, the torque of the shaft dropped drastically, and in one clear-sighted moment I realised what was wrong. I was just in time to let go of the measuring system and fall down behind the wagon before the rotor screwed itself completely loose, executed a fine gyro-pirouette and hit the wagon with a bang. Fortunately nothing hit me on my head, but the rotor never made another run.

9.3 The Price of a Wind turbine Is One Krone per Kilogram

The measured efficiency of the 3 m rotor seemed promising, and backed by my experiments I began to build a real wind turbine in the spring of 1977. This wind turbine should also have sail wings. Although I had still some money left over from my months of work during the previous year, I had far from enough money to build a decent-sized wind turbine from new components, so like so many other do-it-yourself wind turbine builders I had to use recycled parts. That was no sacrifice, by the way, as it is a special pleasure to track down second-hand parts that might be used in your own project. I found most of the components at the local scrap dealers, and I really felt a chill along my spine when suddenly in a corner behind all the trash, I caught sight of something that might be a candidate for part of the many systems of a wind turbine—brake, yaw system, etc. The scrap yard price was almost always fixed, 1 Krone per kilogram, so the price even of a do-it-yourself wind turbine could, as a rule of thumb be determined with great precision by weighing it. Owing to the military service nothing much happened on my project from May 1977 to February 1978. In return, quite a lot happened just after my disbandment.

Figure 9.3   Henrik Stiesdal's 22 kW prototype wind turbine, 1978/1979.

First and foremost I had been invited to give a speech at the Danish Organisation for Renewable Energy (OVE) VindTræf (Wind Workshops) on 18 February 1978 at Snoghøj Folkehøjskole (Folk High School). Along with other do-it-yourself wind turbine builders I was to tell about my project. There had been three other workshops before this one, but this was my first, and it was a very inspiring experience to be with so many like-minded people. I had, of course, been aware that other do-it-yourself builders existed, but it was a great surprise to see such a crowd of enthusiastic pioneers, with so many different projects.

9.4   OVE Safety Guidelines

The VindTræf at Snoghøj became decisive in many ways. There had already been cases of wind turbines being wrecked as a result of runaways, and the wind power community was nervous that if we had too many spectacular wrecks at an early date this might lead to the intervention by the authorities, which might hamper developments. That was the reason that the VindTræf agreed that we should forestall events and lay down our own guidelines for wind turbine safety. To this end an OVE safety committee was appointed, counting Preben Maegaard, Erik Grove-Nielsen, Johannes Grove-Nielsen, Ian Jordan, Askel Krogh, Dorte Arp and myself. During 1978 the safety committee met a number of times, and early in 1979 the committee was able to present a small folder describing the safety guidelines that had been the result of meetings and inspections. Later, the OVE safety guidelines became the basis of the Danish approval rules, and their fundamental principles can be found in most international codes of practice.

Another important event at the VindTræf was a very inspiring speech by Torgny Møller, who had bought the first 22 kW Riisager wind turbine in 1976. At the end of his speech Torgny Møller suggested that an owners' association should be established, something like Association of Danish Wind Electricity, which had organised the owners of the la Cour–type of wind turbines producing power during the years around the First World War. Torgny Møller's association came into being in May 1978 and was named Danske Vindkraftværker, and it soon reached a membership of almost 50. Later it changed its name to Danmarks Vindmølleforening, at almost the same time that it reached 10 000 members. The importance of this association for the development of wind energy cannot be overestimated. Particularly in the area of community wind turbines and subsidies for establishing wind turbines, two factors that are considered by many people to be the most important reasons why Denmark got its lead position in the wind energy sector, the association has been the decisive moving power.

9.5 Æ Dunderværk

Although the VindTræf and what resulted from it was of essential importance for the development in an broader sense it was still more important for myself that a few days later I met the man who was to be my partner in the development of wind turbines during the coming years. I needed to have some big holes machined in the flange of an elastic coupling that was to be mounted on the main shaft of my wind turbine, and the only place in the vicinity that this could be done was in a blacksmith's shop in Herborg near Videbæk. Here I met Karl Erik Jørgensen. He was standing there, messing around with a so-called super wind rose, when I drove up to the door with my clutch flanges. Karl Erik Jørgensen was in many ways the epitome of an entrepreneur. He was born and grew up in the countryside in modest circumstances but had succeeded in becoming the proprietor of a machine pool. When I met him his main activity was a machine workshop that made various subcontracting jobs for enterprises in the area, often assisted by his eldest son, Per, who was at that time an apprentice mechanic.

Some years earlier Karl Erik had suffered from a serious form of cancer, which had left him with some disabilities and a full-disablement pension; but that did not keep him from working full-time and often more. Full of curiosity and energy he tackled various tasks, and it was evident in many ways that he had not forgotten how to play. He had built his own home with direct access to the workshop from the kitchen and bathroom. An old war-time German diesel generator had been reconstructed in such a manner that the heat from cooling and exhaust provided central heating when the generator was producing power for the workshop. All thermal energy of the fuel was used, and the exhaust was smooth and cool—a high-efficiency forerunner of the present decentralised combined heat and power stations. During the heating season the generator, for obvious reasons called by the family "æ dunderværk" (a Jutlandish expression, meaning something like the roaring noisy machine), was more economical than purchased power and heating.

Karl Erik could become choleric when things were not as he wanted. The family used to tell (when he did not hear it) how he had lost his little finger. Once, when he ran the machine pool his little finger got stuck in the V-belt of a combine harvester. The finger broke and consequently got stiff. Later on, in his machine shop his stiff little finger was often in his way when working with the machines, and of course it would happen that one day it was stuck in the lathe and got somewhat battered. Karl Erik drove to the doctor in Videbæk but the doctor was convinced that it was a job for the emergency clinic of the hospital in Herning. Karl Erik said that he was working on an important task, which had to be finished and that he had no time to drive to Herning, sit there and wait. Could he not just have the finger off as it was always in the way anyhow. The Videbæk doctor was less willing to do that; no, he had to go to the emergency section. Karl Erik's answer to that was that it was something he was going to decide for himself, and at a glance he reached across to the doctor's desk for a pair of cutting nippers, ready, perhaps for cutting toenails, and cut the finger off himself. After a hasty patch-up, done by the nurse (the doctor refused) Karl Erik could go back to his lathe.

9.6   Entering the Inventors' Bureau

Karl Erik and I soon discovered that we had common interests. Right from the beginning we got on well together, he was an eminent artisan and wanted to construct real wind turbines, but needed backup on theory, whereas I needed assistance in the practical area. At that time, in spring of 1978, Danish Technological Institute had a special branch, the Inventors' Bureau, which offered consultancy services to inventors, and which could also be of assistance in getting certain kinds of government subsidies. We read in the newspaper about one of these subsidy schemes, Profeo, and we agreed that we ought to apply there, for Karl Erik to get a proper wind turbine in addition to his home-brewed wind rose, which had been somewhat of a disappointment. I sat down and made a rough drawing of a simple wind turbine with a 3-bladed rotor, and with that and a small description of the project we contacted the Inventors' Bureau. We were met with energetic and kindly treatment by the leader, Peter Cordsen, and soon, somewhat surprised and probably with rather stupid expressions on our faces, we were holding in our hands a cheque for DKK 50 000 for our joint project.

The Profeo subsidy was an economic necessity and at the same time a tremendously encouraging pat on our shoulders—someone evidently thought that we had got into something special. Consequently we set to it, I was doing calculations, shopping for components and building permit, Karl Erik put up with the practical side. He made the entire machine himself, including the tower. We got the blades from Erik Grove-Nielsen, whom I knew from the safety committee.

Figure 9.4   Karl Erik Jørgensen working on the 22 kW HVK 1 gear and generator, 1978.

The prototype had a compact design with the rotor installed directly on the output shaft of a gearbox with reinforced bearings. The service brake was an electromagnetic brake mounted on the rear side of the generator. As a novelty for small turbines the turbine has electric yawing, controlled by a wind vane. In Karl Erik's opinion yaw rotors, which at the time were the norm, made a wind turbine look visually disfigured. In the end Karl Erik made the yaw bearing for the yaw system himself. It was a clear proof of his mechanical competencies.

The prototype wind turbine was erected in early summer of 1978, and it soon began to run. In general, it was surprisingly easy, considering how innovative the product was.

However, a sharp break in our delight occurred in early autumn of 1978. While the wind turbine was operating during a gale there was a short grid failure, the turbine ran away and threw off the blades. Afterwards we were analysing what happened. We found out that the violent acceleration of the rotor in the short interval between the disconnection of the generator by power failures and activation of the brake, was sufficient to increase the rotor torque to a level that could not be handled by the mechanical brake. In a few seconds the rotor reached such a speed that its output was multiplied, and the brake became red-hot without being able to stop the rotor. And when the brake had been burned off the rotor really began to run fast and threw off the blades. Two of them hit various objects near the wind turbine whereas the third followed a ballistic curve and hit the ground 475 metres from the wind turbine.

This event made us realise that we needed to design blades with built-in air brakes. For this we liaised with Erik Grove-Nielsen, and he quickly came up with a solution with turning blade tips. Under normal operation the blade tip was kept in place by a spring. However, if the rotor speed exceeded a certain limit the centrifugal force on the tip would cause it to compress the spring and move axially outwards, and with a guide pin system the tip would turn to act as a brake. The first blades with air brakes were completed late in 1978. The set for the prototype had outside springs, because one could not get a standard spring that would work. All production wind turbines had special-design inside springs, but the fundamental principles were unchanged. Nowadays such blades are made with tips that are kept in place with hydraulic cylinders; this allows the air brakes to be used also during normal shutdown.

9.7   From Sail Blades to Wooden Blades

Meanwhile I finished building my own wind turbine, of 15 kW with a 9 m rotor. It was not such a great success as the one we built at Karl Erik's. The machinery did work well, but the sail blades were a bad disappointment in full size. The air inside the sails was forced out towards the tips because of the centrifugal force, and when the wind turbine was operating normally the outboard part of the blade puffed so much out, that in practice it destroyed the aerodynamic profile. At the same time the inmost part of the wink was sucked almost completely flat. In principle opening the blade at the outer and the inner end faces could solve the problem, but the only thing obtained by that would be that the rotor became one big centrifugal pump for air and would not produce any serious power. After many vain attempts I had to swallow the bitter pill and replace the sail blades with some home-brewed wooden blades. With those blades the wind turbine ran until 1991.

Figure 9.5   Installation and testing of the prototype windmill, 1978.

In late spring of 1979 Karl Erik wanted to go on with things. The prototype wind turbine ran well, and with air brakes as well as mechanical brakes, we were convinced that the safety issue had been solved. Consequently we worked together on the development of a production version of the wind turbine. Fundamental principles of the construction from the prototype were maintained, and the wind turbine appeared as a kind of an "archetype" of what for a long time remained the epitome of a modern Danish wind turbine—a 3-bladed rotor with air brakes, electric yaw steered by a wind vane, and two speed operation with generators directly connected to the grid.

Figure 9.6   Henrik Stiesdal designing wooden blades.

9.8   From Herborg to Vestas

The first production wind turbine was erected at Gunhardt Keseler's farm in Dejbjerg near Skjern. Here I had an unfortunate flirtation with fate that left me somewhat shaken. At that time, in 1979, it was not ordinary among pioneers to use personal safety gear such as safety harnesses. We were not protected in any manner during the erection and operation of the wind turbine in Dejbjerg. At a certain stage of installation I was standing at the top of the tower on the platform, mounting the belt that would connect the small generator to the big generator. While

I stood there the electrician was connecting the controls at the bottom of the tower. By a mistake the small generator was suddenly powered up while my hands were still on the belt. As a pure reflex without thinking about where I was standing I jumped back. And it was only because my shirt was partly caught by a fitting that I did not fall down from the height of 18 metres. With a torn shirt and my heart in my mouth I was able to climb back onto the platform having acquired a new bit of experience.

The wind turbine in Dejbjerg was our first one to be delivered to a customer, and a short time after installation I had my first, but far from last experience of lying sleepless during the first night when a wind turbine was running during a gale. Fortunately our wind turbines behaved well, and they suffered only from a few childhood diseases. Later in 1979 Karl Erik erected his next wind turbine, this time for Aase Højmark in Stauning.

Figure 9.7   Installation of the HVK prototype (left); The HVK prototype, Herborg, 1978 (right).

The wind turbines operated well, but it was also evident that commercial business was not Karl Erik's strong point. Consequently we agreed that we should find a licensee who could manage the serial production in return for a royalty. Late in the summer of

1979 we gave the machine factory Vestas in Lem by Ringkøbing a call, having heard that they were experimenting with a Darrieus wind turbine. A few days later the manager, Finn Hansen, and chief engineer Birger T. Madsen came to Herborg in order to see our wind turbine. After a short time of negotiations we reached an agreement in the autumn of 1979. That was the beginning of my wind adventures at Vestas—and in actual fact the end of my own pioneer period. The same autumn I started my university studies, which meant that my time for experiments became much more limited. Most of what I did during the following years was primarily aimed at improving wind turbines that had already become commercial.

Figure 9.8   Stiesdal home wind turbine, 1983.

9.9 Will Power and Drive

Now I need to tell how things worked out with my most important friend and cooperation partner during the pioneering epoch.

In 1978, when I met Karl Erik he was still doing fine, despite the after-effects of his serious cancer. Sadly, the cancer had not been done with; in the spring of 1982 he was attacked by new tumours, which caused visual disturbances and paralysis.

Figure 9.9   Karl Erik with 22 kW HVK-10 turbine.

He continued working at new projects with an incredible willpower and drive, among others on the development of a new type of gear system. As one of many solutions to his growing disablement problems he installed a lift in his most recent wind turbine. The disease, however, was stronger, and Karl Erik died in October 1982.

About the author

Henrik Stiesdal is one of the pioneers of the modern wind industry. In 1976, Stiesdal built his first small wind turbine and, in 1978, designed one of the first commercial wind turbines, licensed to Vestas in 1979. Stiesdal worked as a consultant for Vestas until 1986 while simultaneously studying medicine, physics and biology at the University of Southern Denmark.

Stiesdal joined Bonus Energy (now Siemens Wind Power) in 1987 as a design engineer. In 1988 he was appointed technical manager, and in 2000 chief technology officer.

During his 25 years with the company, Stiesdal has worked with all aspects of wind turbine technology. Besides this he has been engaged with a wide range of other activities, including sales, manufacturing, project implementation, service and quality management.

Stiesdal has made more than 160 inventions and has been awarded more than 200 patents. In 2008 he received the "Siemens Inventor of the Year" award and in 2010 "Siemens Top Innovator" award. In 2011 he was awarded the EWEA Poul la Cour Prize.

During 1979–1988, along with his wind turbine work, Stiesdal studied medicine, physics and biology at Syddansk Universitet in Odense, Denmark. Henrik Stiesdal, 56, is married and has two teenage daughters.

Chapter 10

From Danregn to Bonus

Egon Kristensen
1943–2010

10.1   In Search of Inspiration

In 1980, I was employed as a sales consultant at A/S Danregn, a company owned by Søren Sørensen. By dint of diligence, competence and thrift, he had created an incredibly solid enterprise with a considerable proprietor's capital. The structure of the enterprise was of a seasonal kind—a great number of people were needed from April to June, and fewer for the rest of the year. The son of the proprietor, who also worked in the company, Peter Stubkjær Sørensen and I often discussed how the low-employment periods could be used in the best way. We had a number of ideas that were tried out: community water supply, gear for fire

This chapter has been translated from Danish and was published in *Vedvarende energi i Danmark. En krønike om 25 opvækstår 1975–2000* (2000) OVE's Forlag, Århus.

Wind Power for the World: The Rise of Modern Wind Energy
Edited by Preben Maegaard, Anna Krenz and Wolfgang Palz
Copyright © 2013 Pan Stanford Publishing Pte. Ltd.
ISBN 978-981-4364-93-5 (Hardcover), 978-981-4364-94-2 (eBook)
www.panstanford.com

extinguishing and heat pumps. The first two items were soon dropped, but we reached amazing results with heat pumps and heat recovery.

In order to see what our rivals could offer, we visited HI-messen (industrial fair), which had a hall with alternative energy forms. As we arrived, we saw multitude of people offering heat recovery equipment, and Peter Stubkjær Sørensen turned to me and said quite succinctly: "There are too many of them, we'll drop it." Shortly afterwards we saw a picture of the Nordtank windmill. Peter Stubkjær Sørensen saw the poster and noticed: "That is something that cannot be done just by anybody—it must be the thing for us." We tried to obtain the franchise for the Nordtank wind turbine in Denmark—but it failed. Then we just had to start with our own.

10.2 A Crazy Idea

The situation was that we had already spent a good deal of money on projects that were not realised. Søren Sørensen, therefore, was not willing to take the idea of wind turbines on, as he was not ready to throw more money out through the window for our crazy ideas. Consequently, I was sent home.

Late in the evening of the same day, Peter Stubkjær Sørensen called me and asked if I would come the next day as his father wanted to have a talk with us. The following day at 9.00 in the morning, we were asked to come to Søren Sørensen's office, where he told us that it was a brainless idea. But on the other hand, he did not want us to think he was a close-fisted old man, who had no thought for the future. In the end he told us that an account would be opened in the company, and that DKK 1 million (EUR 150 000) would be deposited on it for us. When we spend it, we were not allowed to come back for more, for then the cash-box will be closed—and he meant it. He did, however, offer to buy the first wind turbine, which was to be set off against the amount we were to owe him.

When we left his office, we sat together—what to do now? We decided to find all the literature about the subject and find some people who knew something about windmills. And a few weeks later we knew a great deal more about windmills, we had also come to know about people who worked with wind power.

10.3 Important Fundamental Rules

Our goal was to make the simplest possible windmill. We visited Risø, Gedser windmill, Tvind, Økær Vind Energi, Hjerm Elektro, Folkecenter, Kuriant, whose owner I knew, and many more. And when someone knew something about windmills—we listened avidly. Folkecenter had a complete concept for the production of a windmill. I asked if they could help us by making drawings and dimensioning for a windmill. The answer was no.

Few days later I called Preben Maegaard and asked him for help with the major issues and about what we should watch out for. We would, of course, pay for these services. He answered "I cannot take a fee, but I am going to Copenhagen next week, then I can look in for a couple of hours." During his visits we learnt about some important fundamental rules, such as don't forget that

- the main shaft should be of good quality and have a thickness of 10 mm for each rotor diameter
- yawing should only be one degree a second
- the gear must be twice as strong as the generator capacity
- all other parts should be able to stand up to a permanent load of 120%
- the shaft should be lined up in such a way that it puts a constant load on the yaw system

10.4 Quite a Good Thing to Be Little Behindhand

At the outset there were only three persons: Stubkjær Sørensen; his wife Alice Sørensen, a draughtswoman; and me. We did things from the end backwards to the beginning, building the wind turbine piece by piece and made the drawings afterwards from what we had made. It was a hard time, as we had to hurry because Nordtank and Vestas were already marketing their wind turbines.

When our prototype was ready, Risø came to inspect the wind turbine. I distinctly remember their remarks: this bearing house will slide, that plate will not stand for the axial loads, etc. Afterwards we set these things right, and upon their next visit they thought that things looked very good.

Figure 10.1   Danregn (Bonus) 22 kW prototype with 5 m Økær blades, 1980 (left); unique colour code on the tower for marking 6 m tower modules depending on the capacity of the turbine. Modular tower was a flexibile solution (right).

Now we embarked on the sales job, which was done in the evenings and on weekends. Luckily we sold a few, and when we installed our first wind turbine, we could not wait to know whether everything would be as we hoped. We were lucky. The wind turbine worked without any major problems. The first windmills were 22 kW, but we soon went on to 30 kW.

In 1981, Danregn Vindkraft A/S was established as an independent enterprise. We were permanently behindhand and felt a strong pressure from our colleagues from other companies. When we were developing a 30 kW wind turbine, they were already at the 55 kW wind turbine. So the only thing was to develop a similar 55 kW wind turbine, and we succeeded comparatively rapidly. Looking back, it may not have been such a bad thing to be behindhand, because we escaped some of the mishaps that others ran into, I am thinking in particular of blade problems that Vestas had.

Figure 10.2 Installation of Bonus 55 kW prototype with Økær blades, 1981, Ydby, Denmark. The prototype was developed with consultancy from Preben Maegaard and belonged to Asger Maegaard Kristensen. Later AeroStar blades were used but after the bankrupcy of the AeroStar manufacturer DCER in 1986, LM Glasfiber blades were used.

On 1 August 1981 we employed the first engineer. We also engaged a part-time consultant to assist us in the Californian market. The first six wind turbines were installed in California in December 1982.

Figure 10.3 Danregn (= Danish rain) headquarters in Brande (Photo: Jens-Chr. Kjaer, 1982).

10.5   When I Grow Old

In 1983 we changed our name to Bonus Energy A/S,[1] as Danregn Vindkraft could not be pronounced in English. In that year we installed 30–40 wind turbines in California, but the great breakthrough came in 1984 when we installed more than 200 wind turbines.

Figure 10.4   Growing in size and changing features: Bonus 95 kW wind turbine with tripod tower, 1985. After Nordtank developed and popularised tubular towers Bonus also employed this concept (left); Bonus 600 kW turbine with tubular tower (right).

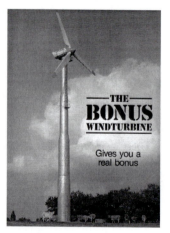

Figure 10.5   Bonus advertising folder, 1980s.

[1]Bonus A/S was taken over by Siemens Wind Power in 2004.

It was also in that year that we employed our new manager, Palle Nørgård. We started with very few employees, but the office soon became 20 staff strong with 40–60 people in the workshops. We got safely through 1984, and since then there has been nothing but progress. In 2000 we employed 350 people with a turnover of DKK 1.5 billion (EUR 200 million). Only a few can experience that kind of growth in a new industry. I have been lucky. And when I grow old, I think that what I shall be looking back to will be my time in the wind power business.

Chapter 11

Vind-Syssel 1985–1990

Flemming Østergaard

ADventure.dk, PR and Marketing, Copenhagen, Denmark

Flemming@ADventure.dk

11.1   Vind-Syssel

Vind-Syssel[1] was one of the 13 major Danish WTG manufacturing companies appearing in the 1980s out of the grassroots' experiment to make the utilisation of wind power from before 1950 come back into public use. During the following 10 years, most of the 13 companies were out of business due to changes in politics in Denmark and the United States and in combination with a too weak economic foundation in the companies. Vind-Syssel was closed down in late 1990, after manufacturing and erecting 50 units, mainly 150 kW. Most of the WTGs were sold to farmers and local owner-groups in Denmark. Three units were sold to Sweden,

---

[1]In the old Danish language, *Vind-Syssel* meant "working with wind". It also indicates the very northwest of Denmark—an area called Vendsyssel—where this enterprise began.

Wind Power for the World: The Rise of Modern Wind Energy
Edited by Preben Maegaard, Anna Krenz and Wolfgang Palz
Copyright © 2013 Pan Stanford Publishing Pte. Ltd.
ISBN 978-981-4364-93-5 (Hardcover), 978-981-4364-94-2 (eBook)
www.panstanford.com

one was erected in the Faroe Islands, and four in China—the first grid-connected WTGs ever in China (now China is leading the world in wind energy). Vind-Syssel spent a lot of effort in the US market, but did not succeed.

Figure 11.1 Vind-Syssel advertisement from the 1970s.

11.2 Background

It was springtime 1985. I felt like a free man, after my decision to leave "Miljøhøjskolen" (Danish Ecology Center), where I had spent my body and soul 24 hours a day as the director since 1974. Yet, my freedom did not last too long, as I was contacted by Preben Maegaard, the director of Nordic Folkecenter for Renewable Energy, who encouraged me to establish a new company for producing WTGs. Six months later, Vind-Syssel was a reality.

A couple of years earlier, the Folkecenter initiated a WTG-project, building on the philosophy of local and clean energy supply, delivered by local technology, and controlled by local democracy. Practically speaking, the project should end up in a WTG-technology, which could be owned by local groups or individuals and manufactured and installed by local blacksmiths. This philosophy was partly workable, because assembly and installation as well as building foundation and infrastructure can be managed locally. Also quite a lot of parts can be used, which are low technology. Many of the parts in a modern WTG are either large or complicated—such as generator, gear, control system, and even the tower. This means that feasible manufacturing takes minimum quantities, and therefore also major investments, which is not possible or affordable for most local blacksmiths. Therefore cooperation was necessary, and was organised by the Folkecenter. There was formed a national group of blacksmiths who should be able to cooperate in a certain degree of specialising and in purchasing of such parts, which they could not manufacture by themselves.

The project was a success, and resulted in developing a 100 kW WTG with an integrated gearbox. Several units were installed in local Danish societies, and also in California. But cooperation became tough, because local blacksmiths are individualists with many ideas, which might not fit to the ideas in the next village. At the same time, increasing the size of the WTG was on demand, and Preben realised that someone else should be in charge of commercialising this major development work, which had been done by the Folkecenter.

11.3   Founding Vind-Syssel

"Hedelund" is the name of the restaurant in Brønderslev, where we used to have all kinds of meetings in my "former life", so why not use this home feeling place for the founding meeting in our new WTG manufacturing cooperation.

At first I told Preben "no thanks". I considered a one-year's travel around the world, and I had also other options in teaching or engineering jobs within ecology and renewable energy. But Preben was never a good listener, especially not when it was something he did not want to hear. First he invited me to an interesting trip to California visiting the impressive wind farms and meeting all kinds of people participating in the WTG industry. Second, he found a small amount of start-up money. Most of all, he offered to put the WTG engineering staff of the Folkecenter at my disposal for final documentation of a commercial 150 kW, and later for a 200 kW WTG, ready for serial production. The group consisted of Peter in charge of the electrical part, Birger, Henrik and Karstensen responsible for everything else, and Lilian, who did the drawings. These five people made Vind-Syssel possible, and so the Folkecenter did a remarkable contribution in developing the modern WTG industry.

Figure 11.2 A Vind-Syssel 150 kW wind turbine.

Seen from a commercial point of view, I still should have said no, because this experience turned out to be hard and expensive for me. But seen from an educational point of view, it was the best and the most interesting thing, which I ever did in my entire life. And the cooperation between Vind-Syssel and the Folkecenter has contributed remarkably in developing the modern WTG industry.

Back to Restaurant Hedelund in Brønderslev, summer 1985. Trying to be loyal to the basic philosophy, as mentioned earlier, we decided to form the company like the old locally owned entities, where your influence is based on "heads of persons" and not "heads of cattle". About 30 people came to the meeting, among them my friends, Tage, the lawyer and Kaj, the auditor, who both offered support with no charge until Vind-Syssel would be on the track—and of course Preben. I also especially remember Bent, an old man with a white beard (like mine is now), who was quite irritating, asking a lot of questions, but who later on appeared to be of a great support to Vind-Syssel.

11.4 The Beginning

Vind-Syssel was a reality, but we had no organisation, except from the board, me and the Folkecenter engineering staff. The money was absolutely limited, so my task now was to work with the Folkecenter engineers on final documentation, negotiate with suppliers for parts, assembly, installation, etc., and to plan the sales strategy and do the sales material. I hired my daughter, Pernille, as my first secretary, and she and I did all of this while I also accepted a part-time job teaching ecology at Nordiska Folkhögskolan in Kungälv, Sweden, because so far, Vind-Syssel could not afford to give me any real salary.

After about six months we were ready to take off by erecting our prototype, the first commercial 150 kW WTG. It was ready for serial production, and the first ever with hydraulic tip brakes, implemented in the first set of the new LM 11 m blades.

Now we were ready to manufacture and sell. But we had no assembly facilities, so we agreed with Stig, the owner of a small machinery factory in Frederikshavn, to do the assembly and we secured all different parts from suppliers in Denmark and abroad. Erection was outsourced to crane entrepreneurs Gert and Donald from GSS PowerMills in Frederikshavn, who was also manufacturing their own wind turbines called the Wind Rose.

Figure 11.3 Installation of the 130/150 kW Vind-Syssel in Sterup, Vendsyssel, 1986.

Figure 11.4 The 130/150 kW Vind-Syssel wind turbine, 1987.

One day a local guy, Kjeld, appeared in my office asking for a job. I told him that everything was outsourced except for the sales work. Even though he had no sales experience, he begged me to give him a chance, and he appeared to be a great seller right from the start. When clients asked him all the technical questions, he replied: "I don't know, but it is a very good windmill". He sold 20 units mainly to farmers in the local area of Vendsyssel. My next salesman spent four months to educate himself in all details, and he sold only one unit.

11.5 The Iron Curtain

I travelled to Finland visiting Kumera and Valmet to buy gearboxes, helped by Olaf from the industrial broker company P. N. Erichsen. Olaf was even more creative, so he introduced the idea to the gearbox manufacturer Penig, situated close to Karl Marx Stadt (now Chemnitz) in the former DDR (East Germany). There could be many stories to tell about purchasing from "die Getriebekombinat", the centralised East German gear industry in Berlin, represented by Frau Weis, but let me leave that for now. The Penig gearboxes became our victory and our defeat.

We also found our tower manufacturer in Eastern Europe—in Poland. And based on these two major components purchased from behind the iron curtain, we were in a very good price-competitive position, because purchase prices in the former communist countries were negotiable in a different way than what we were used to in the West. It was also interesting to experience how workers behind the iron curtain were working and acting. One funny example was the Polish welders, who deliberately build in some welding errors in each tower, for the purpose of getting permission to go to Denmark for repairs.

However, quality in general was in top, and our suppliers were very focused on following our western quality standards and our quality management system. They succeeded in meeting our standards, so we could supply high western-standard quality for competitive prices because of Eastern European purchase. But one of our problems was that we were not aware of the negative image the purchase east of the Iron Curtain could create, caused by the political black-and-white way of thinking of that time and how it could be used by our competitors.

We had some technical problems with the yaw-system, which was purchased in the West, but one of our competitors, Wind World, succeeded in making people believe that these problems were because of the components manufactured in communist countries. This negative campaign gave us some setback.

11.6   The Vind-Syssel WTG

An interesting thing throughout history is that stories of how things began can easily be made into "religion". Just think about the combustion engine for the car industry. Electrical cars came first, but were abandoned. The Wankel engine was also abandoned. Where could we have been today if electrical cars had been given preference since 1900?

The machinery in a modern WTG consists of a main shaft, supported by two main bearings, and the rotor in one end, and a gearbox in the other end, delivering the high speed needed for the electrical power generator. Disk brakes were supposed to be inside the nacelle and tip brakes for emergency only. But the Vind-Syssel technique was different.

Figure 11.5   A 150 kW Vind-Syssel wind turbine.

If the main shaft and main bearings could be integrated into the gearbox, everything would be much more compact and simple, and thereby cheaper. At the gearbox manufacturer Kumera in Finland, Preben from Nordic Folkecenter found an integrated gear-box meant for an industrial mixer. By turning it 90 degrees, replacing the electrical engine with a generator, and replacing the mixer with the WTG rotor, the concept for a simple WTG machinery was in place. The integrated gearbox system became the foundation of the Vind-Syssel technique, together with the hydraulic wing brakes in the tip. Everything else in the Vind-Syssel WTG was more or less ordinary.

Figure 11.6   A Vind-Syssel 150 kW wind turbine (Photo: Arne Jaeger).

While working on the 150 kW documentation, I told Karstensen, one of the Folkecenter engineers, that I did not like the extreme loads on the disc brakes in emergency cases, and that I also did not like the idea of having extra doors in the tower for manually resetting the wing brakes after use. I claimed that the wing brakes should be hydraulic, and be not only for emergency use, but also for relieving the load on the disc brakes. However, people responded with a word which I never understood: "impossible". How would you get the hydraulic pressure from the nacelle into the rotor?

After considering this question, I responded—exactly how hydraulic excavators are functioning. Then things took off. I went with Karstensen to visit the grand old engineer and owner of LM Blades, Mr Skouboe, who, despite his position, preferred working in the workshop rather than at his mahogany desk. We ended up with the first hydraulic wing brakes, which were installed for the first time ever in the Vind-Syssel prototype. This was a revolution in the WTG industry at that time, and it was an important parameter of making LM into the world leading blade manufacturer.

It is of high importance that a WTG is producing maximum, and when it stops due to too high wind speed or due to any failure, it is important to get restarted as soon as possible.. Therefore, I wanted Vind-Syssel to introduce a remote control based on telephone connection to the WTG. I asked Per Gravesen, my friend and electro technician, to make a prototype. He succeeded so well, that he later supplied this system to the Vestas WTGs for several years. Now his company is the leading supplier of remote control systems for Danish water plants. Also in this matter Vind-Syssel was a pioneer.

11.7 Vind-Syssel's First Manufacturing Facilities in Jerslev

Outsourcing of the manufacturing of components was, and still is, appropriate in the WTG industry. Outsourcing of the erection was also natural, as this takes a special expertise and mentality to work with heavy cranes in the field with different weather conditions. After finishing the first four nacelles, outsourcing of the assembly work turned out to be problematic for several reasons. One reason was that it was important to have a close and direct communication between the workshop and the engineering and management. So, we decided to establish our own assembly factory in Jerslev, where our office was situated. We found a former garage for farming machines, which could be changed into an assembly hall within a reasonable budget, and with space for assembling of six nacelles at a time. We also established our own service department, and we were now really ready for takeoff.

The next six units of 150 kW WTGs were sold, assembled and erected, and we continued with the next six units and so on. At the same time, we started developing a 200 kW version, still working with the Folkecenter engineers. We also started developing alternative blades made of wood, and finally we decided to work on entering the US market.

Figure 11.7   A 200 kW Vind-Syssel wind turbine with wooden blades.

Maybe it was because we had too many activities at the same time, and maybe we could have done many things differently, but after two years we ran out of capital like most other WTG manufacturers, and the company went bankrupt. When we knew which way the wind was blowing, we managed to reconstruct in advance by isolating the assets in a new company, owned by the old one (in Denmark called the Hafnia reconstruction model, named after reconstructing a major Danish insurance company). So when the bankruptcy became unavoidable, we were able to keep the concept and idea alive, but only for a short time, unless we succeeded in finding some new capital.

One day by coincidence I met B. O. Jørgensen. He was in charge of attracting new business entities to the Hobro Commune. He claimed to be able to find local investment capital, but on the condition that we would move all our activities to Hobro. After a lot of negotiations, including a major support by Olaf Erichsen (the agent for the gearbox suppliers), some five local investors agreed to participate. The company was now moved to Hobro, where we established our new assembly facilities for the second time in a former storage building at the harbour. And we were now ready for takeoff for the second time.

Figure 11.8 Assembly of the 130/150 kW Vind-Syssel wind turbine in the assembly hall, Hobro, 1988.

11.8   The United States

The Vind-Syssel WTGs were mainly sold and erected in Denmark, but like any other WTG manufacturer at that time we were also looking out to the big world.

When Kockums Shipyard in Malmö, Sweden, was closed down, the buildings were rented to all kinds of new entrepreneurship companies. One of them was supposed to be a WTG manufacturing company working on a new concept. But they received orders from buyers before having any ready technology. Cooperating with them, Vind-Syssel erected three units of 150 kW in Skåne (southern Sweden).

This Swedish company was also organising delivery of four units, 150 kW to China, from our knowledge it should have been the first grid connected WTGs in China ever. Vind-Syssel also erected one unit in the Faroe Islands. But everybody was focusing on California, because this is where the big money was in building wind farms with 20–100 units at the time.

Figure 11.9   A 150 kW Vind-Syssel wind turbine in Skåne, Sweden (Photo: Väsk).

On my first trip to California, I met Steven Smiley from Michigan. Later he agreed to represent Vind-Syssel in the United States, and especially California, which mainly was to introduce Vind-Syssel and our technology to the California developers, and locating projects where we could supply our WTGs.

Travelling with Steven in California in the mid-80s, sometimes by air, and sometimes by Greyhound, but mainly by car, was an experience. This was before mobile phones took over, so one of the great efforts was stopping at payphones on the road for confirming our appointments. We were visiting wind farms, developers, banks, lawyers, authorities, etc., even Indian chiefs, who offered land, and everybody had dollars in their eyes when talking about wind energy. The projects were mainly concentrated in three locations, Palm Springs and Tehachapi delivering power for Los Angeles; and Altamont, sending power to San Francisco.

We targeted two interesting projects, one at Tehachapi, California, and one at Maui, Hawaii. Both projects were supposed to be based on the "sales and lease back" idea, meaning that we should do the whole project, including land lease agreement, power purchase agreement, and building permit, supplying, erecting and connecting the WTGs, and then selling the project to an investor who would lease it back to us for management and maintenance, cashing the money for power sales, and paying the lease to the investor.

We found an investor for the Tehachapi project—the State Street Bank in Boston, who agreed to invest USD 50 million on certain conditions. This investor was targeted by one of the leading Californian leasing providing companies, D'Accord, in San Francisco, where we met one of the directors *via* someone from Steven's personal network. This leasing company was known for arranging aircrafts for American Airlines and trains for the Canadian train company, so we felt that we were in good hands. D'Accord offered us an office, computers and consultancy on a no-cure-no-pay-basis, and we were working there with highly skilled professionals on preparing this leasing deal for approval by the potential investor in Boston. (This is in fact where I was receiving my intensive education in economy, better than four years at school).

When finished, it was mid-December 1987, and I went to Boston on my way back to Denmark, presenting the project with all its details to the director of the investment department of the State Street Bank. I got the approval, on the condition that everything should be finalised latest by 31 December, for tax reasons, and that their lawyer should approve the insurance conditions. I hurried back to Denmark and to my Danish insurance company, who finally agreed with some minor changes demanded by the American lawyer, but on 10 January. And even if approved, the deadline was exceeded, and the project was cancelled. This event was my largest victory and my largest defeat in business ever, and the feeling of having a USD 50 million order was really great, even if it lasted for only 20 days.

11.9 "Middelgrunden Wind Farm", Copenhagen

One thing of which Copenhagen should be proud, is the 20 offshore WTGs outside Copenhagen harbour. This project was initiated by Vind-Syssel in 1987. There is a long story behind it. We were working on feasible foundations, power cables, and investigating the sea bottom, which was kind of a junkyard. To begin with, we asked

12 different authorities for permission (yes, it was 12). All of them said yes, except for the Ministry of Energy, where the ministerial director, Michael Lunn was manipulated by the major electricity companies to manipulate the Minister of Energy Sven Erik Houmand, to send it to a long-term committee for killing the idea. About 10 years later, the project was recalled by the Copenhagen renewable energy organisation, and then the wind was blowing in a more friendly direction. The way this was handled by the ministry was very bad for Vind-Syssel, but because of the technical development it was good for Copenhagen, because Vind-Syssel would have put 200 kW units, and now the size has increased to 2 MW.

11.10 Vind-Syssel Finally Closed in 1990

After losing the California deal, and postponing the Copenhagen project, Vind-Syssel was again in a financially weak position, and we had to go back to the investors and ask for additional capital. This resulted in a conflict between me as the MD, and the investors. I decided to sell my shares for some small money and leave Vind-Syssel. My sales manager was instated as the new MD, but a few months later Vind-Syssel was declared bankrupt, and it was not reconstructed this time.

The technology and the organisation of Vind-Syssel were in many ways new and progressive, and there were many victories during the five years of existing, but the low capital base caused problems right from the beginning.

When Vestas went bankrupt, it was reconstructed with more than DKK 100 million from the major Danish workers organisation foundation, LD. When WindWorld was close to collapse, it was supported by its owners with DKK 50 million. Vind-Syssel was originally established with less than DKK 1 million, and later reconstructed with DKK 4 million, which was not enough for selling and manufacturing units, priced about DKK 1 million and more. However, as they say, the bumblebee cannot fly, but it does it anyway—so we tried to fly with Vind-Syssel. We did 50 beautiful units. By 2013, some of them are still there at the same spot, some have been moved to other destinations.

As for all the pioneers, our road was filled with holes, but the wind is always blowing, and I will never regret.

About the author

Flemming Østergaard, born in 1946, received his degree in mechanical engineering, production, logistics and management from the Danish Technical University in 1969. He did project development and project financing from D'Accord, San Francisco, USA, in 1987 and graphic design and media from Autodidact, Copenhagen, in 1992. Other areas in which he is skilled are farming, carpentering and international trade and economy.

Flemming's job history includes teaching and research at the Danish Technical University in 1969–1970. He was an engineer at Burmeister & Wain, Copenhagen, in 1970–1972, where he was involved in the development of power stations. He was also the founder and manager of the Danish Ecology Center in 1973–1985, founder and board member of the Danish Ecology Farmers School in 1974–1986, founder and director of Nordic Energy Fair in 1979–1984, and managing director and project developer in Vind-Syssel A/S, a Danish wind turbine manufacturing company, in 1985–1990. He became a private business consultant in Szczecin, Poland, in 1990–1992, and then director of Seifert Reklame (marketing), Copenhagen in 1992–1996.

Flemming has initiated several projects, mainly related to environment and energy supply, in Denmark and abroad. He is a frequent lecturer on financial, commercial and political subjects, and is board member in small Danish institutions and business enterprises. He is also member of the social liberal political party ("Radikale Venstre") since 1979. He has gained business experience from Denmark, Sweden, Germany, Poland, Czech Republic, the United States, Russia, Kenya, Uganda, Ghana, and Zimbabwe.

Flemming is the managing director of ADventure (marketing, advertising, publishing) since 1996 and managing director of LLOYD.S Mægler & Finans/ EuropeBrokers (real estate and business consulting) since 2000. All in all, he is currently associated with management, economy, strategic planning; Project and company development, graphic planning and design; production planning, logistics, quality management; and calculation, sales and purchase.

Flemming lives in Copenhagen, Denmark, is single, and has one adult child. His main interests are business development, human behaviour, politics, anti-corruption, anti-violence, economy, environment, modern art, architecture, and music.

Chapter 12

The Story of Dencon

Bent Gregersen

Løkkeshøjvej 2, 5970 Ærøskøbing, Denmark

w-in-gen@hotmail.com

The first oil crisis of October 1973 triggered many various activities in the Danish countryside. How was it possible to maintain the supply of electricity and heating using local resources from the wind, sun and biomass? This debate happened on the Island of Ærø as well, located in the southern part of Denmark facing the Baltic Sea.

12.1 Powerful Island

The Island of Ærø has 7 000 inhabitants living in the two towns Marstal and Ærøskøbing and in the surrounding rural areas. The main trades are agriculture and fishery while the shipping sector provides the basis of shipyards and associated industries. There are four ferry routes to the mainland. By tradition, the island has struggled against involvement from the outside and often demonstrated a strong ability to find its own solutions.

Wind Power for the World: The Rise of Modern Wind Energy
Edited by Preben Maegaard, Anna Krenz and Wolfgang Palz
Copyright © 2013 Pan Stanford Publishing Pte. Ltd.
ISBN 978-981-4364-93-5 (Hardcover), 978-981-4364-94-2 (eBook)
www.panstanford.com

Figure 12.1 The map of Denmark and Ærø Island (red).

By 2013, it is a pioneering community within renewable energy with one of the world's biggest solar thermal installation in Marstal, seasonal hot water storages, several big wind power installations and district heating using solar (Fig. 12.2) and biomass even in rural areas like village of Rise.

Figure 12.2 Marstal solar power plant have an area of 18365 m². It covers a third of Marstal's power consumption (Photo: Erik Christensen).

Change to renewable energy was the future oriented reaction to the serious energy situation in the 1970s. Main actions were taken by a group of involved people from all walks of the local community: the director of the folk high school, representatives of the local industries and shipyards, farmers, businessmen, teachers, politicians and some activists as well. Being an island community we wanted to consolidate the interest in energy security, job creation and new local businesses.

The social democratic government by this time supported such local energy initiatives financially. The local study circle with its 10 to 15 participants employed two urban planning people to organise the activities, starting with an analysis of wind and biomass resources and comparing this with the existing energy demand.

12.2   Denmark's First Wind Farm

Initially this resulted in the construction of Denmark's first wind farm consisting of 11 units 55 kW Vestas turbines.

Figure 12.3   Early wind farm at Ærø with 55 kW Vestas wind turbines.

The National Steering Committee for Renewable Energy supported the project financially on the condition that the company name and logo of the wind turbines should be painted over, and the gearbox, already then a weak component in commercial windmills, was up-sized with the positive consequence that the Ærø Vestas turbines though located close to the rough climate on the Baltic beach, never suffered from gear breakdowns.

Figure 12.4 Dencon wind turbines at the Ærø wind farm.

The wind farm was commissioned in 1985. Biomass utilisation led to the construction of a heating plant based on straw and connected to Ærøskøbing district heating network. The plant was commissioned in 1988. Biogas was abandoned as a solution due to prohibitive costs of transport of the animal waste products.

The community was awarded a prize for the initiative and work of the study group. We were happy to see the Ærøskøbing municipality receive this honour, which helped to open the eyes of some of the citizens and authorities. Eventually renewable energy made Ærø internationally known due to its genuine bottom-up initiatives.

12.3   Innovative Concepts

The study group had good relations with the Nordic Folkecenter for Renewable Energy and in this regard I was invited, along with 15 others, to join a study tour in California to learn from their experiences within wind power. The Danish companies were pioneering in wind turbines and dominated the development.

As a direct result and with inspiration from the wind study tour to California, a wind turbine manufacturing company named Dencon was founded by three people, Leif Gottlieb, John Dreyer and me in 1986. It had company address on Ærø. The production of wind turbines took place at the Ærøskøbing Shipyard. Folkecenter supplied drawings and consultancy.

First a 75 kW turbine on lattice tower was erected at Ærø. The blades were well proven LM 8.5 m with spoiler type air brakes. The integrated gearbox came from Finnish Valmet and had a safety factor of two or so being of 145 kW for a 75 kW wind turbine. ABB delivered the generator, which was flange, connected to the gearbox. The yaw gear motor and the brake calibres were bolted directly to the gearbox according to the design principles that the Folkecenter director, Preben Maegaard, had developed during the early 1980s. The concept made it possible for newcomers and small industries to become manufacturers of state-of-the-art wind turbines even without their own engineering competences within wind power.

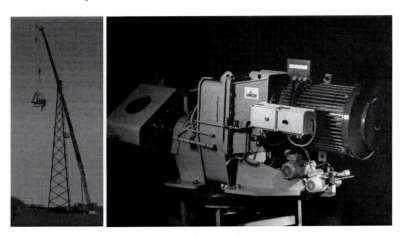

Figure 12.5 Installation of the Dencon 75 kW wind turbine (left); Dencon integrated gearbox with generator (right).

The blacksmiths who dismantled the 75 kW turbine, after 20 years of trouble-free operation, found it as far the best of all Vestas and Nordtank turbines they had ever dismantled. This proves well that in the early 1980s even small companies and shipyards were able to manufacture wind turbines to the highest international standards while using the Danish concept and quality well dimensioned components from specialised sub-suppliers.

12.4 Going Up and Down

The 75 kW wind turbine was installed close to the medieval town of Ærøskøbing near the beach of the Funen archipelago. The annual production figures proved to be significantly better than anticipated which encouraged Dencon to follow the trend of making bigger wind turbines.

Again it was the Folkecenter design team that supported Dencon. This time the integrated gearbox was delivered from Finnish Kumera with a low profile design that gave the nacelle of the new 200 kW wind turbine a very elegant appearance. The prototype with lattice tower was installed on Ærø launching the AeroStar 12 m blade. This was soon followed a series production of Dencon well-designed new wind turbine using conical towers and LM blades. Dencon wind turbines were installed in many parts of Denmark.

Figure 12.6   Dencon 200 kW wind turbine, Vigsoe, 1986 (left); Dencon 250 kW wind turbines, Strandby (right).

When Nordex, that became an important international manufacturer, entered the wind turbine business, their first two models of 150 and 200 kW were almost identical to the Folkecenter derivative that Dencon launched.

Finally Dencon engineered the first 500 kW turbine in Denmark, which was sold to Soenderjyllands electricity utilities. In the design process Dencon obtained a patent of a three-point mechanical suspension in order to reduce the dynamic load going through the drive train, a design principle that soon became industrial standard.

We had financing of the production. In fact I was encouraged by the local bank to use their services in regards of the production of turbines, but not for the development of the company, as we defined it in our business plan. This involved expansions on the international markets. Due to this limitation unfortunately we decided to close the business. An overriding factor was the shipyard's bankruptcy. It was a repair shipyard servicing the fleet of small freight ships, which numbers were strongly reduced during the 1980s.

By 2013, Ærø still maintains its pioneering position as a result of the local initiatives of the 1970s. Phase-out of fossil fuels and transition to the renewables enjoys the full support among the industrial and political leaders of the island that is well safeguarded against damages to the local economy coming from the high future oil prices.

About the author

Bent Gregersen was born in 1935 in Copenhagen. He undergraduated in mathematics in 1956 and did his M.Sc.(Pharm.) in 1962. He has been trained in finance and strategic planning. By profession, Bent is a wind energy planner, but has been the lead manager with various Danish Pharmacies from 1962 to 1985. He cofounded Ærøskøbing Energy Plant, the largest power plant based on combined renewable energy sources, and issued analyses of biomass resources as a part of the planning of the combined power plant. The plant, partly funded by EEC and commissioned in 1988, is in operation in Ærøskøbing as a part of the district heating system.

Bent was invited to California by the Nordic Folkecenter for studying the export potential for wind turbines. This lead to the formation of the wind turbine company, Dencon, in 1986. Bent was associated with Dencon A/S as director from 1986 to 1989, where his responsibility was sales *via* agencies, and with Dangrid Consulting Group Aps as wind energy planner from 1989 to 1992. During this period he also co-edited a report issued by the Danish Department of Energy on the perspective of the Danish wind turbine industry. From 1992 to 1995, he was with Consolidated Investment Ltd. and provided consultancy on various planning issues, such as identifying business opportunities for larger wind energy projects on a global scale; setting up of development companies in selected market (Spain and Scotland); supporting the companies for all technical, administrative, and legal issues; training the staff; controlling and supporting all contracts and agreements such as land lease and power purchase agreement; identifying subsidies; and negotiating with local and governmental entities, financial institutions and investors.

In 1995, Bent described a wind energy project for the Danish island Ærø (Vind energi udnyttelse på Ærø), including owner structure, legal issues, physical assumptions, local economy, etc.

Bent delivered a lecture on *energy price structures* to economists and engineers in Kiev in 1989. He has also delivered two lectures on *wind energy in general* at the University of Lugo (Spain) in 1994.

Bent has two daughters and a son.

Chapter 13

Water Brake Windmills

Jørgen Krogsgaard

Kastrup, Denmark

j.c.krogsgaard@gmail.com

13.1 Background

Almost all wind turbines in Denmark are producing electricity. Electricity produced by an asynchronous generator is sent to the public grid at 50 Hz with the related speed of rotation. The windmill is, thereby, locked by the electricity grid. With constant rotation speed, the blades will stall with increasing wind speed, and thus reduce the rotor effect. With constant speed depending on wind speed, the blade tip speed will remain constant. Average tip speed of older windmills is about 40 m/s and with increasing size of wind turbines the tip speed increases to about 70 m/s. Usually most windmills have three blades. It is called the Danish concept, and has its origins in the Gedser wind turbine that was designed by Johannes Juul.[1]

---

[1] For information about Gedser wind turbine see chapter *History of Danish Wind Power* by Benny Christensen.

Wind Power for the World: The Rise of Modern Wind Energy
Edited by Preben Maegaard, Anna Krenz and Wolfgang Palz
Copyright © 2013 Pan Stanford Publishing Pte. Ltd.
ISBN 978-981-4364-93-5 (Hardcover), 978-981-4364-94-2 (eBook)
www.panstanford.com

13.1.1   Liquid Brakes

In 1937, E. P. Culver from Princeton University,[2] conducted research and experiments with hydraulic dynamometer also known as water brake. He came up with a mechanical design and the basic formula for water brake power: $P = k * v^3 * d^5$, where P is the power, k a factor depending on brake design, v the speed, and d the diameter of the rotor.

Water brakes are also used for purposes other than for wind turbines. They are used in dynamometers, when measuring effects of engines.

13.1.2   Water Brakes and Windmills

Throughout history there have occasionally been attempts to develop wind turbines producing heat (hot water) instead of electricity, the so-called water brake turbines. Water brakes convert mechanical energy of the windmill into heat by friction in the water. Water brake wind turbines can roughly be divided into three groups:

(1) Do-it-yourself turbines
(2) Wind turbines constructed according to a design manual
(3) Industrially manufactured wind turbines

Water brake windmills are used to heat up buildings. They are quite small with rotor diameters of maximum 12 m may be more. Windmills with water brakes operate according to the cube power curve of the wind rotor as well as the rotor of the water brake, which are the ideal operating conditions for both. They are running with constant optimal tip-speed ratio.

When the maximum production is achieved, the efficiency must be limited. This can be done either by letting the water brake change its characteristics and overbrake. When this occurs, the blades will stall. Another way to control the power is by letting the blades pitch and the yield decreases.

The water brake windmill, with its power curve of three and constant tip ratio, runs optimally on a much bigger wind range in comparison with the simple stall regulated electricity-producing

---

[2]Culver, E. P. (1937) *Investigation of a Simple Form of Hydraulic Dynamometer*, in Mechanical Energy, School of Engineering, Princeton University, p. 749.

wind turbine with constant tip speed. That is optimal in a smaller range of operations. Thus, in principle a water brake windmill is producing more than windmills running at constant speed of rotation.

Water brake wind turbines produce hot water, which can be circulated into a nearby house by an insulated pipe network. In the house, there is often a heat storage tank to compensate for the consumption and production.

There are few disadvantages of water brake turbines. Pumps, used for circulating water into the nearby houses usually cause some efficiency losses. Also, transmission loss in the pipes limits the distance to maybe 20 m to 50 m or more but this is enough to enable the turbine to be sheltered by buildings or trees. Noise and too close to neighbours can also be a problem.

13.2 Developing Water Brake Windmills

13.2.1 Institute of Agricultural Engineering

As a result of the energy crisis in 1973 Svend Sonne Kofoed and Ricard Matzen from the Institute of Agricultural Engineering in Taastrup, became interested in wind power. They began to develop an experimental windmill Mark I with a water brake. It was at first a turbine with a blade diameter of 6 m, two blades made of wood, and a 12 m lattice tower. It was downwind rotor and was running anti-clockwise. In the nacelle there was a gear with a shaft down to the water brake at the bottom of the tower. A number of measurements were carried out. These measurements resulted in reports "Wind energy as an opportunity for agriculture" and report no. 36 of the Institute of Agricultural Engineering.

The next windmill Mark II had also a diameter of 6 m and two blades shaped out of one piece of wood and the same as Mark I. The tower consisted of three wooden beams and could rotate according to the wind as the yaw system was on the ground. The blades were placed upwind and held up in the wind by a yaw vane. The rotor was connected to a gear at the top and a long shaft to the water brake at the bottom of the tower. The water brake "k" factor was 1:3. The tip-speed ratio was 1:10, so it was quite a fast-running 2-bladed rotor.

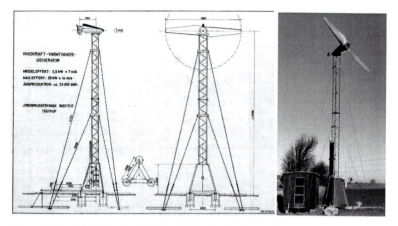

Figure 13.1   Mark I water brake windmill developed by Svend Sonne Kofoed and Richard Matzen.

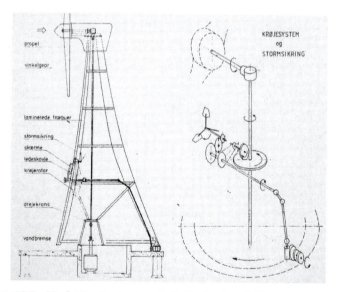

Figure 13.2   Mark II water brake windmill: Section (left); Yaw system and power security (right).

The measurements were recorded in the report "Wind Energy-Heat Generation" and presented in 1978 at the Second International Symposium on Wind Systems Energy in Amsterdam. Ricard Matzen's many experiments and measurements with water brakes resulted in mentioned reports. These, in large part, formed the basis for the development of water brake turbines in Denmark.

13.2.2   Gunnar Broe and Værløse Group

Gunnar Broe was the main member of a group of experimenting engineers from Værløse, active from around 1970 onwards. The group was developing wind-powered water pumps for developing countries. They were made very simple and consisted of simple parts that were produced in big numbers. Parts came from the automobile industry, and were used or new components at a reasonable price.

Broe developed a simple turbine with a three-bladed rotor. The rotor diameter was 5 m and tower height was 8 m. The nacelle was made of second hand car parts. The tower was bolted lattice. The blades were upwind of the tower and were held up against the wind by yaw blades.

The windmill was later converted into a heat-producing windmill with an open water brake, based partly on results from the experiments held at the Institute of Agricultural Engineering.

13.2.3   Windmill Group ECO-RA

The windmill group ECO-RA continued the development of Gunnar Broe's ideas with water brake windmill experiments. The group consisted of Gert Ottosen and Jørgen Krogsgaard. The prototype wind turbine used parts from a car and the open water brake that looked like the one from the Agricultural Technical Institute.

The entire windmill was of a very simple construction. After taking some measurements and testing of the water brake windmill, a design manual was created in 1976. It was very brief with three pages of text and drawings, which was sold in 200 copies.

Then in 1977, Jørgen Krogsgaard designed a much better water brake windmill. It had three blades with the profile from the Gedser wind turbine. The rotor diameter was 6 m. For the transmission in the nacelle, second-hand car parts were used. The tower was a tube held up by four wires. The blades were placed downwind. The power was limited by mechanical pitch regulation. The speed of rotation was variable, which gave a constant tip-speed ratio until the pitch regulation set in and the tip speed was constant. The water brakes were of the closed Culver-type and mounted in a special lattice tower top in such a way that they were easily removable. A design manual for the windmill was created and 220 copies were sold.

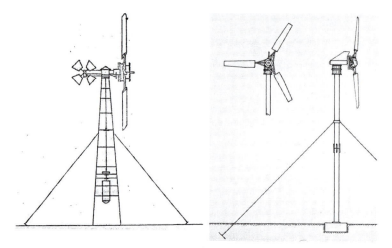

Figure 13.3 Prototype 2-bladed ECO-RA water brake windmill (left); 3-bladed ECO-RA version (right).

Figure 13.4 The 3-bladed ECO-RA water brake windmill (left); ECO-RA water brake windmill built by Ib Grosser, Odder (right).

In 1978, Ib Grosser from Odder used this construction manual and built a water brake turbine with 6.4 m rotor and a tower height of 12 m (Fig. 13.4 right). It ran very well and the pitch regulation could keep the rotation speed nearly constant. After four years of operation it was taken down for revision and was not put up again. Also other windmills were built in accordance with the same construction instructions and worked very well.

13.2.4   The Calorius Windmill

Hans Henrik Ekner from Slagelse developed a 2-bladed house-hold windmill with a water brake, the Calorius windmill. It was approved by the Risø test station for windmills. The windmill was mounted on a 9 m tube tower held up by four wires. The rotor diameter was 5 m. It was upwind and held by a wind vane. The water brake was of the closed Culver-type.

The power limit was set in the way that the water brake changed characteristics in a very clever way. When the maximum effect was achieved, the rotor stalled. The windmill had variable rotation conditions with constant tip-speed ratio until the power limit stall takes over and the tip speed was held constant.

This wind turbine—Calorius type 37, was the only water brake windmill approved and tested at Risø test station and documented in the reports "I-986" and "I-1205". The measurements showed that the maximum capacity was 3.5 kW at about 11 m/s. The company Acoustica conducted the noise measurements: The Calorius windmill was very quiet.

The safe-guard against run away was a mechanical brake activated by too high speed. However, it was difficult to adjust and this caused some problems.

In the period from 1993 to 2000, the firm Westrup from Slagelse built a total of 34 Calorius wind turbines. By 2012, there were still 17 in operation.

Figure 13.5   Calorius windmill.

13.2.5   The Svaneborg Windmill

In 1982, in Stagstrup, two brothers Hans and Bent Svaneborg each built their own windmill. Hans built a water brake windmill and Bent a wind turbine with a generator for electricity production. The water brake windmill was a downwind type, with a 17 m tube tower held up by four wires, rotor diameter of 7.5 m and three fibreglass blades. For the transmission in the nacelle a Mercedes small truck rear shaft was used. From the top of the wind turbine there was a shaft down to the bottom of the tower where the water brake was installed.

At 9 m/s wind speed, a power limitation was applied by using a simple mechanical pitch regulation; the windmill changed from running with a constant tip-speed ratio to a constant tip speed and hence to a constant rotation speed.

The water brake was of the same type, as used at the Institute of Agricultural Engineering. The turbine operated for 14 years until it was taken down for revision. Folkecenter conducted measurements that showed a surprisingly high efficiency and an output of 8 kW. Hans Svaneborg's windmill for heating can be found in the collection of historical turbines at the Folkecenter.

Figure 13.6   The Svaneborg water brake windmill.

13.2.6   LO-FA Heat-Producing Windmill

In 1980, Knud Berthou from Sakskøbing built a heat-producing windmill LO-FA. It was a 3-bladed turbine with wooden blades, rotor diameter of 12 m, and 20 m tubular tower with an insulated storage tank inside. Heat was generated by directly coupled adjustable fluid brake with hydraulic oil as the liquid. The power was estimated to provide 90 kW at 14 m/s wind speed.

The Institute of Agricultural engineering in Taastrup did testing and measurements of the fluid brake. Ricard Matzen had a great influence on the entire process of developing the LO-FA windmill. Another involved in the project was the engineering student Thomas Krag Nielsen. Unfortunately, the main initiator Knud Berthou passed away. Then the project was moved to the island of Fejø. The idea was to put it in operation again, however, it did not succeed to get a local approval and the LO-FA windmill was scraped.

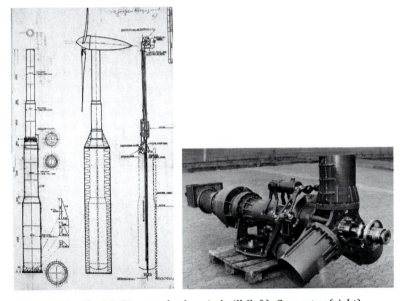

Figure 13.7 The LO-FA water brake windmill (left); Generator (right).

13.2.7 Ørn Helgason

In 1982, Ørn Helgason from the Science Institute of the University of Island started a heat-producing windmill project. The water brake windmill was built and tested on the most northern island of Iceland, Grimsey, at the Arctic Circle, where wind speeds reach on average 8.5 m/s. The 2-bladed windmill had rotor diameter of

5.7 m and a 9 m lattice tower. The windmill was of the upwind type with a wind vane. The water brake and storage tank were at the bottom of the tower, connected to the turbine hub with a transmission shaft. Power limit was achieved by changing the water brake characteristics, which then stalled the blades.

Experiments and measurements were conducted on the turbine and the tip-speed ratio was estimated to be 3.5 to 4 times the velocity of the wind. The windmill is documented in the report "Exploiting wind power of high wind speed for house heating city water brake with variable load in Iceland" from University of Island.

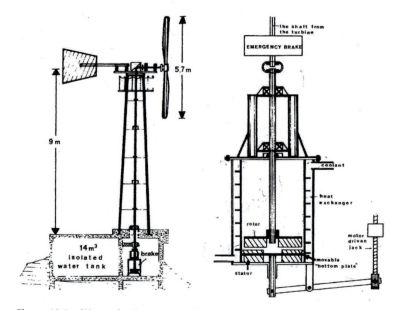

Figure 13.8   Water brake windmill developed by Ørn Helgason (left); water brake with variable load system (right) (Images: Ørn Helgason[3]).

13.2.8   Gerlev

In 1981, Gerlev Sports Folk High School built a water brake windmill for heating. The designer was Jørgen Krogsgaard. It was a downwind rotor with a diameter of 12 m. The blades were made

---

[3]Helgason, Ø., Sigurdson, A. S. *Test at very high wind speed of a windmill controlled by a waterbrake*, Science Institute, University of Iceland.

of solid wood, the tower was an 18 m tubular tower held up by four wires. The water brake was a closed Culver-type.

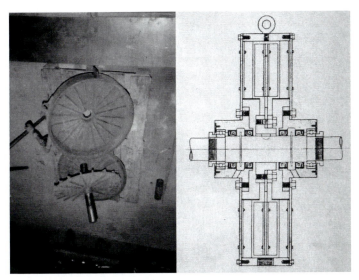

Figure 13.9 Water brake used on Gerlev windmill (left); Hydraulic dynamometer of the bladed type (Culver-type[4]) (right).

Figure 13.10 Water brake windmill in Gerlev.

---

[4]Culver, E.vP. (1937) *Investigation of a Simple Form of Hydraulic Dynamometer*, in "Mechanical Energy", School of Engineering, Princeton University, p. 749.

The power was limited by a simple mechanical pitch regulation. At design wind speed the pitch regulation took over. The blade speed changes from constant tip ratio to constant speed. After installation, it worked very well for some time. The project was stopped because of a mechanical failure.

13.3 Conclusion

Windmills with water brakes work very well. They are cheap to build and have high efficiency. They consist mostly of simple and same mechanical technology. No electricity and wire are involved. For heating purposes a combination of a water brake windmill and a solar heating system is perfect. In spite of all advantages, only a few windmills with water brakes were built in the period 1980–1990 but they never entered mass production. The electricity-producing windmills took over.

About the author

Jørgen Krogsgaard, born in 1937, grew up in Korinth Fyn, and is now living in Kastrup, Denmark. He completed a four-year apprenticeship as a mechanic at Thomas B. Thrige, manufacturer of electric motors and equipments in Odense, Denmark. He then completed his engineering education in Helsingborg, Sweden.

Following his education, Jørgen worked on new machine room installations for the shipping industry in Germany and Sweden, and was involved in planning the installation of stationary diesel power plants for B&W, a former Danish diesel motor factory which manufactured large diesel motors. From 1979 he was employed at the test station for Windmills, Research Centre, Risø, Denmark, for more than 25 years and from where he retired in 2004. In Risø, he was involved in setting up the new test station to develop wind energy.

Jørgen is also member of a number of ecology groups, including OVE, ØKO RA and their windmill sub-group, and other eco groups and has worked with the ØKO RA windmill sub-group, where he designed two windmills with water brakes and published construction reports for windmill builders. He has developed a solar furnace for use in developing countries and developed and patented two different types of the micro hydro JK turbine. The first of these turbines project was sold to Vestas.

Jørgen participated in the wave energy project, Wave Dragon, and is still active. He is one of the founding members of the governing board in ESHA, European Small Hydropower Association.

As a pensioner, he is still active in a number of different innovative activities related to renewable energy and other forward-looking ideas.

Chapter 14

Cooperative Energy Movement in Copenhagen

Jens Larsen

Copenhagen Environment and Energy Office, Blegdamsvej 4B,
2200 Copenhagen, Denmark

jenshmlarsen@gmail.com

After several attempts to establish offshore wind farms in various places in Denmark, the Middelgrunden wind farm became the first offshore wind project based on the sale of shares. Before that, an attempt in the Århus Bay had failed, and when we presented a proposal for a project of 20 offshore wind turbines at Middelgrunden in 1996, it seemed an almost unattainable vision. A fortunate combination of political will in the Danish parliament and a strong local engagement in Copenhagen carried the idea through.

14.1 No More Space on Land

In 1993, twelve Bonus wind turbines each of 300 kW were set up in Avedøre Holme, and in 1996, seven Bonus 600 kW windmills

Wind Power for the World: The Rise of Modern Wind Energy
Edited by Preben Maegaard, Anna Krenz and Wolfgang Palz
Copyright © 2013 Pan Stanford Publishing Pte. Ltd.
ISBN 978-981-4364-93-5 (Hardcover), 978-981-4364-94-2 (eBook)
www.panstanford.com

were installed on Lynetten. Both projects came into being through a successful cooperation between local windmill cooperatives and the power utility.

Figure 14.1 Middelgrunden wind farm (40 MW) (Photo: Kim Hansen).

Conflicts of interests created problems to locate new sites for wind turbines in the capital area, therefore interest centred on areas at the sea. The mapping of the ministry pointed to Middelgrunden as a possible site, but the area had low priority among power utilities and civil servants. The wind power cooperative was formed, and with the approval of the Danish parliament, the Danish Energy Agency granted funding for the pilot study. At first, some rivalry with Copenhagen Lighting Company arose, but Middelgrunden Cooperative was soon established in order to develop further the project.

14.2 Middelgrunden Offshore Wind Farm

Before the project could be authorised many problems had to be solved. The situation just outside the capital of Copenhagen caused a heavy debate. The humour column "At Tænke Sig" in the daily *Politiken* carried an advertisement for Middelgrunden Wind Farm: "Offered: Resistance to wind. Wanted: Wind resistance."

A turning point was reached when an overwhelming majority of representatives of Danish Society for Nature Conservation supported the project. A few interest groups, such as the leisure sailors associations and the suburban municipalities of Gentofte and Lyngby-Taarbæk would not budge from resisting. On the other side the municipality of Copenhagen and a big number of green organisations as well as the 8 500 members of the windmill cooperative favoured the project.

At the same time the pilot study showed that Middelgrunden was well suited for an offshore wind farm with the existing wind power technology and regulation. The final approval from the Danish Energy Agency was given in December 1999, and at the same time all contracts were signed just before a new energy reform was to come into force on 1 January 2000. All this was necessary in order to make the finances of the project work for the owner cooperative as well as the power utility.

Figure 14.2   Jens Larsen working on Middelgrunden, repairing damage on the fence on top of foundation (left); repairing damage on the tower (right).

Now Middelgrunden has been running for more than 10 years producing more than 100 GWh. It delivers about 4% of the demand for electricity in Copenhagen. The site has been a popular icon for wind enthusiasts and developers around the world. Copenhagen City and people of Copenhagen are very proud of the wind farm. Former environmental minister Connie Hedegaard called it "a green landmark" for Copenhagen. Former President Bill Clinton said flying over Middelgrunden "You are smart—we're not". What he might not know was that this beautiful site is a living result of a unique local engagement and a big success for the cooperative movement.

Figure. 14.3 Stavros Dimas (left), former EU Commissioner of Environment, and Connie Hedegaard (middle), EU Commissioner of Environment in 2012, visiting Middelgrunden Wind Farm in 2005, with Jens Larsen (right).

14.3 Cooperative Movement—Important Player in the Future

Already in the 1980s there were many projects based on cooperatives, and after Middelgrunden, many new energy projects have been implemented on the same basis all over Denmark. Just to make a short list of projects with close relations to our group in Copenhagen: Samsø Offshore Wind Farm, Copenhagen Photovoltaic Cooperative, Hvidovre Wind Turbine Cooperative, and Bio Gasification. Also five near-shore wind farms are ready to take off in Denmark, all with local engagement and support.

The wind industry of today is a big player on the global energy market, however, local protests against wind turbines on land are getting stronger. Both in Denmark and abroad. In this perspective politicans and opinion makers in Denmark tend to forget how important local cooperatives were when it comes to the history and development of the unique Danish success on wind energy. It is crucial to point that cooperatives and the cooperative movement have played a unique and important role in this development.

Middelgrunden is one good example were 8500 people have done something which was a mission impossible in year 2000. The educational effect of this cooperative has been and is still outstanding.

The Danes are beginning to understand the job we have ahead of us. And it is not an unattainable vision anymore. Today more traditional energy companies and private entrepreneurs have the biggest influence on the Danish wind energy scene anno 2013. I accept that as a fact. But it is my strong belief that local engagement and the cooperative understanding should have a more central and essential role for a sustainable energy future. The ordinary players in the energy market will not be able to deliver a sustainable energy future without a stronger local engagement. Or the market will deliver it too late—and I do not think we have the time to wait.

Figure 14.4   Members of the Middelgrunden cooperative participating in the annual meeting.

Important Web Sites

www.middelgrunden.dk
www.solcellelauget.dk
www.hvidovrevindmollelaug.dk
www.bioforgasning.dk

About the author

 **Jens Larsen** has been a strong supporter of cooperative actions and local engagement for 25 years. He developed and supported Middelgrunden Cooperative and many other local energy projects in Denmark and all over the world. He is currently working as a climate coordinator for a municipality in Denmark.

Chapter 15

The Danish Small Wind Power

Jane Kruse

Nordic Folkecenter for Renewable Energy , Kammersgaardsvej 16, Sdr. Ydby,
DK-7760 Hurup, Thy, Denmark

jk@folkecenter.dk

15.1   Overview

Within two years, from 2010 to 2012, small-size wind turbines have re-created broad public interest in land-based wind power, awareness of the importance of renewable energy and enabled citizens to make energy self-supply a life-style issue.

More and more individual users prefer to produce their own electricity from small windmills, photovoltaic installations, or a combination of both, rather than buying energy from the national grid. In terms of investments, many choose a small wind turbine instead of a second car, a boat or a new kitchen. People living in rural areas consider an ecological lifestyle as beneficial. Also, development in modern small wind power enables people to choose an innovative technology at reasonable costs (payback time usually

Wind Power for the World: The Rise of Modern Wind Energy
Edited by Preben Maegaard, Anna Krenz and Wolfgang Palz
Copyright © 2013 Pan Stanford Publishing Pte. Ltd.
ISBN 978-981-4364-93-5 (Hardcover), 978-981-4364-94-2 (eBook)
www.panstanford.com

6 to 10 years). Small wind turbine owners become community advocates and energy savers, bringing revenues for the benefit of local economy.

The main benefit of small power installations is the self-sufficiency for the individual consumer and independence from national supplies (coming both from renewable energy or fossil fuels, Denmark has no atomic energy). Small-scale windmills are visible in landscape—they are installed in close distance to houses and farms helping to create public awareness.

15.2 The Market Potential

There are many types of small-scale windmills ranging from 100 W up till 10 kW, some of them using advanced carbon fibre blades and permanent magnet generators (PMG). This type of equipment will have great opportunities for paving the way for wind energy in the many unserved areas of the world. Off-grid small windmills are often used in developing countries, mainly for a wide range of rural energy applications. A new and unexpected application for small windmills (often self-made) called the Home Power[1] has emerged in developed countries, for example, in the United States. In some countries small windmills are integrated into architecture, especially in densely built cities.

Small wind power development varies in different countries, due to climate, politics, technology development, market or governmental policies:

China

China has many years of experiences within small wind power, for off-grid applications for consumers that do not have access to the national power grid. There is a growing and important development of a wide-range of wind turbines up to 50 kW, gearless, PMG, with inverters for grid connection. In the period from 2005 till 2009 there were about 400 000 small wind turbines, produced in China, installed in the country and 120 000 Chinese wind turbines exported to other countries.

---

[1]Named after *Home Power* magazine which promotes small wind power and DIY small wind turbines

Great Britain

In the period 2005–2009, 13 514 small wind turbines up to 20 kW were installed. The British Wind Energy Association (BWEA) estimates that 600 000 windmills of max. 50 kW will be installed by 2020. Only in 2011, 23 MW of small wind turbines were installed in Britain, making Britain ahead of the United States in terms of installed capacity (in the United States it was 19 MW in 2011). New feed-in tariff program was implemented in April 2010 Microgeneration Certification Scheme) with price of 0.21 GBP/kWh (0.241 EUR/kWh) for wind turbines up to 100 kW.

However, new feed-in tariff revisions introduced in December 2012 reduce existing tariffs to:

- 0.152 GBP/kWh (0.174 EUR/kWh), 42% reduction of price for wind turbines up to 1.5 kW
- 0.070 GBP/kWh (0.080 EUR/kWh), 25% reduction for wind turbines of 1.5 kW to 15 kW
- 0.043 GBP/kWh (0.049 EUR/kWh), 17% reduction for turbines from 15 kW to 100 kW

The United States

In 2011, more than 19 MW capacity small wind turbines were installed in the United States.

It is expected that by 2013 there will be 650 MW small windmills installed (26 000 units, average capacity 25 kW) with eight-year favourable tax-credit incentives.

According to AWEA,[2] the US small wind turbine market (up to 100 kW) decreased by 26% (in MW) in 2011. However, domestic sales and export of wind turbines manufactured in the United States increased by 13.4% (to 33 MW).

There is a growing interest for do-it-yourself small wind turbines of 1 kW or less capacity. Construction of home-made open-source wind turbines has been made popular through construction guides, magazines, web sites and TV series.

Japan

After the Fukushima disaster in March 2011, Japan strives to increase the share of renewable energy (wind and solar) in power

---

[2]AWEA, *2011 U.S. Small Wind Turbine Market Report,* published in June 2012.

generation by increasing investments in the sector and introduction of a new feed-in tariff in July 2012. The feed-in tariff is applied to small-scale wind power and solar energy installations and requires power utilities to buy back electricity generated from wind and solar energy at prices set by the government.

Small-scale wind turbines (of less than 20 kW capacity) will be subsidised at least JPY 57.75/kWh (about USD 0.74/kWh).

Figure 15.1   Examples of modern small wind turbines in Japan. Professor Izumi Ushiyama established demonstration and testing facility for small wind turbines at the Ashikaga Institute of Technology in Japan.

Denmark

There are over 200 000 homes in the open land the Danish that depend on heating from oil. It is the government's policy to phase out oil. With small windmills installed at 10% of the 300 000 homes in the open land the Danish market potential is 30 000 small windmills. Net-metering for small wind power up to 6 kW has been introduced in June 2010.

Danish small windmills: approximately 300 units; in 2011, one year after new legislation, there were about 300 units installed. In the country there are 20 manufacturers and suppliers of small wind turbines.

In Sweden approximately 1 100 small windmills are in operation. There are 12 manufacturers of small wind turbines in the country. In Finland there are slightly less than thousand small wind turbines operating, whereas there are only two manufacturers.

Table 15.1   Overview of small wind power characteristics for Scandinavian countries (in EUR)

	Sweden	Denmark	Norway	Finland
Remuneration Model	Market + Certificate	Net-metering + Feed-in	Market (Certificate 2012)	Market(Feed-in May 2010)
Average Price	0.05 + 0.03	0.28 + 0.08	0.05	0.08
Production value 4.000 kWh/year	320	1 100	200	320
Production value 10.000 kWh/year	800	1 580	500	800

15.3   The Transition to Decentralised Power in Denmark

The difference between fossil energy and renewable energy forms is that the latter are decentralised by nature. Thus, a small number of large-scale wind turbines for energy production may be replaced by a large number of small ones. In principle, any building, house or farm can produce energy from renewable sources, which turns out to be cheaper than production from fossil forms of energy. Wherever wind resources are available, costs of electricity generated by small-wind windmills will be higher than from large windmills, however, generally lower compared to photovoltaics.

In Denmark small wind turbines (household wind turbines) are defined by the Danish Energy Agency (DEA): maximum height from base to the blade tip is 25 m; rotor diameter up to 13 m; maximal size of the generator is 25 kW; maximum capacity is 6 kW and 40 m^2 swept area are principles for net-metering (electric meter reverses). Small wind turbines are to be installed in private households, nearby and connected to existing buildings for self-supply in rural areas.

Figure 15.2 Calorius 6 kW windmill supporting a household with other renewable energy applications, Thy, Denmark.

Figure 15.3 Simple assembly installation of blades on a 5 kW wind turbine can be done easily and quickly.

15.3.1 Danish Legal Framework of Household Renewable Energy Supply

With the new law from June 2010,[3] the Danish parliament decided that single household should be allowed to have self-supply of electricity using renewable energy sources.

Figure 15.4 Examples of small-scale wind turbines.

The main background of the law was family self-supply for the rural population as compensation to obtain energy services at costs similar to urban residents that are supplied by cheap and efficient CHP and district heating (70% of population). Small power supply is defined as maximum 6 kW output and this rule applies to wind, solar, hydrogen and biomass as single source of supply or with a combined capacity of max 6 kW. Installations larger than 6 kW are considered as production units. In self-supply installations (wind turbines up to 6 kW) the user is also the owner—home power producer owns his wind turbine just like he owns his car or swimming pool.

Net-metering means that electricity purchase meter runs backwards. In this way autonomous power production from solar or wind energy has the value of power purchase prices, inclusive taxes and VAT. The right to connect to the grid and to deliver power from the household to the grid according to the net-metering principle means that every kWh produced has the value of 0.28 EUR/kWh. Excessive power is remunerated by the company responsible for the grid at 0.08 EUR/kWh for a 10-year period.

---

[3]On 4 June 2010 the Danish parliament passed net-metering legislation that applied to small power supply (less than 6 kW).

Figure 15.5   A 6 kW Thymøllen small-scale wind turbine with 40 m² swept area.

For the permission to install a small wind turbine the following documentation is required by the local authorities: Power performance test, durability test, and acoustic sound test (noise) as well as some requirements of proximity to existing buildings, taking into consideration the neighbours.

Danish Legislation of 4 June 2010 defines following categories of household windmills:

- Maximum 25 m to blade tip;
- Maximum 6 kW for net-metering (electric meter reverses) and up to 40 m² swept area;
- Maximum 25 kW generator size and to 200 m² swept area.

In Denmark there are various types of small wind turbines approved by the Danish Energy Agency, fulfilling requirements such as safety, duration test, power performance test, acoustic sound test, maintenance manual in Danish language.

Approved small wind turbines in Denmark by 2012:

(1)   Up to 5 m² swept area there are ca. 10 different types
(2)   In the second category, with 5 to 40 m² swept area eight types got approved:

- Evance Wind (English) 5 kW
- Windspot (Spanish) 3,5 kW
- Vindby (English) 1 to 2 kW
- Easy Wind (German) 6 kW
- Eco Wind (Scottish) 6 kW

- Thymøllen (Danish) 6 kW
- Energytech (German) 5.5 kW
- ZE Wind (Danish) 6 kW

(3) In the third category, from 40 m^2 to 200 m^2 swept area four types got approved:

- Gaia Wind (Danish) 11 kW
- Eco Wind (Scottish) 15 kW
- HS Viking (Danish) 25 kW
- ZE Wind (Danish) 25 kW

15.3.2 Danish Institutional Framework for Small Wind Power

Small wind power in Denmark is supported by favourable legislation but also many institutions and the industry, accredited inspectors for approval (private companies) which are part of the institutional framework.

The Small Wind Turbine Association (SWTA) had a leading role in the launch of small-scale wind power at the political and general public level. The SWTA was founded in June 2009 to promote the interest of wind power for the supply of individual homes, SMEs and farms with a capacity up to 25 kW. It is a manufacturers' association that represents and promotes small size wind power in Denmark by various strategies:

- Create public awareness—political level, consumer level, industry
- Lobbying
- Establish technical standards, including safety
- Streamline planning/permit process
- Network with other committees, public bodies, insurance companies, planning authorities, etc.
- Pave the way for financial support to new test station for small wind turbines
- Develop new industrial sector, creates new jobs in manufacturing, installation, approval, testing, maintenance and export
- Benefits from Danish wind turbines' favourable reputation for quality

Other institutions are Danish Small Wind Turbine Association or State Approval Office (Danish Energy Agency).

15.4 Small Wind Turbines at the Folkecenter

The Folkecenter for Renewable Energy is an independent, non-governmental organisation. It was established in 1983 to pave the way for renewable energy by developing, testing, and demonstrating technologies, which are designed for manufacturing in small and medium scale industries.

15.4.1 Development and Testing

At the Folkecenter's test station for small wind turbines 10–12 small-scale windmills are permanently being tested, measured various standards or demonstrated for national and international clients. The windmills deliver more electricity than required for Folkecenter's own power needs. A 75 kW and a 525 kW windmills are the centre's own property.

Figure 15.6   Folkecenter test station for small wind turbines.

Manufacturers of small wind turbines can have their products tested at the Folkecenter test station. It has platforms and foundations for testing of electricity producing windmills of 1 kW up to 30 KW and mechanical wind pumps.

The test station is equipped with data loggers, wind measurement masts, towers for installation of wind turbines and water wells where the performance of small windmills for electricity and water pumping can be measured as per international standards.

Measurements have been conducted for a variety of renewable energy manufacturers in Denmark.

Figure 15.7   Installation testing of wind turbines at the Folkecenter.

Figure 15.8   Wind test station at the Folkecenter for renewable energy.

Folkecenter also provides general information about small wind power to the public. There are 20–100 personal visits per week from the industry, potential customers, authorities, students, etc. The Folkecenter web site[4] has 4 000 visits per day.

Figure 15.9 EasyWind small wind turbine for installation (above); Being installed by two men during the day from 09:00 till 17:00 (below), ready for supplying to the grid!

15.4.2 Application of Small Wind Power at the Folkecenter

In 1998, Folkecenter built the first straw-bale house in Denmark. The experimental construction of the straw-bale house is based on wooden structure filled with 842 weed straw-bales. The building

---

[4]www.folkecenter.net

has two floors and is perfectly insulated against temperature changes. From the inside the walls are plastered with clay and painted with yoghurt. Outside walls are plastered with lime mortar. Floor foundation is made of seashells. Advantages of using natural and renewable materials like straw include low costs, availability of materials, high insulation and moist regulation. The straw-bale house has the living area of 225 m^2.

Figure 15.10 Folkecenter's straw-bale house—autonomous wind power installation. In operation since 1996 without replacement of any of the components.

The house has its own supply of electricity from a 2.2 kW Proven windmill. The straw-bale house is not connected to the public grid. The wind turbine delivers AC power with fluctuating voltage and frequency. The rectifier (top left of Fig. 15.10) delivers DC power to eight Surrette batteries, each of 250 Ah. The Trace inverter (bottom left, Fig. 15.10) delivers 230 V, 50 Hz, AC power of grid quality to the building.

Figure 15.11 The 2 kW Proven windmill installed at the Folkecenter. By 2012, it has been in operation for over 15 years without repair and maintenance works, except for grease of the two bearings every three years.

Figure 15.12 Procure 400 W: Installation on the roof of the office building in 2007 (left); Nacelle, blades and control are delivered in a box, ready for installation. The mast is produced locally (right).

SkibstedFjord is a 670 m² training centre and assembly house built in 1996, which features experimental low-energy underground architecture. The building has photovoltaic cells integrated in window panes along the glass façade. Solar cells of 10 cm × 10 cm size are placed in various patterns and demonstrate the possibilities to use the cells as shades and architectural effect. On the building there are also photovoltaic cells integrated in the façade.

Can wind turbines be placed on buildings? On the SkibstedFjord training centre, three 400 W Procure windmills have been operating since 2007. The tower is a 2″ pipe bolted to the concrete structure of the roof. Small windmills are very quiet—audible noise from the windmills or vibrations are not registered inside the building.

Figure 15.13   SkibstedFjord underground building with PV façade and 400 W wind turbines, Folkecenter.

15.5   Future of Small Wind Turbines

It is assumed that 150 000 families in the United States are self-reliant in terms of energy supply using renewable energy this has formed the background for a significant new industry for home power equipment. Because small size windmills are ideal for mass

Figure 15.14 New generation of vertical-axis wind turbines exhibited at the New Energy exhibition in Husum, Germany, the leading trade fair for autonomous energy technologies.

Figure 15.15 Modern DELA wind turbines with low or no sound in Mecklenburg, Vorpommern, Germany.

production similar to bicycles and various kinds of appliances, prices may in the future when mass-produced become very low compared to conventional energy supply technologies.

About the author

Jane Kruse was born in 1946 in Faaborg, Fyn, Denmark. From 1976, Jane was involved in grassroots work for sustainable develop-ment as opposed to nuclear power and traditional fossil energy supply. In 1988, she became head of a private wind turbine corporation including 50 families. She also co-owns an ecological farm of 12 hectares, and lives and works at the Nordic Folkecenter for Renewable Energy, Denmark. She is one of the founding members of Folkecenter and secretary of the board for 11 years. Jane is also the head of the Information & Training Programs Department of Folkecenter. Since 1999, she has managed various projects at Folkecenter, including establishment of test facility for wave power machines, testing of wave power machines, and testing of small windmills from 400 W to 25 kW. Jane has been the editor, author and photographer of many Folkecenter publications.

Besides being an indispensable member of Folkecenter, Jane has also been board member for the Committee for Green Technology (under the Danish Engineer Association) and Centre for Rural Area Development (Aalborg University under Minister of the Interior). She is closely attached to many NGOs working on renewable energy. Her many years of experience ranges from coordinating European and international renewable energy information tours to handling projects concerning developing countries and training courses; advising the Danish Energy Agency and the Home Secretary of Danish Ministry on their renewable energy projects; and managing various local and international projects within renewable energy, including those funded by the European Union, Danish Ministry of Energy, Danish Ministry of Home Affairs, Danish Ministry of Foreign Affairs, and Danish Ministry of Housing. Jane is also actively involved in local politics, and since January 2011 she is member of the Thisted Municipality Council.

Chapter 16

Consigned to Oblivion

Preben Maegaard

*Nordic Folkecenter for Renewable Energy, Kammersgaardsvej 16, Sdr. Ydby,
DK-7760 Hurup, Thy, Denmark*

pm@folkecenter.dk

16.1   Introduction

The modern wind turbine as we know it, was not developed overnight in a big industrial company—its birth and development was marked by hard work of many inventors, engineers, pioneers and amateurs. It is not possible to point to any single enterprise or institute that came up with an epoch-making invention, neither is there a Steve Jobs or Bill Gates, who makes a breakthrough. It was a result of a long process of making; it came through as a result of the efforts of numerous engaged and gifted people, with their success and failure stories. This chapter presents the history of projects and inventions, that were an important, though often forgotten, contribution to the development of the modern wind power as we know it today.

Wind Power for the World: The Rise of Modern Wind Energy
Edited by Preben Maegaard, Anna Krenz and Wolfgang Palz
Copyright © 2013 Pan Stanford Publishing Pte. Ltd.
ISBN 978-981-4364-93-5 (Hardcover), 978-981-4364-94-2 (eBook)
www.panstanford.com

In Denmark during the 10 year period between 1974 and 1984 a great number of enterprises and inventors emerged, who worked on many different versions and concepts of windmills. This bottom up process is part of the one of the flourishing periods of innovation and dynamics that led to the foundations for a new Danish industrial sector, which soon acquired importance on the international scale. While other countries also had their inventors and pioneers with designs and constructions that did not become commercial, in Denmark several dozens had the dream to develop the technologically ultimate solution. Only some of them are mentioned in this chapter.

16.2   Home-Made Inventions

"I planted the trees, I cut them down, and now I have carved the wood into blades for my windmill," says **Jacob Overgaard** from Jelstrup, Thy, in Per Mannstaedt's film, "Dansk Energi" ("Danish Energy"). It is a one hour documentation of the bottom-up efforts by the people to make the vision of a Denmark without nuclear power come true. The film was made in 1977, following another film, "Flere Atomkraftværker" ("More Nuclear Power Plants") in which Per Mannstaedt had shown us what is going on behind the metres of thick walls in a nuclear power plant, something that was not at all reassuring.

The contrast between the world of atomic energy and the inauguration of Jacob Overgaard's windmill was overwhelming. When the neighbours, most of them over 60, saw the green blades on the red and white tower do their first turn, they proceeded to the farm house where the tables in the best parlour had been laid with coffee, cakes and aquavit. Jacob and his windmill were celebrated, and debates on energy policies took place. My son who was four at the time also appeared in the film along with the group of men at their afternoon coffee. The people had taken things in their own hands. Jacob had delivered a manifest statement that we could manage without nuclear power.

During the process of making the film, **Per Mannstaedt** had met so many ordinary people that knew and demonstrated the craft to build a well-functioning windmill for supply of their own energy needs and even sale of power, that he got captivated by the idea of building his own windmill in Northern Zealand. I advised him to use a well-known concept and to forget many of the exotic and imaginative designs he had encountered whether where they came from amateurs or professional as well. His wind turbine was to be a variant of the master blacksmith windmill with 5 m Økær blades. Special heat controls were to be made by Professor Ulrik Krabbe, as the windmill was not going to be connected to the grid. The tower was second hand from a building crane and had guy wires with anchors in the ground.

Figure 16.1 Jacob Overgaard's windmill (left); Per Mannstaedt's self-made wind turbine of NIVE design (right).

Per Mannstaedt's windmill functioned and produced well. Had it not been installed in Denmark but in Sweden or Germany it may well by that time have been a technological sensation and seen as a prototype of an inexpensive and reliable wind turbine, almost ready for a serial production. In Denmark, however, it was just one of the many, and was soon forgotten.

The art of designing and building wind turbines at that time did not belong to high-level laboratories with enough finances, technological resources, and public relations, but to the people.

In Øsløs, artist **Rud Ingemann Petersen** set to work in a professional manner. He built an 11 kW windmill according to the Johannes Juul's template with stall-regulated blades, a well-dimensioned Nord gearbox and asynchronous generator.

Figure 16.2 Painter Rud Ingemann Petersen's 11 kW wind turbine of the Juul concept (left); Nacelle of Ingemann's wind turbine (right).

He did endeavour not to reinvent a better concept by using the simple and reliable 45 kW Bogø-type technology from 1953 that preceded the later Gedser wind turbines. He constructed and built every bit by himself, including the blades and the controls. For several decades the windmill delivered power for lightening and heating for the family. It possessed qualities that might have made Rud become one of the well-known windmill designers. But that was not his ambition.

16.3   The Quite, Quite Different Ones

In Denmark not only conventional wind turbines were developed—quite peculiar models appeared as well. One of them was the 4-bladed Vendelbo windmill placed on the roof of the Aurion Bakery in Vendsyssel. It was constructed by manufacturer Arne Brogaard especially for the operation of the grain (wheat, rye, spelt) grinders at the organic bakery. For Aurion organic bread was more than assuring the consumers that the cereals were cultivated without using chemical fertilisers and pesticides. Also, mechanical energy for grinding should not come from fossil fuels but from wind to complete the products' organic origin.

The Greenland Tele Administration had some fifty small 1 kW or 2 kW turbines for the supply of electricity to their transmitter stations in the ice desert. Wind power was a supplement to the propane gas tanks which had to be brought there hanging below a helicopter.

Figure 16.3 Small 2 kW wind turbine prototype for telecommunication in the ice fields of Greenland (left); Testing of the small wind turbine for Greenland (right).

This form of transport of gas is so expensive, that it was viable to pay a very high price for a small wind turbine to supply most of the energy to the transmitter stations so that a container of propane would last for three years instead of one year and thus save two costly helicopter transports of gas. However, the wind turbine had to able to survive a temperature of minus 60 C and winds of 90 m/s. This is uniquely tough weather conditions.

In the beginning French Aerowatt turbines were used, but around 1984 a Danish development project was initiated in which Transmotor furnished the generators and the Folkecenter developed wooden blades.

The blades were tried out in three types of wood (laminated spruce, laminated ash tree, special Kerto plywood) and three kinds of surface treatment in order to harvest experience on what would work best in Greenland conditions. The result was blades that were more solid and far cheaper than the French ones—another important aspect concerning wind turbines blades in Greenland's icy mountains.

In special weather conditions ice builds up on the tower behind the rotor caused by condensation of ice crystals. When it happens, the rotor blades are chopped in two pieces and must be replaced by new ones. So, the costs mattered as new blades were constantly needed.

Figure 16.4 Development of blades for arctic applications of different kinds of wood and with three types of surface coating.

Worthwhile to remember from the 1970s as well is the small Veflinge windmill, built by a group with some of them former Tvind Windmill team members: Sanne Wittrup, Iben Østergaard, Martin Winther Jensen, Lars Peter Riishøjgaard and Hans Peter Ravn. Their small windmill was to be sophisticated and optimised with specially hinged sail wings in order to reduce the rotor loads. The young windmill builders had to give up their efforts. However, their time was not wasted.

Later on, Sanne Wittrup became a leading energy journalist on energy for the weekly *Ingeniøren*, and Martin Winther Jensen had an education as engineer. For some years he was employed at Risø National Laboratory and later at Siemens as a member of Henrik Stiesdal's very successful design team. The Veflinge team obtained decisive influence; they had learned about things the hard way.

16.4 A Very Personal Experience

In 1974, I was myself an amateur that was caught in one of the many traps of early wind energy. I fell a victim to the Swede Bengt Södergaard's book *Vindkraftboken*. It was the only accessible windmill design book by that time. I expected that Södergaard was an expert whose recipe could have been trusted. I did not want to experiment, but to build a windmill that could supply the house with power.

Figure 16.5 Cover of Bengt Södergaard's book *Vindkraftboken*, personal copy of Erik Grove-Nielsen with the Smiling Sun sticker, spirit of the times.

Two aerodynamic blades were made in one piece from an 80 mm steel pipe, ribs of plywood and a thin aluminium plate cover.

The generator was an asynchronous electric 11 kW motor. For gear I found an Allgaier farm tractor rear-thigh at a scrap dealer's shop in Sundby.

In the same place I found the tower—a rusty 8 m steel tube that had been used as a drain pipe in a gravel pit, which had worn it paper thin at one side and considerably weakened it.

Figure 16.6 Preben Maegaard's first windmill design from 1974–1975. The first version had two airfoil-type blades of 8 m diameter. Later it had three sail-wing blades. The drive train was an Allgaier farm tractor rear-thigh.

However, it was not the tower that failed. It turned out that the blades yielded scaring power during its maiden run on a not very windy day. I wanted to stop the windmill. However, I found that I could not turn the self-yawning wind turbine out of the wind with my hands. While the windmill roared on, I was just in time to fetch my small Ferguson tractor and fortunately it was possible to fasten a wire to the nacelle and pull it away from the wind before something went seriously wrong.

I was relieved and thankful and got my first real-life lesson within wind turbine design and operation. The windmill was never started again in the 2-bladed version. According to the fashion of the times, it was later furnished with three sail-wings and a rudder and stood there until the area had to be used as a parking lot.

16.5   Turning Wind into Heating

In the middle of the 1970s, a lively debate took place on whether windmills should be used for the production of power or for heating. In a temperate climate, three times as much energy is used fo heating as for electricity, and back then both were produced by importing oil. This made a good reason for developing heating windmills.[1]

In summer of 1975, many people were rigging up their first homemade wind turbines in their backyards and gardens. One of them is master blacksmith **Jørgen Andersen** from Serritslev near Brønderslev. He had built a 3-blade fast runner windmill which heated water by means of a water brake. The height of his windmill was 16 m with a blade span of 10 m. It looked quite modern with three blades pulling the water brake, placed on the ground. The wind turbine was placed by the main road 14, and many cars stopped to have a talk with the blacksmith on his heating windmill. By using a water brake, you save an expensive generator costing DKK 6 000–7 000 (EUR 900), the blacksmith pointed out.

Figure 16.7   Windmill for heat production developed by Jørgen Andersen.

---

[1]For more information on water brake windmills, see chapter *Water Brake Windmills* by Jørgen Krogsgaard.

At the same time a windmill of the same size as the Serritslev Blacksmith's was developed by a teacher, **Georg Pedersen** in Karby, on the island of Mors. He enjoyed the advantage of an excellent location close to the waters of the Limfjord. As soon as there was some wind, the turbine could have kept the entire house warm.

Georg Pedersen built the windmill all by himself, using parts from a lorry as well as new parts. "I used old records and my own calculations," he said to local daily newspaper *Morsø Folkeblad* in November 1975. It is not easy to find any new data on how to build a windmill; but if only you are able to read and do sums it can be done.

Everything concerning the construction and capacity has been meticulously calculated," said Georg Pedersen, who had already built his first heating windmill in 1974, during the energy crisis. In some respect it worked well, but it had a wooden tower, which was not sufficiently dimensioned. "One day it simply tipped over."

An American-type wind rose, from the plumber's enterprise **Peter Jensen & Sønner** in Frederikshavn, satisfied the needs of the market. They were practical people who made a wind turbine at a price level that sold to hundreds of people who wanted to make their own wind-generated heating. It was a simple and sensible household windmill. The rotor was mounted directly on the integrated gear from the Aalborg enterprise DESMI, while the generator was a 10 kW British synchronous engine.

Figure 16.8 SJ Windpower wind rose from Peter Jensen og Sønner.

Unfortunately, the makers promised a too high production of 10 000 to 12 000 kWh/year. It was twice as much as this windmill could do, no matter how you bent the laws of physics. Things became worse when, in an attempt to placate unrest, the manufacturers sought assistance from Risø, with a view to increase the energy production.

The method used was to enlarge the diameter of the rotor; but the tower simply could not carry that extra load. It broke down in windy weather, which led to a bitter realisation for the person responsible, when Erik Grove-Nielsen, windmill blade pioneer, presented pictures of the crumpled-up little windmills that were passed around during a contact group meeting at the Risø test station in the late 1970s.

Later on they succeeded in making the handy windmill from Frederikshavn work well, and for many years many of them were seen in the Danish landscape producing heating for their owners. But inflated production figures had dented the reputation of the product. The owners of windmills now had their own association and journal; its editor, Torgny Møller, demanded honest production figures, a reasonable thing. The owners were an easy prey for sellers who promised too much.

Figure 16.9 Davidsen's Lolland wind turbine in California.

Heating windmills never became a real success, but it would be unjust to mention this type of windmill without telling about the LO-FA windmill. It was part of a windmill culture had came into being on the island of Lolland, including master blacksmith Anders Davidsen's Lolland windmills in different sizes which were produced in nice quantities and exported to California as well. The LO-FA heating windmill developed by Knud Berthou as well as the PADEMO windmill did not have a market breakthrough; the island of Lolland did not get the new industry, which was so badly needed with its long tradition within mechanical industry.

Activities around the LO-FA windmill were established in the old dairy factory at Slemminge by Sakskøbing. Here the inventor and industrialist *in spe*, **Knud Berthou**, had gathered a group of ten persons who developed a windmill which was technically different from anything else at the time, apart from the blades made of laminated wood.

The windmill had three blades with a diameter of 12 m and was designated as a high-tech windmill with variable pitch blades, planetary gearbox and a very refined liquid heat-brake system. The heating medium was not water but hydraulic oil. Whereas the other heat-brake windmills had their mechanical parts close to the ground Berthou placed everything inside the nacelle, and this opened for new potentials. This is how the bottom 10 m of the steel tube type tower were filled with 15 tons of water and thus made up the heat storage which was insulated. Besides serving as friction liquid in the heat-brake hydraulic oil transferred heat to the water storage in the tower and lubricated all moving parts in the gear, bearings, etc.

However, technological merits did not lead to commercialisation. Even if the need for heating is three times higher than for electricity, windmills connected to the power grid were favoured by the state of subsidies. With days of strong winds the excess energy was not lost but could be fed into the grid and finally, an energy supply chain supported the power producing windmills with a variety of industrially produced components. With too many odds to overcome the pioneers behind the LO-FA and other heat-brake wind turbines disappeared into the oblivion. However, their achievements and innovations have to be remembered by the history of technology.

Probably the most successful water brake windmill in Denmark was produced by the brothers **Bent and Hans Svaneborg,** two intelligent and creative artisans. The windmill was erected at Stagstrup in Thy, where it heated Hans Svaneborg's house for years. The two brothers had gathered all accessible knowledge on that kind of windmill and gave it three fibreglass blades. The angle gearbox came from a Mercedes vehicle; the home-made liquid brake was placed at the bottom of the tower. From here the pipes led to a water storage container inside the house. The Svaneborg windmill was demolished but not shredded. When restored it will be part of the Folkecenter Renewable Energy Museum where the blades are already exhibited.

16.6 The "Folke" Windmill: One Blade Suffices

"From the point of aerodynamics more than one blade is really almost wastefulness," according to civil engineer Burmand Jensen. Together with Finn Jensen and Søren Olsen, who had harvested inspiration at a sail-blade-windmill-course at Kolding Folk High School during the spring of 1977, they wanted to design an ideal windmill for the people. It was super light, of 6.5 kW and had an enormous swept area for its size of 95 m^2. It was calculated to be able to produce more than 20 000 kWh annually. Other people also worked on one-bladed windmills, among those the German weapon giant MBB. They shared similar problems and fate: the fewer the blades, the faster they have to rotate in order to utilise the wind efficiently. And that creates noise and problems of counter-weights for proper balancing of the rotor.

The group was good at getting money grants, Burmand was an excellent theorist, the others could turn theory into practice. This resulted in several prototypes which did not, however, get past the experimental stage. Burmand recounts that the last type was erected next to the old Harreby Co-op outside the town of Ribe. This place also had a workshop where the pioneers could work. But one windy day things went wrong. Some of the ingenious mechanical bits in the hub had got stuck, and the windmill began to race. The long laminated wooden blade hit the tower and was splintered so that the bits of the blade flew as far away as 250 m!

Of the three pioneers, Søren Olsen later founded his own blade factory, Olsen Wings, which became a leading supplier of blades for 3 kW to 30 kW wind turbines.

Figure 16.10 Burmand with one-bladed wind turbine called "nice weather turbine" (Let vejrs mølle) 1977 (left); one-bladed wind turbine prototype (right) (Photos: Flemming Hagensen).

16.7 The Kramsbjerg Windmill: A Technical Innovation

Small windmills for individual homes was the marked segment that Jørgen Kramsbjerg Hansen went for with his 15 kW Kramsbjerg wind turbine, which emerged in 1983–1984. Hansen was technically knowledgeable and had during some years been developing his construction of a three blade downwind turbine with KJ blades. His windmill used an industrial gear with a safety factor of almost three or far more than those of the established factories. The windmill had a great deal of technical refinements compared to the general standards within the business. Thus the generator was not a traditional asynchronous engine connected to the grid.

Jørgen Hansen used a Danish Transmotor synchronous generator normally used as a generator in smaller ships. It offered the advantage that it was possible to adjust the generator to particular operational conditions by regulating the energising current. During gales when other windmills were brought to a standstill the anemometer sent a signal that changed the energising current and reduce the velocity of the rotor to a non-critical rotation speed. Today this principle is being developed on the advanced multi-polar ring generator wind turbines, whereas the ordinary types with asynchronous generators just stop when the winds are really strong, over 25 m/s or so.

The Kramsbjerg windmill is an instance that shows that even with a great personal involvement, technical quality and a sensible price the market will not just descend to you from above. At that time people were so engaged in what was going on within the established windmill trade and export adventure that the Kramsbjerg and the like were not given adequate commercial conditions.

Figure 16.11   The Kramsbjerg windmill (Photo: Flemming Hagensen).

16.8   The Ring Generator Turns Up

It is almost possible to repeat the story just told about the Kramsbjerg windmill when we describe the Odder windmill. It was at 18.5 kW with three blades, downwind. It did not become a commercial success, either, when it was launched in 1983–1984, but not because of lack of technical qualities. The necessary legal and planning framework was absent, and as renewable energy is dependent on such framework, the market did not emerge.

Developer of the wind turbine, Werner Kastrup Pedersen observed from his earth excavators that the hydraulic tubes got very hot due to friction heat in the hydraulic system. He concluded that a windmill would be able to agitate the oil in a heat brake designed for the purpose. But this idea was abandoned, and instead he began to make power-producing windmills with variable pitch blades and a 58-pole synchronous generator from Transmotor.

And this was certainly a real innovation. Now it would be possible to do without the gearbox which has caused so many problems for the windmills. If the Odder people had continued along this track, then, we in Denmark might have had ring generator wind turbines on the market ten years before Enercon in Germany, who introduced the multi-pole generator wind turbine in 1993 and since then has been among the leading and biggest players in the world.

In Odder, however, they had problems with adjusting the pitching of the blades to ensure the necessary constant speed of the rotor and this made them drop the multi-pole generator principle. Subsequently it has become evident that the speed of rotation should be regulated electronically and not mechanically in the way they tried to do it on the Odder windmill.

But there was a stage in between: In the "1983 Nordisk Vindmøllekatalog" (Nordic Windmill Catalogue) the Odder windmill is seen with two blades of their own making but with a Hansen gearbox and a generator from the DDR.

Figure 16.12   Odder windmills: 2-bladed (left); 3-bladed version (middle, right) (Photos: Flemming Hagensen).

And thus, after another disappointment the company employed engineer *Jens Thillerup*, who created a windmill, which was pleasant to look at, robust and conventional with its three blades, Sala gear and a French Leroy asynchronous generator. But unfortunately there was no market for single-family wind turbines at the time. After the erection of half a dozen, nothing more was heard about the Odder windmill.

16.9   An Organic Windmill

The Thy windmill emerged around 1984 as a somewhat different product. Flemming Grønkjær of Grønkjær Maskinværksted, later Thymøllen, and architect Niels Helmer Christensen from Tygstrup, jointly developed a simple and robust 7.5 kW windmill. Potential customers were homes standing on their own grounds in North-western Jutland. What made this windmill special were its solid, equilateral and un-twisted wooden blades. The inventors made use of the experience harvested from engineer Jacob Bugge's experiments with this type of blades at the Folkecenter.

Figure 16.13   Thymøllen—Thy windmills, successful design from the 1980s. Thy 13 kW wind turbine with KJ blades at the Folkecenter (left); form redesigned in 2010 for the new energy market (right).

On locations with good wind this small windmill could produce up to 17 000 kW, and that was very suitable for heating and lighting an ordinary home in the countryside.

Thymøllen—the Thy windmill—became available as well with fibreglass blades from KJ. A special version of the Thy windmill gained an award in the windmill competition arranged by the Danish Design Council in 1984. Niels Helmer Christensen, the windmill team architect, had developed a tripod laminated tower and re-designed the nacelle so that the blades, tower and nacelle were all made from wood. This windmill is probably the most Danish and the most ecological windmill seen so far.

In June 2010, Denmark introduced a new legal framework to promote single family renewable energy installations for self supply. The criteria for such windmills were: maximum 6 kW, 40 m^2 swept area and a total height of 25 m to the blade tip. In 2010, Thymøllen, now involving windmill veteran Leif Pinholt, redesigned their windmill. It got wooden blades designed by engineer Viggo Öhlenschläger and a new appealing white-green livery.

Figure 16.14   Ecological windmill in Tygstrup: tower, blades and nacelle cover are entirely made of wood.

Thymøllen has obtained important shares on the emerging household windmill market for their pretty and robust design. House owners in the countryside place such windmills in the corner of their gardens, about 25 m from the buildings. The windmills are grid-connected and benefit from the net-metering principle where each kWh of self supply has the value of the purchase price of electricity. In Denmark power costs are rather high as 65% of the kWh price are various kinds of taxes.

16.10   From Grassroot to Producer

Ole and Kenny Hansen from Kongsted on Zealand passed from active OVE, a leading Danish NGO, grassroots to manufacturers. They actually made some of the best windmills of the period.

Figure 16.15   The Kongsted windmill (Photo: Flemming Hagensen).

The type though of a conventional design was in some aspects ahead of its time with a heavy duty Stephan gear of the integrated type, well dimensioned and with a legendary longevity and good performance.

They stopped in 1982 after a score of windmills and three years in the business. The only reason for that was that they could not bear the kind of business world where you cannot trust one another and where disagreements end up in a law court. A sub-contractor had several times used non-authorised materials for the blades which consequently had to be replaced. They were also tired of windmill customers who would call in the middle of the night and complain about the windmill, even when the problem most likely was lack of wind, says Ole Hansen.

But how did Ole Hansen become a manufacturer? In 1976 there were only two commercial manufacturers to choose among: Riisager and Borre. Now it turned out, at an exhibition in Skive, that the Borre windmill could not turn itself. An electric motor made it go round, revealed Ole Hansen. His next move was to buy a Riisager windmill, "the perfect windmill", as claimed in a book about wind energy. But it was a disappointment.

Figure 16.16   Kongsted nacelle displayed at the Folkecenter Collection of Wind Energy.

Ole Hansen found that the gearbox was under-dimensioned, the fibreglass hybrid blades disintegrated, the tower was too low, and the light flashed off and on. Improvements led to the Kongsted windmill which obtained system approval and consequently became eligible for subsidy. This did not please Riisager, says Ole Hansen, but he wanted to develop a well functioning windmill. "That can best be done by building on experience. Riisager did the same by building on the experience from the Gedser windmill," says Ole Hansen. It was quite true that the concept had been taken from J. Juul, who designed the Gedser windmill.

16.11   The Fascinating Egg Beater

Special interest was shown in vertical-axis windmills, VAWTs, which for some years attracted and fascinated a great number of developers in Denmark and abroad, and resulted in many experiments. The belief in the VAWT concept as the rational, simple, cheap and reliable windmill, was so strong that it was often overlooked that the wind is more turbulent, and the energy contents lower, when the rotor was placed down on the ground and not on a high tower like on a horizontal type windmill.

The designers and inventors of the many varieties of VAWTs often were focused on having the generator placed on the ground, in order to facilitate servicing. Meanwhile, they forgot that the component that needed to be developed was not the generator, as the technology was already well-known, but the vertical rotor technology, which was new and yet unknown.

After 40 years of experiments and testing the situation by 2012 is almost the same. The VAWT concept still has to demonstrate reliability and efficiency. That was the reason that windmills of the Darrieus type have in fact been abandoned as commercially irrelevant.

The preliminary successful research, testing and practical experiences of Johannes Juul and Ulrich Hütter were available for the designers of horizontal-axis wind turbines. However, people involved in VAWT design had only the idea and theory of an ideal wind turbine but no real operational experiences. And that caused great problems, particularly because of the vibrations and durability of the various configurations of VAWT rotors. Yet, the dream of the "egg beater" as the ideal wind turbine survived, and many projects materialised.

Civil engineer Jean Fischer from cement machinery giant F. L. Smidth & Co. (FLS) that created the Gedser wind turbine in 1956, became well known at an early stage for his dashing concrete column with its six egg beaters. Also other vertical-axis types were presented by Jean Fischer.

Until contact was obtained to Karl Erik Jørgensen's and Henrik Stiesdal's successful 22 kW windmill project, at Vestas the very earliest experiments 1976–1978 with wind energy were also aimed at the Darrieus principle. They worked on a kind of 15 KW "bi-plane Darrieus", developed by engineer Leon Bjervig. The tower was elegant, tapered, had the look of a professional windmill, but except the prototype it was never manufactured. Vestas preferred to produce the HVK blacksmith's type 3-blade windmill on a license.

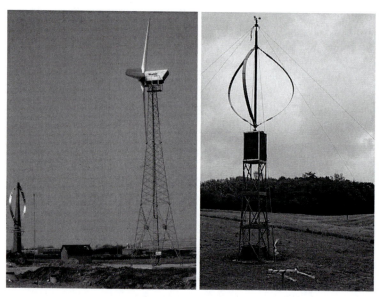

Figure 16.17   To the right, 55 kW HVK Vestas in Lem, 1980. To the far left, an elegant windmill of the egg beater type (Darrieus) developed by Leon Bjervig in 1978. It never entered production (left); The exam project Darrieus wind turbine, developed by engineers Troels Friis Petersen and Flemming Rasmussen just outside their windows, Risø test station (right).

Let us mention some more people who spent much energy on developing the vertical-axis-type of wind turbine during the 1970s. Aluminord, a division of the biggest Danish manufacturer of cables and wires, Nordisk Kabel og Tråd, made an offer to extrude aluminium profiles for Darrieus wind turbines. At the new test station at Risø, engineers Troels Friis Petersen and Flemming Rasmussen, could study their exam project Darrieus just outside their windows.

At Nørgaards Højskole in Bjerringbro Arne Bech was working at the egg beater, as the type was called among those in the know; engineers V. Lassen Jordan and Per Ove Christensen were making test windmills of this type as well. OVE veteran Ole Højland and Søren Arthur Jensen of the Danish Maritime Institute were doing examination projects on Darrieus windmills; Søren Arthur continued to develop them far into the 1980s together with Morten T. Jensen. At a convivial summer camp in 1976 in Hjerk, Salling, engineer John Kvint led a project and built a 3-blade Darrieus windmill, with wooden blades which were shaped during the camp.

But all this did not lead anywhere. Despite the fact that actually more well-educated, "real" engineers worked with Darrieus designs than with the 3-bladed Danish concept, in a retrospective it can be stated that it was not money, zeal or arithmetic competence that decided the result. The 3-bladed upwind windmill with the asynchronous generator was a better concept, technically and economically.

16.12   The Phenomenon Borre

Engineer Niels Borre made the entire development and became a legend. He developed a derivative of the Darrieus windmill with a vertical axis and four perpendicular blades. Dansk Vindkraftindustri in Slangerup was to produce it. The blades were straight as on the old Persian windmills. The first Borre windmill was erected with the electric utility EFFO in the autumn of 1977. Invitations for a grand opening ceremony had been sent out, but the windmill would not really get started in spite of an acceptable wind.

Still, the egg beater had an unconditional appeal to the public. The Borre windmill was placed as well at one of the low-energy demonstration houses in Skive for which it was meant to supply heating. It just did not work here either; in fact it was turned around by an electric motor to give some illusion of being a wind turbine!

The Borre windmill was erected and tested at the Risø test station. During testing things went wrong. When the wind was strong, the windmill speeded up; an attempt was made to stop it with a lasso to throw ropes up into the rotor when the windmill started racing, a method which had been practised at runaways of the thousands of la Cour windmills on Danish farms in the period 1905 till 1950. The local blacksmith was called in; it was his task to stop the windmill, employing the lasso and ropes that were to get entangled in the blades. However, the run-away drama did not always have a happy ending—the windmill disintegrated.

Already before the windmill had been tested and documented, Borre was so certain of the commercial future of his project, that he had a designer see to the aesthetic appearance. After thorough studies, a dark blue colour was chosen for the tower. At the end the Borre windmill never came into industrial production, although several customers were interested.

After the accident Niels Borre dedicated himself to propeller-type windmills. About 1980 sail wings had become a fashion. And Borre was ready for prompt delivery. He found good customers in Bente and Erik Johannessen at Gyrup in Thy, who ordered two windmills from Borre to be on the safe side. If one could not provide heating for their big house it was only logical that the other would be able to do so. It was unthinkable that both should break down at the same time.

But things turned out still worse. Erik Johannessen can tell of terrible experiences with the Borre windmills. It is well known that winds may be strong in Thy, and one day things went wrong. The mechanical parts had grown warm, and suddenly red hot parts of the machinery fell to the ground and made the wet grass seethe. Soon after they could look out at the windows and see that one of the towers was swaying. Then the entire windmill rose from the ground and fell down with a thud, and everything was smashed to smithereens. Electric cables had grown so hot that they melted, and soon nothing was left.

The entire investment was wasted, and there was nothing to be got from Borre. He had taken off to Canada where he was working as "Mr Windpower". Danish customers no longer meant anything to Niels Borre.

Figure 16.18   Windmill designed and built by Niels Borre, installed on low-energy house in Skive (about 1976). It could not turn by the wind, it needed an electric motor.

The Johannessens had been enriched by a wind power experience, but they did not give up wind power because of the Borre scandal. Soon after, they ordered a 22 kW Smedemester windmill from H. Poulsen & Søn at Lyngs, a Folkecenter/NIVE design. It provided power for them for many years. All through the years the windmill was looked after by *Bendy Poulsen*; but when a blade was broken after many years of service it was given up for lost. Erik Johannessen would have still preferred to buy a bigger windmill but the local planning authority would not permit it.

16.13   Wind Matic and Tellus

Wind Matic was represented in the "1983 Nordic Windmill Catalogue" with a veritable cavalcade of types. Here one can find something that would appeal to all tastes.

Figure 16.19 Riisager-type Wind Matic wind turbines at the Altamont Wind Farm, California (Photo: David J. Laporte from Vancouver, Canada).

Figure 16.20 The first modern wind turbine in Brazil: Folkecenter 75 kW installed at Fernando de Noronha Island by the Folkecenter in cooperation with Brazilian Wind Energy Centre (Recife, Brazil) (left); Wind Matic 120 kW 19S with LM blades (right).

In the beginning Riisager was part of the company, so we find these creations of his: the 22 kW 10S, the 30 kW 12S and the 55 kW 14S. The numeral before the S signifies the rotor diameter. But we also find some interesting novelties: a small 7.5 kW Wind Matic Folkemølle with LM blades aimed at the same market as the Thy windmill described above.

However, the 1983 catalogue also marks the significant transition to a conventional 55 kW windmill with three LM blades, the evidence of the complete breakaway from Riisager. Wind Matic now wanted to enter the market where Bonus, Nordtank and Vestas had been ruling. And it was quite successful. They made as well a 75 kW windmill and took part in the California boom.

During this process Wind Matic was sold to an American developer, very imaginatively called "Electricity". The rumour about this was the major subject of a debate during a cruise on a luxury yacht below the Golden Gate Bridge by San Francisco, where some American businessmen had invited a Danish study group. I was the leader of the group together with Steve Smiley from Michigan. The Americans thought that we had Wind Matic people in the group, which was not the case. But it turned out, however, that among us we had the founding managing directors of future windmill manufacturers such as WindWorld, Vind-Syssel, Dencon and Wincon West, just to mention some of the people who must have harvested much experience from this study tour.

When the cruise was over, everybody was confused, the Danes and not less the Americans who absolutely got no share in the posh repast. We got a sample of the notorious Californian "business ethics" which also killed Wind Matic because their American partner was unable to pay the millions of dollars owing to them, some told. After the Californian experience the old Wind Matic gave up. The company was reconstructed by one of the main creditors Alfred Priess from a company in Vinderup with a long tradition in steel pylons and substations, that took over Wind Matic.

Windmills of 150 kW to 200 kW with LM blades and a new designed gearbox from the Jens Fisker Maskinfabrik were developed. All of a sudden Danish industry was able to furnish Danish produced gearboxes of a design which mostly smaller Danish windmill manufacturers cooperating with the Folkecenter had used till then.

Only seven big Wind Matic windmills of the new generation were made, and some of them had accidents. One was wrecked during gales in early year 2000 in Zealand. The air brakes had failed. They were of the front spoiler type, that is, a solution which the father of the Danish windmill concept, Johannes Juul, would never have used, but which the authorities had given their system approval.

Figure 16.21 Wind Matic brochure for the WM 15S model with two generators of 13 kW and 66 kW output. "Money's in the Air" as the slogan in the brochure advertises the new wind turbine with six variants of towers (lattice/tubular; different sizes).

Thirteen former Wind Matic people created Tellus after Wind Matic's collapse when the wind rush in California ended. They were experienced people such as engineers Rio Ordell and Poul Højholt who in cooperation with a group of colleagues did in record time make themselves manifest in the market with a well-designed 80 kW windmill with 8 m LM blades. The windmill had several similarities to the Wind Matic design, that is the same LM blades. Thirty-five of these were produced and they performed well.

Later on this wind turbine was followed by an up-scaled version with blade extenders and a 95 kW generator. Nine of these were erected on a wind farm by Dragør near the Copenhagen airport where they saluted guests landing at the airport. 1987 was the year of disaster. Three out of four windmill producers vanished. Tellus was one of them.

Figure 16.22   Tellus T-1995, 95 kW wind turbine/Risø LiDAR Measurements/ Wind Matic (Photo: Lucas Bauer/wind-turbine-models.com).

16.14   A Genuine Pioneer: Claus Nybroe

Many more than those mentioned in this chapter worked with windmills. Some other work of history must catch up with what I have neglected or overlooked and do it before it is too late and the key persons have passed away. My experience has been that whenever I dug one spit deeper in the Danish wind history 1974–1980 a wealth of new sources were revealed for future researchers to bring forward.

But let this overview be finished with a far too short mention of a pioneer who has for almost 40 years left his strong mark on the development of contemporary wind power.

Architect, writer, developer, missionary, inventor, and manufacturer Claus Nybroe has been described as one of the most brilliant designers of windmills in the twentieth century. One of the first stamps that he placed together with Carl Herforth was incredibly useful and well thought-out: the book *Sun and Wind*[2] a small state-of-the-art handbook. It became the Bible of many windmill builders in the 1970s. It was the right book at the right time.

Figure 16.23   Cover of the book *Sol og Vind* by Carl Herforth and Claus Nybroe. Personal copy of Erik Grove-Nielsen with the Smiling Sun sticker (windsofchange.dk).

In the late 1970s, Claus Nybroe together with Rio Ordell started the Dana Vindkraft on the island of Endelave. They developed the Holger Danske windmills of sizes including 22 kW. They set new standards for technique and good design. Later on Nybroe paused for a while but reappeared in 1983 with the report "Californiske Vindmølleparker" (Wind Farms in California) on a Folkecenter arranged study tour to the Californian wind farms. In 1984 he was co-author of the three-volume publication "Manual for Wind Farms" published by the Folkecenter.

When the Danish Design Council arranged a competition on new windmill types design in 1984, Claus Nybroe got the first

---

[2]Herforth, C., Nybroe, C. (1976) *Sol og Vind*, Information Publishers, Copenhagen; sold in more than 15 000 copies.

prize. After that he devoted his time to develop the quite small windmills from Ole Windflower, Rasmus Windflower, etc. His own experimental test station attracted many interested people from Denmark and abroad. Claus Nybroe's rotors, the Windflowers had a design of their own—created by his own hands, they were and still are elegant, noiseless, efficient and durable. He has also taken time to create quality design especially of nacelles for Bonus and since 2004 for Siemens in Brande.

Figure 16.24 Claus Nybroe's Ole Windflower.

About the author

See chapter *From Energy Crisis to Industrial Adventure: A Chronicle* by Preben Maegaard.

Chapter 17

Hütter's Heritage: The Stuttgart School

Bernward Janzing[a] and Jan Oelker[b]

[a]Freiburg i.Br./Baden Würtemberg, Germany
[b]Radebeul/Sachsen, Germany

Bernward.Janzing@t-online.de, jan.oelker@gmx.de

The development of modern wind power in Germany began in the research institutes for aviation in Stuttgart. After World War II and after the oil crisis of the 1970s, the vital impulses for the use of wind energy came from the community around Professor Ulrich Hütter. With their research on the field of rotor blade aerodynamics and the composite construction principle Hütter with his students and associates at the University of Stuttgart and at the German Test and Research Institute for Aviation and Space Flight (Deutsche Forschungs- und Versuchsanstalt für Luft- und Raumfahrt DFVLR) built the foundation for design and construction of rotor blades.

17.1 The Lack of Energy

It came as a shock wave. Heiner Dörner, lecturer at the Institute of Aircraft Construction of the University of Stuttgart remembers the

---

The chapter is an abbreviated excerpt from the first chapter of the book *Windgesichter—Aufbruch der Windenergie in Deutschland*, Sonnenbuchverlag, 2005.

Wind Power for the World: The Rise of Modern Wind Energy
Edited by Preben Maegaard, Anna Krenz and Wolfgang Palz
Copyright © 2013 Pan Stanford Publishing Pte. Ltd.
ISBN 978-981-4364-93-5 (Hardcover), 978-981-4364-94-2 (eBook)
www.panstanford.com

autumn of the year 1973: "On Sundays we took walks on the highways or took out our bikes." Oil had become scarce on the markets of Europe and North America. Therefore many governments saw the need to take unpopular action, for instance the ban on driving private cars on Sundays. From one day to another, the public noticed how strongly industrialised countries depended on imports of oil and gas.

The situation was triggered by oil drilling Arab states. That autumn they had reduced the crude oil production by up to 25%, and had set-up embargos from time to time. In January 1974 they started a new pricing policy and in consequence the listed prices for crude oil were almost four times as high. The Western world faced a serious oil price crisis.

Under the impression of rising energy prices the use of alternative energy sources, as for instance wind power, came back into the focus of public attention. "Wind power always comes over mankind in waves. People remember it always in times of crisis," Dörner resumes, "after World War I, when coal was getting scarce, and after World War II, when the energy supply lay waste, and once more after the oil price shock in the early seventies."

Dörner had directly experienced the renaissance of the use of wind energy after the first oil crisis at the University of Stuttgart. Although he had worked at this institute of research for more than 35 years, he did not see himself as a pioneer. The rebirth of wind energy had a previous history, and this was connected especially to the name of the man whose academic heritage Dörner has administrated, and had passed on to students for more than three decades, the name of his teacher, mentor and role model—Professor Ulrich Hütter.

Figure 17.1 Professor Ulrich Hütter (Photo: Heiner Dörner's Archive).

17.2 The Dream of Flying and First Theoretical Works for Wind Energy

The academic work on wind energy for Hütter, as for many other pioneers, was closely related to the dream of flying. Hütter, born in Pilsen in 1910, became a great fan of gliding with unpowered aircraft. Together with his brother he had designed a few light-weight gliders in the thirties. The company Sportflugzeugbau Schempp–Hirth in Göppingen built them in series. During the war, he was the head of the construction department of the aviation research institute "Graf Zeppelin" near Stuttgart. In summer 1944 he took over a chair for fluid mechanics and flight mechanics at the Technical University of Stuttgart.

After the end of World War II the allied forces prohibited aircraft construction in Germany. This was a hard blow for many engineers who had dedicated their entire professional career to aviation. For the aircraft designer Ulrich Hütter this ban was reason to return to another subject he had pursued with great attention several years before: wind energy.

Hütter had completed his doctorate proceedings already in 1942 at the University of Vienna with the thesis "Beitrag zur Schaffung von Gestaltungsgrundlagen für Windkraftwerke" ("*A contribution to creating the design principles for wind turbines*"). Back then he worked as a lecturer at the college of Weimar. On a test field of the local company Ventimotor GmbH he collected valuable experience with small wind turbines working according to the aerodynamic principle. The design of their rotor blades used aerodynamic uplift to drive the rotors instead of air resistance as in old type windmills.

The benefits of such machines have been undisputed ever since the fundamental analyses of the scientist Albert Betz in the 1920s. At that time, Betz had analysed wind roses with different numbers of blades in the wind tunnel of the Aerodynamic test station in Göttingen. From these tests he derived the physical principles of the conversion of wind energy. He also determined the upper limit for the theoretical maximum value of wind power performance, known today as "Betz's factor", which is 16/27. According to this factor, a maximum of 59.3% of the total perfor-mance of airflow meeting the rotor of a wind turbine can be converted. Betz's theoretical works have been the base for the calculation of wind turbines ever since.

Hütter's doctorate thesis also builds on Betz's theory. Hütter was the first to transfer the principle of airfoil aerodynamics of airplanes to rotors of wind turbines. Until today, the design principles Hütter derived for the calculation of wind turbines widely subsist.

The fact that Hütter's philosophy was rooted in aircraft construction had effects on all his constructions and in particular on rotor blade development. "All things that rotate should be as light as possible but of course also as solid as necessary," was his creed. Even for the smallest systems Hütter took all efforts to minimise the loads affecting the rotor blade by reducing its weight. It follows that his basic principle was a consequent lightweight construction.

As an aircraft designer Hütter paid special attention to the profiles of the rotor blades and worked hard on their aerodynamic optimisation. He opted for fast-running machines and high lift coefficients through the blade profile. "From the aerodynamic point of view, the profiling is important for high-speed blades, not the number of blades," Dörner cites his teacher. "In theory even one blade alone might be enough for a wind turbine to come close to the theoretically possible energy yield from the wind." From today's perspective he adds: "Two blades were chosen for reasons of weight symmetry, but because of the alternating loads, this was not too beneficial for the service life. Therefore in the end three blades have won." However there was a long way to go towards this finding.

17.3 Allgaier's Investments

Hütter's search for the perfect concept began shortly after World War II with the construction of small systems. Together with company Schempp–Hirth he had developed a small single-blade rotor in 1946. After further experiments, he was able to build a 3-blade 1.3 kW unit for the power supply of a chicken farm in Ohmden near Kirchheim/Teck only one year later.

Thereupon the entrepreneur Erwin Allgaier from Uhingen near Göppingen in Wurttemberg recognised the potential of this new technology for his medium-range mechanical engineering company Allgaier–Werke. The company that once had been founded as a workshop for cutting and stamping tools by then was producing a wide range of metal products, from pots and pans to pumps for manure.

Wind power somehow seemed to fit in the mix. Allgaier immediately employed Hütter as the senior construction designer for his company. Not far from the company site, Allgaier in 1948 set-up a test field where Hütter could research systematically. The first pilot wind turbine had a diameter of 8 m and 1.3 kW output.

Hütter, a passionate designer developed the 3-blade machine further to realise 7.2 kW output with a rotor diameter of 11.28 m. As these systems were primarily designed for island operation, and hence were not controlled by the grid, Hütter opted for performance control through adjustment of the angle of airflow hitting the rotating rotor blades called pitch regulation. He accepted that this solution required complicated construction even for the smallest systems. On the other hand it was possible to adjust the perfect blade angle for each wind speed. This led to a reduction of wind loads and suited well Hütter's lightweight construction philosophy.

From 1950 the wind turbine was produced under the name Allgaier WE-10. It was the first serial produced wind turbine in Germany that worked according to the aerodynamic principle.

Figure 17.2   Allgaier WE-10 turbine, Bonn (Photo: Jan Oelker, 2000).

Approximately 200 wind turbines of this type were produced in the following decade with outputs between 6 and 10 kW, and they sold quite nicely in Germany and abroad.

Figure17.3 Allgaier WE-10 turbine, on the roof of Klöckner–Möller GmbH company, Bonn (Photo: Jan Oelker, 2000).

Allgaier also developed wind turbines for the first wind park in Germany. In November 1953 the Water Administration Office in Meppen put eight wind turbines of the type WE-10 into operation near Papenburg in the Emsland in order to create power for a pump station. For 10 years, the 3-bladed wind turbines at the "Nenndorfer Hammrich" pumped out groundwater and rainwater which due to a dyke no longer had natural drainage towards the river Ems.

Figure 17.4 The first wind farm in Germany with Allgaier wind turbines in Nenndorfer Hammrich (Papenburg), 1954 (Photo: Sepp Armbrust, Collection of Jan Oelker).

Upon the initiative of the regional power utility in Schwaben, Ulrich Hütter in 1952, for the first time, tested parallel grid operation, and with success. The WE-10 marked the beginnings of professional use of wind energy. With this system, Hütter had created a perfect reference for his work. Whoever was interested in the use of wind could not bypass the aircraft designer from Stuttgart.

17.4 The 100 kW Challenge

The academic wind power association Studiengesellschaft Windkraft had been set up in December 1949 upon the initiative of the regional commercial office in Stuttgart. This association contracted the pioneer in October 1953 for the design of a wind turbine with at that time very daring dimensions: it was meant to deliver 100 kW. Hütter had been the winner of a design competition over his competitor Richard Bauer. Bauer, also a learned aircraft designer, opted for the single-blade rotor as it could realise the highest speeds. Practical implementation however failed due to problems with control and aerodynamic stability.

Hütter had more trumps in hands with his design and his experience. He opted for a 2-blade wind turbine with a teetering hub. This hub was necessary as in the normal height profile of wind on a 2-blade system, the upper blade takes in more flow than the lower one. Hence the loads affecting the rotor have to be absorbed either through stability of the blades or by a hub giving in. Hütter chose the second version in order not to have to give up his lightweight construction principle.

The Allgaier Company was not capable of financing the development of the 100 kW system, which the Studiengesellschaft preferred. Therefore a development joint venture was set-up for the purpose in July 1954 under the name Windkraft Entwicklungsgemeinschaft (WEG). The venture united seven public energy suppliers and five companies from the electro-mechanical industry and fluid-mechanical segment. This proves the interest the industry had in wind energy in these days.

As an aircraft expert Hütter used his comprehensive knowledge of aerodynamic blade profiles. For the mechanical part, his team relied on proven systems and found the support of acknowledged manufacturers. The list of suppliers reads like the "who is who" of the famous mechanical industry of South Germany: Voith Company from Heidenheim delivered the gearbox, Escher/Wyss from Ravensburg delivered the yaw bearings, Mannesmann delivered the tower, Porsche parts of the nacelle, and AEG added the generator plus switch systems. Hütter's team at company Allgaier held the reigns in hands.

In order to enhance his team for development of the system Hütter upon recommendation from his team member Eugen Hänle in February 1955 recruited the construction designer Sepp Armbrust. Armbrust, born in 1930 in Pecs in Hungary, had studied at the Technology College in Esslingen. Like Hütter and Hänle he was a passionate glider pilot. He took part in several German Championships and later was a member of the German national gliding team.

Armbrust, who virtually worked all his professional life on the field of wind energy until his retirement in 1993 after all this time today still remembers his early days at Allgaier in Uhingen: "When Erwin Allgaier had announced his intention to visit our office, Hütter came to Hänle or to me to see who had the most interesting construction drawings on the table. He took the most interesting ones and put them on the board in his room to make sure he had a pretty drawing in the background." He says Hütter loved beautiful drawings and was a very capable sketch-drawer himself. "He was an aesthete, very friendly and charming" Armbrust characterises the man from Austria. "But he was also very ambitious".

This ambition had been awakened by the very challenging 100 kW system. Realising the tenfold of the output of the Allgaier wind turbines was a giant leap to take. The system was to have a rotor diameter of 34 m and hence was called W-34. It was meant to become a giant innovation not only for the wind power industry but also for the materials technology.

The two rotor blades were each 17 m long, and made of glass-fibre reinforced plastic. Hütter took over this technology from glider construction. Ambrust and Hänle tested several glass fibres and resins for the new blade. They constructed a new resin dispensing equipment for precise dosage of the resin quantities. In February 1957 both blades were complete. The first glass-fibre reinforced plastic blades of this size in the world were a novelty not only on the wind energy market. At that time, they were the biggest parts ever made of the new material.

For testing the W-34 a test field was set-up in 1956 at "Schnittlinger Berg" between Schnittlingen and Stötten on the Swabian Alb. Sepp Armbrust was in charge of the test field. Until his retirement, it has been his second work place with but a few interruptions.

Figure 17.5 Sepp Armbrust on the test field in Stötten/Schnittlingen (Photo: Jan Oelker, 1999).

Figure 17.6 Assembly of the 100 kW turbine W-34 on the test field in Stötten/Schnittlingen, August 1957 (Photo: Sepp Armbrust, Collection of Jan Oelker).

At the end of 1956 in Schnittlingen, an Allgaier wind turbine was set-up as the foundation and it was re-furnished with a teetering hub and two glass-fibre reinforced plastic blades. It had 8.8 kW output. It was tested to gather experience with the material and the modified construction. Assembly of the W-34 begun on the test field in August 1957 and on 4 September it first went into operation. The first grid connection took place on 11 December 1957. One day later the system already realised the first full load.

Figure 17.7 Wind power test field in Stötten—W-34 wind turbine (in front) and 2-bladed Allgaier WEC 10 turbine (behind), ca. 1957 (Photo: Sepp Armbrust, Collection of Jan Oelker).

The W-34 was only a pilot wind turbine, and it realised the highest output coefficient measured until that date on a wind turbine. It did not attain high power production. Still the experience gained with this pilot plant, and the basic concept of the lightweight construction and blade adjustment Hütter had favoured were considered in later developments. Hütter had the leading role of pioneer for the development of the modern use of wind power in Germany, and later also in America, comparable to his Danish colleague Johannes Juul in the northern neighbour country. As different as their concepts may have been, with their research after World War II, Juul and Hütter developed the principles of engineering and technology of the modern wind power stations we know today.

17.5   Changes of Interest

The W-34 was to be the preliminary highlight of Hütter's work on the field of wind energy use. When he took over the chair for aircraft construction in 1959 at what was then the Technical College of Stuttgart, he made sure the tests in Stötten were continued, but they were only of academic value. At that time the energy industry and politics had lost interest in further research or even serial production of wind power plants. The power generation costs from the use of wind could no longer compete with electricity based on fossil fuels after the victorious march of cheap oil had begun in the late fifties. The power grids were continuously improved and the capacities of power stations fired with fossil fuels were continuously enhanced. The main reason however was that many politicians and energy managers saw nuclear power as the perfect energy source for the future.

Under these premises, company Allgaier no longer could see any economic potential in the construction of wind turbines and in 1959 fully withdraw from this business segment. WEG too cancelled financing of further tests. After the Studiengesellschaft Windkraft had decided to unwind the association early in October 1964 the board of the association transferred the test field in Stötten as its heritage to the German Test and Research Institute for Aviation and Space Flight (DFVLR) in Stuttgart, where Hütter had been appointed the vice president recently. Soon, however, DFVLR too lost interest in the test field. Research was costly, and the institute was no longer willing to pay the rent for the land. "Considering that one litre of oil costs only four Pfennig, wind power was not really attractive at that time," Armbrust remembers the premature end of wind power research.

Therefore the W-34 was scraped in 1968. The village smith from the near-by village of Schnittlingen set out to disassemble the system. It was a sad moment for Sepp Armbrust, but at least he was able to save one blade of the W-34. Armbrust organised a long timber transporter, paid it out of his travel cost budget, and had the blade transported to Stuttgart. For some years he stored it in the plastics laboratory at DFVLR which he managed from 1968 to 1973. Hütter and his team dedicated the following years mainly to basic research on the field of composite materials.

Thirteen years later, Heiner Dörner set up the rotor blade as "Kunstwerk am Bau" (art on a building) in front of the faculty at Stuttgart to honour Hütter's achievements. Standing in front of the five-storey campus building, this 17 m long blade today still impresses people—not only by its size, but also because of its particular aesthetics, its slender and elegant appearance. As a "warning energy finger", Dörner says, it also points towards the sparing handling of natural resources. The shortage of oil at the beginning of the 1970s was to be the trigger for the renaissance of wind energy.

Figure 17.8 Heiner Dörner and the W-34 blade in front of the building of the Institute of Aircraft Design, University of Stuttgart, 1999 (Photo: Jan Oelker).

17.6 The Birth of GROWIAN

After a few years during which not much happened in wind energy research Americans were the first to feel the energy crisis. Searching for alternative energies they remembered Hütter's earlier works on this field. Dörner, who since 1968 had been working as Hütter's assistant at the Institute of Aircraft Design (IFB) in Stuttgart, remembers that an inquiry from the US aerospace agency NASA to buy Hütter's good old W-34 reached the institute in 1972.

Together with the National Science Foundation (NSF) NASA developed the US Federal Wind Energy Program which was adopted in 1973. NASA was responsible for development and testing of large wind energy plants. Unfortunately, their interest in Hütter's system came too late—the W-34 was no longer available. At least Sepp Armbrust was able to unearth the plans and construction documentation and they were sold to NASA for USD 55 000.

It was no coincidence that Hütter's system served as the model not only for the first wind turbine developed by Westinghouse under the American wind energy program, a 100 kW system with the name MOD-0 that went into operation in 1975. All other big American wind turbines built in the 1970s and 1980s by companies such as General Electric and Boeing were clearly influenced by Hütter's fundamental philosophy.

Hütter had some consulting contracts with NASA and later also with the aircraft manufacturer Boeing. However, soon to retire, he felt less and less desire to teach the basics to engineers on the other side of the Atlantic Ocean. "I'm not going to nurse the kindergarten of wind energy by myself," his former assistant Dörner remembers Hütter saying. Instead Hütter sent Dörner "across the ocean", who thus experienced the wind energy revival close-up.

Finally when the oil price shock had reached Europe the University of Stuttgart too resumed activities for wind energy research. This new phase was triggered in 1974 by a call from the German Ministry of Research and Technology (BMFT) to the wind power experts in Stuttgart. At BMFT people were in the process of preparing the "General Energy Research Program" for the coming four years, which was meant to analyse non-nuclear energy sources so far unused. One item to be subsidised was the "Exploration of new energy sources for large-scale technological application" and one of these was wind energy.

The Ministry invited the engineers to a meeting. So Hütter and Dörner went to Bonn, then the capital of the Federal Republic of Germany, to answer the questions of the government official Alois Ziegler. "How many megawatts can you deliver with that wind power of yours," was the first question the civil servant asked. Hütter reported on his experience with the W-34 and proposed to take a leap into the scale of one hectare rotor surface. He had a liking for grand figures. This swept area equalled 112.8 m rotor diameter and Hütter thought technologically this was taking it to the limits but still feasible. Taking into account the known yield rates, this would lead to approximately three megawatts output, he calculated. He said this dimension should be interesting for the power industry.

"What, so little?" The response of the officer showed the real dilemma between what was technically feasible and what was politically desired. While wind power use so far had experiences only up to a scale of 100 kW, politicians and power providers thought only in terms of hundreds and thousands of megawatts for the national economy.

Although the visions had been very different, this meeting gave birth to GROWIAN, an acronym for the German words for giant wind power station which was to be the core project of state-subsidised wind energy research in Germany after the oil price crisis. However this birth took place under an unlucky star as the figures went to the press the next day, where as Dörner says they were "cemented". So the dimensions of GROWIAN were defined *a priori*.

Hütter had been fully convinced that rotor blades could be produced on this scale by composite construction but he had to face a lot of critical responses to this proposal. He openly said that further research was needed before trying to technically realise these dimensions. He down-scaled his proposal within the program study "Energy sources for tomorrow? Non-fossil–non-nuclear primary energy sources" financed by BMFT. Under this study which, and this is quite savoury, had a question mark in the title—his team at the University of Stuttgart and at the DFVLR tackled Part III, wind energy.

With this study, Hütter's team estimated the potentials of wind energy in Germany and analysed the economic profitability as well as defined a draft concept for development of a large wind turbine essentially based on Hütter's experience with W-34. In the first step, the team proposed the development of a wind turbine with 80 m rotor diameter and approximately 1 MW output, and building on the experience with this design, to move forward to bigger systems later on.

The BMFT did not want to hear about this. Even if this program study served as a guideline for the development of new energy technologies, and as the foundation of the research programs of the coming years, the officials in Bonn did not want to back up the intended dimension of the prototype. At the one hand, this giant leap was considered the only way to create business interest in this project. The low density of energy and the unsteady wind conditions did not match the hard-set concepts of energy suppliers. The establishment in politics and industry was fixed on nuclear power, and did not see wind energy as a serious form of alternative energy. BMFT was convinced this sceptic attitude could be overcome only with a very big system.

There was a lot of political prestige on stake with GROWIAN. The Ministry knew of the plans of Boeing for building a system of 91 m rotor diameter, and wanted to exceed it. The goal of BMFT seemed to be proving the international competitiveness of the German industry. The biggest wind power station in the world was to be built in Germany by all means. This outlined the road to GROWIAN. BMFT appointed the Nuclear Research Centre Jülich GmbH (KFA) as general manager. The realisation of the project and the coordination of all involved companies and institutions were contracted to MAN Neue Technologien GmbH in Munich.

The research institutes in Stuttgart also were on board. The Research Institute for Wind Energy Technology (FWE) at the University of Stuttgart had been set-up for Hütter's wind energy studies and it took over the calculation of aerodynamic performance. The Institute of Structure and Design (DLR) at DFVLR, also under his direction, was responsible for the construction of rotor blades. Hütter's colleague Franz-Xaver Wortmann, Professor at the Institute of Aerodynamics and Gasdynamics (IAG) at the University of Stuttgart, took over the calculation and optimisation of the blade profiles.

However, during the project phase to deliver construction drawings to build GROWIAN there was a collision of interests between the client BMFT, Hütter as the brain giving ideas, and engineers of company MAN. Hütter's engineers favoured rotor blades in composite construction made of glass-fibre compound or carbon-fibre compound. In the last 15 years before the project they had gathered a lot of experience with composite materials, and they appreciated the benefits of what was then still quite a new material. They proposed to reduce the rotor diameter, or at least to design a test rotor blade. BMFT refused both for reasons of time and budget, and insisted on a rotor diameter of at least 100 m. The mechanical engineers from MAN in Munich shied away from this step as they had too little experience with fibre-compound materials.

Client and contractor agreed on a compromise which was to build rotor blades of 50 m length in hybrid construction. A hexagonal steel frame bar takes-up the loads. It is shrouded with a shell of glass-fibre compound providing the aerodynamic form. This solution suited the engineers from MAN but it was much heavier than a self-supporting blade of glass-fibre compound. Therefore the entire system design had to be revised.

This meant the principle of consequent lightweight construction Hütter had in mind was thrown overboard. He was no longer deeply involved in the practical realisation of GROWIAN as he had retired from the chair of the Institute of Structure and Design at DFVLR already in 1976. The Institute of Aircraft Design (IFB), where he was still acting director at that time, was no longer involved in the GROWIAN project.

17.7 Limit of Feasibility

Hütter saw better opportunities to realise his academic ideas through cooperation with Voith GmbH from Heidenheim. Wolfgang Weber, one of the last to complete his doctorate proceedings under Hütter's reign, was already working there. Together with Weber, who later was appointed Professor of Business Engineering at the University of Applied Sciences in Aalen, Hütter designed an extreme high-speed 265 kW wind turbine trying to push his light-weight construction philosophy to the maximum. The system was a 2-blade downwind turbine. The 52 m rotor diameter gave it the name WEC-52. In order to reduce weight of the nacelle, the energy was transmitted through a tapered gearbox and a fast shaft in the tower to a second gearbox, which just as the generator was easily accessible at the bottom of the tower.

When the system was put into operation in October 1981 on the test field in Schnittlingen, which had been re-opened two years before, it was the biggest wind turbine built in Germany to that date. However the test run lasted only a few months. Hütter had tried to maximise the performance coefficient of the rotors with an extremely high speed at the blade tip of more than 100 m/s. The aerodynamics of such speeds requires very slim blade profiles, and that stretched the issue of rigidity to its limits.

The extremely slim rotor blades were bending during certain critical operation modes. When braking, torsion forces affected the blades so they whipped out up to 3 m wide. When it was decided to cut back each blade by 5.5 m this was the end of the ambitious big project WEC-52. Even though the Voith WEC-52 was the star of the revived test field in Schnittlingen for only a short time, it was another gem in Hütter's crown of academic achievements. Hütter's philosophy was extended by the Voith wind turbine to the limits of what was feasible with regards to high-speed and weight reduction.

Figure 17.9 Voith WEC-52 blade and Debra 25 wind turbine on the test field in Stötten (Photo: Jan Oelker, 2000).

After having been acting director of IFB for five years after his retirement, he finally withdrew in 1980 and worked only as consultant. The research work of this visionary thinker had laid the essential foundations for the use of wind energy. From then on it was up to his many students to go farther along the road he had opened with their practice-oriented work.

17.8 Hütter's Heritage

Hütter's blade design principle was already in the late seventies adopted for the Tvind turbine in Denmark. After 1980 it became a standard for blade technology in the Danish wind industry, with thousands of blades fuelling the wind power boom in California.

When the wind energy boom started also in the north of Germany, it was the end of the great times of wind energy research in the region of Swabia. Ulrich Hütter, whose name was inseparably tied to it, and whose philosophy vitally influenced the development of wind turbines in Germany, unfortunately did not live to see the industrial breakthrough of wind energy in his home country. He died in 1990.

For Heiner Dörner the time to pass on the relay baton to younger hands came in the autumn of 2004. He retired after having introduced more than 1 000 students through three decades to the scientific principles of the design of wind turbines and the aerodynamics of rotor blades.

When Dörner looks back on his professional career he also recaptures the development on the field of wind energy. The technological breakthrough, Dörner resumes, came with the serial production of wind turbines. The Stuttgart scientists and engineers however had built the foundations on which, thirty years after the oil price shock, the big modern wind turbines were an alternative technology for power generation to counteract the once more rising oil and gas prices. The vital impulses for system development are coming from the industry. Still Heiner Dörner proudly reminds us: "The academic roots for the modern use of wind energy in Germany are embedded in the research institutes for aviation in Stuttgart."

About the authors

Bernward Janzing, born in 1965 in Furtwangen/ Schwarzwald, is a freelance journalist based in Freiburg. After having studied geography, geology and biology in Freiburg and Glasgow, he took to freelance work since 1995. During his studies Bernward had developed a growing interest in the subject of climate conservation, and renewable energies became the central issue of his work as a journalist. In 2002 he published his book *Baden unter Strom*, which documents 125 years of the history of electric power in the region of Baden where he has his roots. For his book *Störfall mit Charme— Die Schönauer Stromrebellen im Widerstand gegen die Atomkraft*, published in 2009, he was awarded the Umwelt-Medienpreis der Deutschen Umwelthilfe (the Environmental Media Award of the German Environmental Aid Association). In 2011 he published his latest book *Solare Zeiten—Die Karriere der Sonnenenergie*.

Jan Oelker, born in 1960 in Dresden, is a freelance photojournalist living in Radebeul. Ever since the first wind turbines went into operation in the German region of Saxony in 1992, Jan has been documenting the development of wind energy with his camera. He works for different magazines, companies and publishers. In 2005 he published the book *Windgesichter— Aufbruch der Windenergie in Deutschland*. For his photos on the GEO-report "Strategiespiel um die Windkraft" in 2006 he was awarded the journalists' prize "Unendlich Viel Energie" (*Endless Energy*). Photos by Jan were published in the books *Störfall mit Charme* (by Bernward Janzing, 2009), *The Nature of Wind Power* (by Frode Birk Nielsen, 2009) and *Alpha Ventus—Unternehmen Offshore* (Stiftung Offshore-Windenergie, 2010).

Chapter 18

Overview of German Wind Industry Roots

Arne Jaeger

Angerbenden 23, 40489 Düsseldorf, Germany

ajwind@web.de

18.1 Introduction

After the world economy was hit hard by the oil crisis in 1973–74 alternative ways of producing energy were sought—and found. Wind energy was one of them. In the 1970s Denmark and the United States quickly became pioneering countries concerning wind turbine research and development. Federally funded projects as well as a bright private engagement gave birth to a new industry. These very early activities made the two countries play a major role when the wind industry came to rise in the 1980s. Apart from Denmark and the United States, the Dutch wind industry had also progressed significantly during the 1980s.

In Germany main attention was paid to federal large-scale wind turbines until the early 1990s. These machines, however, all failed in the end. Federal support for private projects was low or non-existing. There was a deep lack of experience in design, engineering and operation of wind turbines. Thus, Germany had

Wind Power for the World: The Rise of Modern Wind Energy
Edited by Preben Maegaard, Anna Krenz and Wolfgang Palz
Copyright © 2013 Pan Stanford Publishing Pte. Ltd.
ISBN 978-981-4364-93-5 (Hardcover), 978-981-4364-94-2 (eBook)
www.panstanford.com

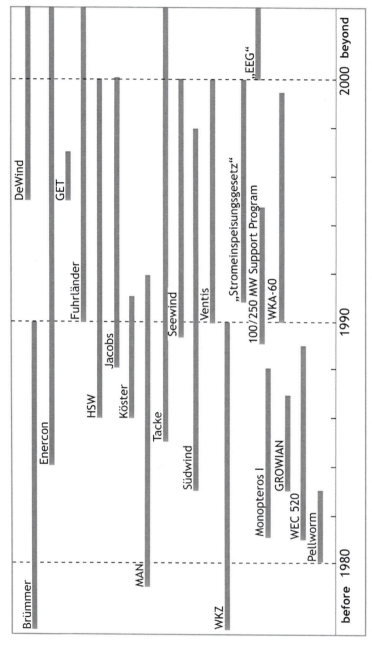

Figure 18.1 Timeline of the development and existence of federal and private initiatives in Germany in 1980–2000.

serious problems in keeping up with the other pioneering countries. In the mid-1980s nobody believed that Germany would ever rise to that extent as we know it by 2012.

Nevertheless, there were several early initiatives in Germany of which few turned out to become global companies. The combination of the private sector and federal support backed the late rise of wind power in Germany and the countries' strong position in the global wind industry of the recent years. The number of installed wind turbines in Germany exceeded 22 000 in 2011. Installed capacity grew to approximately 29 GW while the percentage of electricity produced by wind turbines approaches the 10% mark. And when looking at employment, the wind industry provided a job for roughly 100 000 people in Germany alone.

This chapter looks back to the 1980s and 1990s, and presents various early companies that made the German wind power scene. Manufacturers, wind turbine types and the market development are examined to give an overview of the origin of what would become one of the largest and important wind turbine markets ever. Though many of the names and types mentioned in the following pages are dated far back in the 1980s, it is important to avoid focusing solely on global companies but to honour smaller attempts, too.

18.2 Early Activities in the 1980s

18.2.1 Summary of German 1980s Wind Power Activities

18.2.1.1 General overview

In the early 1980s the German wind power scene consisted of two opposite parties: On the one hand the government that funded large-scale wind turbines ranging from 265–3000 kW. The main intention was to quickly develop large wind turbines for the country's supply and for the international market. At that time some European neighbours and especially the United States, where large-scale prototype machines already operated, quickly made progress in this new technology. Thus, large wind turbines based on German know-how and manufactured by German companies had to be realised in order not to drop out of the international competition.

On the other hand, there was a rising number of people who did not believe in the government's "megalomania" and heavily criticised these plans throughout the whole 1980s. For the critics, a popular argument was the lack of technological know-how to build and operate such giant wind turbines. Another argument was neighbouring countries like Denmark or the Netherlands, where small wind turbines were successfully produced in large numbers and up-scaled step by step. Many Germans believed this was the right way to succeed in wind technology. But for many years they were left disappointed: Contrary to the federal wind projects, that received enormous sums of money, all private initiatives were poorly funded, if they received external funding at all. However, few German companies like Brümmer or Windkraftzentrale (WKZ) managed to sell a considerable number of wind turbines in the first half of the decade. At the same time, a handful of Danish and Dutch machines were installed making the "wind idea" spread successfully. Like the few German machines, they created and accelerated private interest and engagement.

Around the mid-1980s the young German wind industry slowly moved towards larger serial production. The 20 kW Aeroman turbine, developed and produced by MAN, successfully found its way to many foreign countries around the world. In 1985 Tacke Getriebe entered the wind industry by producing a 150 kW machine designed by Wagner. The same year Enercon installed its first 55 kW machine with dozens to follow the years after.

While the federal large-scale projects like GROWIAN or Monopteros failed one after the other the government realised the need for funding and research of small wind turbines. Hence, a development program was set up in 1989 by the Ministry of Research and Technology. This program lasted until the mid-1990s. It created an incentive for more companies to enter the wind industry as well as the installation of single wind turbines and wind farms as pilot projects.

In the late 1980s a considerable number of German manufacturers made up the industry. Some of them designed wind turbines up to 300 kW, which was a big step for German companies. Parallel to that, more and more foreign companies like Bonus or Vestas, decided to settle in Germany not only because of its vast potential area but a rising market and a large federal subsidy program for the 1990s [1].

Figure 18.2   One of Enercon's first turbines: 55 kW Enercon-15 located in Norden, northern Germany, 2006. Note the Danish AeroStar 7.5 m blades used for the first Enercons.

Governmental activities were ceased in the early 1990s due to lack of success and a rising private sector. Despite all the negative aspects related to the federal projects it should not be left without notice that these projects still were beneficial to the German wind industry and its technological progress, respectively: The amount of experience in design, manufacture and operation of large wind turbines was very big and influential for the years to come. The insights gained during these activities were very important for the private sector. Finally, Germany impressively demonstrated the will to utilise wind energy.

18.2.1.2   Overview of federal wind projects

The following text presents an overview of greater German federal research wind projects.

Pellworm Test Site: 1980–1983

The test field on the island Pellworm in northern Germany was one of the very few federal research projects using small wind turbines. The Society for Utilising Atomic Energy in Shipbuilding and Shipping (GKSS) was asked to carry out tests on 10 different wind turbines with rated power between 5.5 kW and 22 kW. The inauguration was in 1980 and the testing period was from 1980 until 1983. Many of the German designs failed catastrophically leading the press to make fun of this test project, the announcement of GROWIAN and the non-sense of wind energy in general. After a series of crashes, modifications and repair, only two machines, Brümmer and MAN, managed slightly to succeed. After the test project ended, some companies continued development, some stopped and others sold their designs to foreign (larger) companies. A remarkable fact was the test of a Danish Wind Matic 22 kW that was included in the project—with no surprise it succeeded and operated "out of competition".

Voith: 1981–1988

The German gear manufacturer Voith was ordered to develop and build a medium-size wind turbine around 200–300 kW. Ulrich Hütter was involved in Voith's design team. Thus, Voith decided to build upon the advanced and successful W-34 (100 kW) machine from 1954, also developed by Hütter. WEC 520, as it was called, had two pitchable blades, a 52 m downwind rotor and both gear and generator integrated inside the tower on the ground. Final rated power was 270 kW.

The progressive turbine was placed at Schnittlingen, southern Germany. Inauguration was in mid-1981. Once, during test operation, the rotor was fully stopped—mistakenly. The consequence was that both tips were splintered and the rotor was shortened to 45 m diameter. Test operation was continued. However, in the mid-1980s, general interest in wind turbines was low in Germany, what made Voith exit the wind energy sector [2]. WEC 520 was in operation until 1989. Very short before its scrapping, the Mannheim Museum for Engineering and Work decided to take over and keep the machine for museum purpose.

Monopteros I: 1984–1988

Monopteros I (Greek, *monopteros:* single-arm) was the first single-bladed wind turbine in the medium-size 400 kW-class. It was placed in Bremerhaven–Weddewarden, northern Germany. Test operation began in 1984. Federally funded, this machine was designed and built by Messerschmidt Bölko Blohm (MBB) of Munich. The 30 m rotor had a pitchable blade running downwind. Yawing was active. In 1985 a massive failure of the whole machine occurred that ended with the runaway of a part of the blade. The wind turbine was never put back into operation and the test program was ceased in 1988.

Like GROWIAN, Monopteros I badly damaged public perception and image of wind energy in Germany and led to a rising disinterest in this new technology.

GROWIAN: 1983–1988

The 3 MW GRoße WIndenergie ANlage (GROWIAN) was the most famous, most discussed and criticised German federal research project in the wind energy sector.

GROWIAN was actually supposed to be a test machine for later serial production. On the one hand, Germany should have had a strong symbol for its will to support renewable energy. On the other hand, a competitive product was to be created to make Germany play a major role in the aspiring new wind industry.

The 100 m two-bladed downwind machine, placed at Kaiser–Wilhelm–Koog, northern Germany, was inaugurated in October 1983. But in 1984, the teetered hub construction proofed wrong and forced the turbine to remain inoperative, which was the case for most of the time. For uncountable times engineers had to carry out repairs, modifications or other attempts to bring GROWIAN back to operation.

Meanwhile GROWIAN became a symbol for failing German technology. It was heavily criticised by various industry experts, right from the beginning, for its size and overall concept. The German press constantly reported about the "event" and the "big wind failure".

Showing that wind turbines are non-sense was the hidden intention of the government, some argued. Whatever engineers decided to do, they were unable to keep the turbine running.

Hence, for a couple of years it seemed doubtless that wind does not work and the focus should be on conventional energy resources.

Figure 18.3   GROWIAN wind turbine (Photo: Flemming Hagensen).

The loss of face for the German wind industry was enormous. But when the test project was ceased in 1988, GROWIAN supporters had few but strong arguments: Firstly, lots of valuable information and experience was won regarding production, installation, operation and grid-integration. Secondly, GROWIAN proved that large wind turbines could be built. And thirdly, it was clear for everyone that unknown technical territory was entered. Thus, failures and problems were no disgrace but necessary.

Future research turbines, for example, WKA-60, would incorporate the know-how gained in the GROWIAN project.

GROWIAN II/WKA-60

One of the last greater federal research projects was a three-bladed, pitchable upwind turbine rated at 1.2 MW—WindKraftAnlage-60 (WKA-60). It was the direct successor of GROWIAN, and included the know-how gained with GROWIAN.

Three units were built of this type. The first one was inaugurated in 1990 on the famous island Helgoland, located in the German North Sea. It lasted until 1995, due to various lighting strikes and other technical problems.

The second unit was installed on the GROWIAN foundation in Kaiser–Wilhelm–Koog in the year 1992. This machine operated until 1997. Since 1999 it served as an attraction for EXPO 2000 visitors and other people interested in renewable energy technology. Nacelle and rotor were placed on the ground and prepared for regular viewing.

Figure 18.4 One of the three GROWIAN II/WKA-60 units for research application rated at 1200 kW. The turbine located at Kaiser–Wilhelm–Koog, northern Germany, 1997. Note the working platform and the voluminous design (Photo: Arne Jaeger).

A third unit was placed at Cabo Vilano, Galicia, northern Spain, together with Spanish partners. The overall aim of these machines was to gain more experience on large-scale wind turbines operating in grid-connection.

18.2.1.3 Influence of Ulrich Hütter

Apart from many other persons who did important and influential research in Germany, the wind energy pioneer Ulrich Hütter should be emphasised for his remarkable technological influence. In the 1950s he developed a highly progressive wind turbine that pointed the way. It was a two-bladed downwind machine rated at 100 kW. Equipped with pitchable blades and a guy-wired tower, the W-34, as it was called, served as inspiration for many commercial applications that came to life eventually. Hütter defined an approach to wind turbine development that lasted for decades: Designing wind turbines on the basis of aerospace know-how. This approach seemed to be practical for his developments as well as for many of the federal research projects of the 1970s and 1980s.

Especially in the United States, the aerospace approach was practiced intensively. Hütter's W-34 found successors in the federal MOD-0 research turbines. In the 1980s American manufacturers like ESI or Carter produced thousands of two-bladed downwind turbines that strongly followed the W-34. In Germany federal projects like WEC 520 or GROWIAN, all incorporated what Hütter did first.

However, it should be noticed that the aerospace approach failed right away. All wind turbines, be it federal projects or commercial applications, suffered massive technical problems or ended up in a total disaster for one particular reason: It is not possible to transfer know-how from aerospace designs onto wind turbine designs. The technical and physical differences were too big. One of the biggest lessons learned in wind turbine technology was that wind turbines need their own specific designs based on appropriate, intensive long-term research.

A second more successful Hütter development found its way to commercial production transferred *via* Tvind. In the mid-1970s, the famous Tvind international school started getting experience in blade manufacturing. They decided to use a blade root designed and applied by Ulrich Hütter in the 1950s. The blades of Tvind's first 15 kW experimental machine incorporated this root construction.

Some Danish blade manufacturers (Økær and Alternegy) continued using that blade root system. In detail the root system consisted of rovings that came from the inner blade root, revolved around the bushings and went back to the blade root. Several thousands of blades using the original and an advanced version of the Hütter root were produced in the 1980s. After the 1980s biggest blade manufacturer, Alternegy, went bankrupt there was no other blade or wind turbine producer who continued using the Hütter system. Despite, after more than two decades, still thousands of these blades are running successfully all around the world.

18.2.1.4   Early foreign influence from Denmark

By 2012, Germany is a global major player concerning wind turbine know-how and technology, and points the way for future developments. Thirty years ago the situation was totally different. Neighbouring countries like Denmark or the Netherlands, were far more advanced and, hence, their influence on German wind energy activities was just a matter of time. Especially the Danish played quite a role. Since 1981, Danish blade manufacturers like Økær, later licensed to Alternegy, delivered their blades to several self-builders or small companies in Germany. A famous example was Aloys Wobben, founder of Enercon, who in 1981 bought a set of three Danish Økær blades (5 m length) for his 15 kW homemade turbine. When Enercon produced first 55 kW machines (type: Enercon-15) between 1985 and 1986, Danish Alternegy blades of 7.5 m length were chosen. By 2012, these first Enercon turbines are still running in northern Germany.

Another example of early Danish influence was the cooperation between German Windkraftzentrale (WKZ) and Danish S. J. Windpower, who serially produced wind roses for heating purpose. WKZ sold these wind roses under the name "Elektromat 12 kW". Since 1985, WKZ used Alternegy blades for its Elektromat 20/25 kW and 90 kW model.

Another highly famous example of early Danish wind turbines in Germany is the Vestas V15-55 located at Cecilienkoog, northern Germany. The Danish 55 kW turbine was set up in 1982 by the farmer Karl Hansen and has since become a very strong symbol for early pioneers in Germany. A further case were three Danish 18.5 kW Kuriant turbines that were placed at Grohnwoldhusen, northern Germany. They represented one of the first wind farms in Germany. These early applications were followed by dozens of Danish units since 1985 and thousands to be erected in the 1990s and beyond.

Figure 18.5 An early WKZ 12 kW Elektromat located near Bad Oyenhausen, Germany, 2003. The design was taken over from Danish S. J. Windpower who went bankrupt in 1980 (left); A WKZ 25 kW Elektromat unit near Büsum, northern Germany, 2000. Note the tail-vane and yellow tips. The blades were manufactured by Danish Alternegy (right) (Photo: Arne Jaeger).

18.2.2 German Pioneers

This chapter deals with pioneer companies in Germany of which most came up in the 1980s. For many the 1973 oil crisis and a greater independence of conventional energy sources were the main motivation to found a company selling wind turbines. For already existing firms diversification was the decisive reason to seek for new industries. When wind energy slowly started to rise in the second half of the 1980s some companies realised the big opportunity that was on the horizon. Parallel small wind turbines became more and more popular since the large-scale machines proved wrong.

Brümmer: 1974–2000s

Herman Brümmer can truly be called a "real pioneer". In 1961, Brümmer started repairing small wind turbines in the neighbour-hood. Soon he developed his own wind turbine concept that featured very simple, three-bladed downwind machine, yawing passively. Rated powers varied between few kilowatts and 15 kW. Throughout the 1970s and early 1980s Brümmer sold some hundred units, most of them to German customers. In 1980 a 15 kW Brümmer turbine successfully underwent tests at the Pellworm site. Larger machines up to 200 kW were developed. However, with the advent of grid-connected wind turbines in the 1980s, Brümmer lost market shares, since his machines only operated as stand-alone units while no grid-connected machines were ever designed. In 1990 the last Brümmer unit was sold. Despite, the turbines were marketed until the 2000s.

Figure 18.6   Early Brümmer turbine near Altenbeken, Germany, 2006. Note the simple blades of welded steel and two sections, the guyed-pipe and the downwind rotor (Photo: Arne Jaeger).

Enercon: 1984–2012

In 1984, Aloys Wobben founded a company that would become one of world's largest and the most influential ever in the wind industry. After building two self-made turbines (a wind rose and a 15 kW three-bladed machine), the company Enercon was founded, that specialised in frequency converters and wind turbines. In 1985, the first 55 kW Enercon-15 (15 m diameter) was set up at Aurich, northern Germany. Very quickly, this turbine became a focus of attention, for its excellent design and production values as well as its high grade of technological maturity. These features led to a fast rise of Enercon turbines. In 1986 the first wind farm, comprising four Enercon-15 on the island Norderney, was realised followed by another three wind farms the year after. While the first Enercon-15 machines still used Danish Alternegy blades, the successor, Enercon-16, already employed the company's own blades—a tradition that is still being practiced. Enercon quickly became a popular name in the circles of the German wind turbine industry and abroad. In 1988 a 300 kW prototype (Enercon-32) was erected at Pilsum, northern Germany. It was a big step for German conditions at that time. Enercon constantly managed to turn old well-proven ideas into serial production. Be it the serialisation of ring generators in 1993, when the E-40 clearly revolutionised the development of wind turbines. Or the adaptation of the old 1920s blade designs since 2003 that were developed by Albert Betz.

Figure 18.7 Enercon's 500 kW E-40 near Leer, northern Germany, 2008. Thousands of E-40 working with a ring generator were sold (Photo: Arne Jaeger).

Figure 18.8 Panoramic view of Enercon-32 turbines (300 kW) at Wremen, northern Germany, 2011. Note the stepped greens painted on foundation and lower tower sections for improved integration with the landscape (Photo: Arne Jaeger).

Enercon offers an unconventional wide range of wind turbines between a few kilowatts and 7.5 MW. The company symbolises the success story of German wind technology and belongs to the group of global wind turbine players. More than 10000 units have been sold worldwide.

Husumer Schiffswerft (HSW): 1986–2000

HSW was an old shipyard from Husum, northern Germany, with huge experience in both craftsmanship and handling the powers of nature. In 1986 the company decided to enter a new business field, because of a serious crisis in the shipyard industry. Wind energy was recognised as a rising renewable energy sector with a large future potential. As the first step, a 30 kW two-bladed machine was developed together with an engineering company [3].

The second step was a 200 kW wind turbine equipped with a 25 m upwind rotor. HSW adopted the Danish line concept that featured a stall-regulated rotor with three blades, and a heavyweight machine construction. In 1988, HSW-250, the successor was introduced and serial production began. Of this type 100 units were sold, 50 of which were chosen for the famous "Nordfriesland Windpark" close to the Danish–German border on the west side. This wind farm became a very strong and popular symbol for successful wind energy utilisation in Germany.

In 1993, a 750 kW pitch-regulated model was set up at the test station Kaiser–Wilhelm–Koog. It created huge attention, since it was seen as a pre-stage to megawatt wind turbines. This became reality when two years later HSW built a wind farm of four HSW 1 000 at Bosbüll, northern Germany.

The shipyard was said to have filed large orders, when in 2 000, it suddenly declared bankruptcy and vanished into REpower in 2001, a large merger of the manufacturers HSW, Jacobs and BWU and the engineering firm pro+pro.

Figure 18.9 Husumer Schiffwerft HSW-250, Burg/Fehmarn, 2011. Note the coloured blade tips typical for HSW turbines (Photo: Arne Jaeger).

Köster: 1986–1991

Obviously there is no other German company having such a long tradition in wind energy like Köster. Originally specialised in agriculture and farm equipment, Köster included wind turbines in the product line around 1900. The Adler wind turbines had a very good reputation in Germany and abroad for their reliability, productivity and long-lasting design. The Adler machines mainly drove saw mills or pumped water. When northern Germany underwent an electrification program after the Second World War, Köster ceased wind turbine production. However, in the 1980s the company rediscovered wind as an opportunity. In 1986, Köster signed an agreement with the Deutsche Luft- & Raumfahrttechnik (DLR), Stuttgart, to produce a wind turbine prototype developed by DLR. This wind turbine was called Debra-25 and was rated at 100 kW. Ulrich Hütter was responsible for its development.

Köster continued improving this machine and renamed it Adler-25 of which a dozen were sold in northern Germany. Once again, Köster ceased production in the early 1990 due to bankruptcy of the blade supplier.

Figure 18.10   Köster 165 kW Adler-25 wind turbine located at Barlt, Schleswig–Holstein, 2007. Note the downwind rotor and the maintenance platform outside the nacelle (Photo: Arne Jaeger).

MAN: 1979–1992

In the late 1970s The Federal Ministry of Technology and Research (FMTR) hired MAN to design a small wind turbine simultaneously to the government's ambitious GROWIAN project carried out by MAN, too. The small wind turbine was called "Aeroman" and the very first unit consisted of several design aspects typical for Hütter: Two blades, downwind, pitch-regulation. In 1980, a unit was sent to Pellworm for undergoing tests. What followed was a series of failures, complete destructions and modifications. Nevertheless, the first type, Aeroman 11/11, still performed as one of the best at Pellworm. MAN developed the turbine further and changed various details. A wind farm on the island Kythnos, Greece, was equipped with five Aeroman in 1982.

Figure 18.11 MAN 20 kW Aeroman wind turbine located at Rödemis Hallig, Husum, 2011. Note the tail-vane (Photo: Arne Jaeger).

The second generation, Aeroman 20, had an upwind rotor and a 20 kW generator. The original lattice tower was replaced by an 8-edged tubular tower. A tail-vane yawed the nacelle.

In 1984, the FMTR started a distribution program for 20 Aeroman units to be spread across northern Germany. For this demonstration program incentives for private were offered. The Aeroman machines generally performed well. Some of these 20 can still be seen in operation. By the mid-1980s, more than 300 Aeroman machines were shipped to Alaska, California and many other locations worldwide. For American locations the rated power was increased to 40 kW. The turbine proved very successful. Aeroman units can still be seen in California.

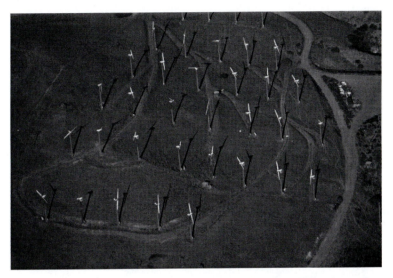

Figure 18.12   Aerial view of an Aeroman wind farm at Tehachapi Pass, California, 2010. More than 300 machines were shipped to California in the mid-1980s (Photo: Arne Jaeger).

MAN continued production until the late 1980s when the company decided to give up the wind energy sector. Through daughter company Renk, MAN had acquired Tacke Getriebe. The license was transferred to Jacobs Energie, Heide, northern Germany. Jacobs up-scaled the turbine to 33 kW and further modified it. Of the new 33 kW turbine some dozens were produced, mainly for customers in northern Germany. Like many of the 1980s Aeroman, the Jacobs turbines are still operative today.

Tacke/GE: 1985–2012

The story of well-known GE Wind Energy begins in the mid-1980s when Tacke was offered a license production by Dr Günter Wagner, who had developed advanced wind turbines for some years. Wagner built a handful of prototypes and was seriously interested in serial production of his wind turbines and claimed to have contacts to American developers. Tacke then produced a prototype installed at Tehachapi that eventually failed. Fifteen improved units were then built for a wind farm at Tehachapi Pass, California. However, these units failed the years later due to poor design. They were re-bladed with Danish LM blades.

Figure 18.13   Tacke 150 kW and 600 kW turbines in the Tehachapi Pass, California, 2010 (Photo: Arne Jaeger).

Tacke stopped cooperation with Wagner and decided to develop its own wind turbines. Parts of the company Tacke, including the gear division, merged with Renk, a subsidiary of MAN. The new RenkTacke developed the 150 kW design further. By 1989 Renk-Tacke had two types serially produced: A 45 kW and a 150 kW turbine. Of both a handful were manufactured. In 2011 some of them still produced electricity.

In 1990 a new company called Tacke Windtechnik was founded. The Former Renk was sold to MAN who was no longer interested in supporting RenkTacke's wind energy activities. The new Tacke successfully build up a line of wind turbines ranging from 60 kW to 1.5 MW. From 1989 to 1992 Tacke cooperated with the Danish Folkecenter for Renewable Energy who delivered the know-how for an integrated drive train, that is gear and generator make up a closed unit in the nacelle. This construction was first applied in 1992 in the TW 500 (500 kW) prototype installed on Borkum Riff, and one year later in the very successful TW 600 (600 kW). Tacke sold several hundreds of its TW 600 worldwide. For many years Tacke was the second largest German manufacturer. Although the firm was able to compete in the megawatt-range it went bankrupt in 1997. The same year business was taken over by American Enron. A "new Tacke" came up, called Tacke Windenergie. After struggling with financial problems, two years later it was renamed Enron Wind which collapsed, and in the end, was taken over by General Electric in 2002. The wind division is called GE Wind Energy [4].

Figure 18.14   Tacke TW 600 units at Utgast, northern Germany, 2009. This wind farm was Germany's biggest in 1996 (Photo: Arne Jaeger).

Südwind: 1983–1998

Südwind was founded in 1983 in West Berlin and represents the first wind energy activities in Eastern Germany. Designed and built by a handful of young engineers a first 15 kW turbine was put up in the western part of Berlin in 1983. This was a three-bladed downwind machine with passive yawing, a guyed pipe and a "schlaggelenk" rotor that was rarely used on wind turbines. In the late 1980s, Südwind managed to sell a couple of single wind turbines to various owners in northern and western Germany. Those were commonly 10, 15 or 30 kW turbines of which some were funded, for example, within the 1989 support program. The 15 kW design, called E710, also underwent tests at the Kaiser–Wilhelm–Koog test site. In the late 1980s a 30 kW machine was built that was based on the previous 15 kW design. This larger machine was also tested at the DEWI test site near Wilhelmshaven. Some dozens of the 30 kW and 45 kW, an up-scaled version, were sold. In the early 1990s Südwind was searching for new designs for larger wind turbines in the 200/300 kW class. It came from the Vind-Syssel, which manufactured various sizes designed by the Danish Folkecenter for Renewable Energy. The design used had an integrated drive train. Südwind applied this design and developed a 270 kW stall-regulated wind turbine that was marketed since 1993. Later a 330 kW up-scaled version followed. Of both the 270 kW and 330 kW some dozens of units were sold to German customers. In the mid-1990s, Südwind sold licenses to Indian Suzlon, which produced several hundreds of the 330 kW turbines for Indian wind projects.

Figure 18.15 A 30 kW Südwind turbine near Vreden, Germany, 2008. Note the guyed pipe and the downwind rotor (left); A Südwind S-46 600 kW turbine located near Lichtenau, Germany, 2008 (Photo: Arne Jaeger).

During the 1990s, Südwind failed to keep up with the ever-growing demand for larger wind turbines. However, a 600 kW design followed in 1996. It was a pitch-regulated machine with a 46 m rotor, called S-46, and turned out successfully. Despite that, in 1998, Südwind declared bankruptcy. The company had a 750 kW up-scaled prototype of the S-46 (S-50) running at the Grevenbroich test site. But rights were sold to Nordex Balcke-Dürr in north-eastern Germany. Few years later Südwind just served as a label in the product line of Nordex.

Windkraftzentrale: 1976–1990

Apart from Brümmer, Windkraftzentrale (WKZ), founded by Horst Frees, was the second pioneer whose roots go back to 1976. At that time, Frees marketed Danish wind pumps in Germany and developed own small battery charger with two blades, which was very successful and counted some hundreds of units sold. Around 1980, the firm began producing wind roses on the basis of a Danish design. These had a rated power of 12 kW and were called Elektromat.

In 1985, WKZ introduced its own 20 kW Elektromat. A 3-blader, which strongly followed the Danish design philosophy. This machine was quickly up-scaled to 25 kW. A dozen of Elektromat 25 kW was sold. Around 1989, WKZ managed to build a larger machine rated at 90 kW of which few were put up [5]. In 1990, Frees suddenly died, leading to the disappearance of WKZ. The Elektromat design was acquired and developed on by Fuhrländer, a well-known German wind turbine manufacturer in 2012.

Horst Frees was very influential for the German wind energy development. He constantly pushed wind projects, fought authorities and tried to improve the image of wind energy throughout the 1970s and 1980s.

18.3   The Wind Boom of the 1990s

18.3.1   Summary of German 1990s Wind Power Activities

After a series of failed federal wind projects and an ever-growing market for small wind turbines, the Federal Ministry of Research and Technology (FMRT) decided to start a support program for 100 MW installed capacity across the whole federal republic—privat wind projects, that is, single turbines and wind farms were receiving subsidies. This program was started in 1989. The scheduled capacity of 100 MW was quickly reached due to a very large demand for small and medium wind turbines up to 300 kW. Hence, the support program was extended to 250 MW in the years to come. A second even more important event was the 1991 "Stromeinspeisungsgesetz" ("Law on feeding electricity into the grid"). It secured a minimum pay for produced electricity, enabled smaller producers to connect to the grid and sell their power to larger utilities. The law and various state incentives strongly accelerated wind development.

Motivated by a growing market, lucrative subsidies and a rising acceptance of wind energy, more companies entered the market. Wind energy slowly lost the negative image from which it badly suffered in the 1980s. For some areas, especially in the economically weaker northern parts of Germany, the new industry created jobs and gave additional source of income. For politicians the new source of energy became more and more interesting to deal with, what partially resulted in a strong political backing.

In the 1990s, competition seriously rose on the market. Now the leading factors were rated power, reliability and economics. Thus, wind turbines grew faster and faster. In 1993, all major companies had a serial machine around 500/600 kW on the market. HSW, for example, was the first to even introduce a 750 kW model the same year. Only three years later the fight for the MW-class began! The first Germans to introduce a 1.5 MW prototype in 1995–1996 were Enercon and Tacke. The following year the first wind farms were equipped with MW-turbines.

The high speed of ever-growing wind turbines had a "market consolidation" as a consequence. Since 1993, a couple of companies either left the wind business or went bankrupt. They were not able to compete with the other established firms in terms of financial and technological power. Furthermore, technological development crystallised into one technical concept: three blades, upwind, pitch-regulated, tubular tower. With few exceptions and modifications, this technical concept survived until the recent years.

A supply chain was also created. It involved mainly blade manufacturers but other component producers as well. Hence, wind turbine companies enjoyed a great benefit. They could rely on proven components that were easier to get and did not have to risk the development of own components—at least of most important and sensible components like blades or gearboxes.

The 1990s also saw Germans doing business in foreign countries. Export became a valuable factor. German wind turbines spread to neighbouring countries in Europe or to the Asian area. Selling complete licences became a common practice, too.

Around the millenium, Germany aspired to a leading country for wind energy. More than 7 500 units were installed around 1999 with an installed capacity of roughly 4 400 MW.

18.3.2   New Manufactures

In the following part newer companies from the 1990s are briefly described. In contrast to the 1980s, a supply chain of reliable and proven components existed that newer companies could make use of. The other side of the coin was a rapid development and the demand for bigger turbines. Lack of capital and know-how and the big advance established firms enjoyed, resulted in many companies going bankrupt.

DeWind: 1995–2012

DeWind was founded in 1995 by a couple of former Ventis employees. A first 500 kW machine was set up in 1996. This machine, called D4, was then up-scaled to 600 kW and became a huge success. Two years later, a 1 MW machine (D6) with a 62 m rotor diameter was presented to the market but soon was succeeded by an up-scaled version with a rated power of 1.25 MW. Like the D4 machine, the D6 sold hundreds of times. In 2002, DeWind installed a prototype of its new D8, a 2 MW turbine, designed by the famous car manufacturer Porsche. In 2006, DeWind started selling wind turbines with various frequencies. Hence, a 50 Hz and a 60 Hz version were offered and new markets could be entered. A new 2 MW turbine, D9, was presented in 2011. This machine has still a rated power of 2 MW but uses a 93 m rotor. Thus, it makes operator's benefit with a considerably higher yield. A 3 MW turbine is under development and supposed to be installed in 2012.

In the story of DeWind, the financial situation was troubled for a couple of times. DeWind was listed on the stock market. In 2002, the British company FKI bought all DeWind shares and strongly influenced the further development. Few years later, FKI lost its interest and DeWind was sold to another British/Indian company called EU Energy/Shiram. In mid-2006, the American Composite Technology Corporation took over the DeWind turbine business. However, another change was made when in 2009, the South Korean Daewoo Shipbuilding & Marine Engineering (DSME) acquired 100% of DeWind.

Figure 18.16   Looking up at 1.25 MW DeWind D6 turbine at Lindchen, Germany, 2008 (Photo: Arne Jaeger).

Fuhrländer: 1990–2012

Fuhrländer originally was founded by a local smith Theo Fuhrländer in the early 1960s and was located at Waigandshain, central Germany. The son, Joachim Fuhrländer, got in contact with wind energy in the 1980s in northern Germany. After discovering wind turbines manufactured by Windkraftzentrale (WKZ), Fuhrländer seriously thought about entering the young wind energy market that seemed to offer a good future prospect. When WKZ stopped all activities in 1990, Fuhrländer took over what was left. The 25 kW Elektromat turbine, a WKZ development, was developed further.

Additionally service for the remaining WKZ units was carried out. Meanwhile, a 100 kW turbine was designed and represented the company's step into its own wind turbine development. In 1994, a 250 kW model followed. Both types together sold about 40 times. In 1996, the FL 800 rated at 800 kW was introduced. It actually was seen as the step into the MW-class, that was realised in 1998, with a 1 MW turbine called FL 1000.

Until 1999 Fuhrländer preferred using stall-regulated machines. That changed when the engineering firm pro+pro was looking for potential companies to serially produce its own 1.5 MW turbine called MD 70. Fuhrländer started a license production of pro+pro's MD 70 and later MD 77. In 2000, the company was changed to a stock corporation but without being listed on the stock market. A further step was taken when Fuhrländer took over the business of Pfleiderer, who developed pitch-regulated 600 kW and 1.5 MW turbines but never realised serial production. The former Pfleiderer turbines were developed further and are still part of the Fuhrländer turbine portfolio. These types have been made on license in central countries, China, Brasil, the United States, etc. In 2005, Fuhrländer introduced another innovative turbine: The FL 2500 (2.5 MW) incorporated a new drive train construction with a three-row roller bearing for the rotor instead of two bearings on a single shaft. In the recent years Fuhrländer moved its main production facilities to a larger industrial area close to the airport of Siegerland, since the Waigandshain location became too small for manufacturing multi-megawatt turbines. The production of the new FL 3000, planned for 2012, already benefits from this new location.

Seewind: 1989–2012

Seewind was a company from Walzbachtal–Jöhlingen, southern Germany. Gerd Seel, the founder, developed a wind turbine for lower wind speed areas. By 1990, he had completed a prototype, set up near Walzbachtal. It was a 110 kW three-bladed upwind machine equipped with feathered blade tips to increase rotor output, which was unique at that time. Seewind produced almost all components on its own. The 110 kW machine was the company's most successful design with more than 60 units sold. By 1997, a 750 kW model was developed but few prototypes were erected.

Soon after Seewind started cooperation with Danish WindWorld on a 750 kW machine WindWorld had developed. Some dozens of this type were sold. When WindWorld was merged into NEG-Micon this work came to an end. Furthermore Seewind was unable to design larger wind turbines. As a consequence, the company changed to a service supplier responsible for creating road access to wind farms or servicing wind turbines.

Figure 18.17 A 110 kW Seewind turbine (Photo: Arne Jaeger).

Jacobs: 1988–2001

Jacobs Energie was located in Heide, northern Germany. In the late 1980s Jacobs produced Aeroman turbines licensed by MAN. In 1992, when MAN ceased all wind energy activities, Jacobs developed the Aeroman turbine further, increased the rated power from 20 kW to 33 kW and offered 31 m concrete towers. Of this model some 20 units were sold in Germany. Plans came up for development of an own wind turbine. In 1994, a 500 kW prototype turbine with a 37 m rotor was installed followed by a dozen of units the years later. The successor was a 600 kW machine of which more than 30 were sold in Germany. In 1998 a 1.5 MW prototype, called MD 70 [5], was installed at Kaiser–Wilhelm–Koog for test purpose. When HSW went bankrupt in 2000, Jacobs took over all wind energy activities of HSW. In 2001 Jacobs vanished into REpower.

Figure 18.18   Jacobs Aeroman 14.8-33, successor of the famous Aeroman 20 kW, Schleswig–Holstein, 2007 (Photo: Arne Jaeger).

Figure 18.19   A 225 kW GET Danwin machine near Aurich, northern Germany, 2009 (Photo: Arne Jaeger).

Gesellschaft für Energietechnik: 1992–1997

GET was founded in 1992 and first concentrated on development and sales of the Danish DanWin-27 (225 kW), of which a dozen was sold in Germany. In 1995 a 600 kW machine was presented— followed by another dozen of units sold. GET cooperated with the shipyard HDW-Nobiskrug, located at Rendsburg. In 1996 GET took over all wind energy activities of Autoflug, a German company involved in developing two-bladed wind turbines [5]. But in 1997, the company went bankrupt and disappeared from the wind industry.

Ventis: 1988–2000

In the late 1980s, Schubert, a company for electronics, decided to establish a subsidiary for wind energy. This new company was called Ventis Energietechnik, located in Braunschweig.

Figure 18.20 Ventis V20-100 near Nordenham, northern Germany, 2011 (Photo: Arne Jaeger).

Ventis' turbine was an upwind, pitch-regulated 2-blader, rated at 100 kW. Some dozens of the V20-100 were sold. Ventis equipped the first wind farm in Egypt with its V20-100. In 1995, the company went bankrupt. Though, the 500 kW turbine was already undergoing tests at the DEWI test site in Wilhelmshaven, the firm had to fight with blade and electronic regulation problems in numerous units of the 100 kW model. Few years later, Ventis Energy was founded, but again went bankrupt shortly afterwards. A last attempt was made in 2000, with Ventis AG, listed on the stock market, but without any success at all.

18.4 The Success Story of Wind in Germany

The unstoppable development of wind energy in Germany was not pure coincidence. There are a number of factors that contributed to the rise of that new industry. The most important factors (the offshore aspect, last point, was added concerning future developments) are as follows:

18.4.1 Constant Research and Innovation

Intensive research and tests play a leading role in the German wind energy development. Since the early 1980s, there were huge actions undertaken in wind energy research. Most of these federal research projects failed catastrophically, but still they demonstrated which technical approaches were right and which were not.

The market for wind turbines was filled with various technical concepts. The famous test sites in Pellworm and Schnittlingen during the 1980s and Kaiser–Wilhelm–Koog, since 1990, constantly made the private sector learn lessons and led the industry find out about those concepts existing by 2012. In Kaiser–Wilhelm–Koog alone, more than 20 different wind turbines underwent tests under though conditions.

In the mid-1990s, another test site was opened at Greven-broich, for large megawatt and multi-megawatt turbines. Until 2012 more than 10 machines (including foreign ones) have been tested.

Further work was done by the German Wind Energy Institute (DEWI), which in the 1990s, operated an own test field at Wilhelmshaven for machines up to 1.5 MW. In the last years, DEWI also set up another test site close to Cuxhaven, for multi-megawatt machines. Additional tests were carried out on a couple of wind turbines at various sites.

Figure 18.21 Various large-scale megawatt machines undergoing tests at the Grevenbroich test site, 2010 (left); The test site of the German Wind Energy Institute (DEWI), near Wilhelmshaven, 1999. Various machines, from left to right: AN Bonus 2 MW, Nordwind 750 kW, Vestas 1.65 MW, HSW 30 kW (right) (Photo: Arne Jaeger).

Another aspect, typically associated with the German industry, is innovation. Many technical trends were set in Germany. Hütter's wind turbine W-34 of the 1950s might reach back a very long time, but represents the first trend. His concept often inspired manufacturers from all around the world, for example, American companies like Carter or ESI.

A typical innovative company was and still is Enercon. When the E-40 was commercialised in 1993, the ring generator concept clearly revolutionised both technological development and the market. Though wind turbines with a ring generator were nothing new, the company managed to produce such machines serially. More than 10 000 Enercon machines have been set up globally. There were more firms trying to succeed with a ring generator. In 1999, a company called Genesys erected a 600 kW machine in southwestern Germany. However, the turbine remained a prototype. In the more recent years, Vensys built a 1.5 MW turbine with a ring generator, of which a prototype was set up at Grevenbroich.[1] This new design proved very convincing to Chinese

---

[1]For more information see chapter *Direct Drive Wind Turbines* by Friedrich Klinger.

company Goldwind, which eventually took over Vensys and now manufactures this turbine serially.

Figure 18.22 The 1.5 MW Vensys turbine at the test site in Grevenbroich, 2010. Note the clearly visible ring generator (Photo: Arne Jaeger).

The previously mentioned examples are just few out of many innovations. Generally, German wind turbines are very popular among foreign investors. They are praised for their reliability, economic viability, safety and pleasant design. All of which are a result of permanent research of the past and visions of the future.

18.4.2 A Strong Support Program

The 100 MW federal support program of 1989 was the first very successful step into a larger utilisation of wind energy. Hence, it was extended to 250 MW. Some German states offered further incentives and subsidies to accelerate wind development. Such typical states were the northern federal states Schleswig–Holstein or Lower Saxony. Without surprise, wind development was most intense there.

Above all, "Stromeinspeisungsgesetz"—the law that secured a minimum pay for feeding the grid with electricity and easier grid-connection became active in 1991. The wind sector has benefited most, since almost all costs could be covered. Since that time, a tumultuous development took place especially in northern Germany where additional incentives from the states were guaranteed.

In 2000, the "Erneuerbare–Energien–Gesetz" (Renewable Energy Law) was passed. It secured minimum pays for other alternative energies and generally helped all renewable energies to rise in Germany.

18.4.3   Political Support

Wind energy always received political backing. On the one hand, this was and still is a real opportunity for the wind industry. On the other hand, this dependency on politics is one of the biggest problems.

In the 1980s it was the government's decision to make a first step towards wind energy utilisation and a package of large-scale wind turbines was realised. Though, many criticised that there was no real political will, but the desire to show that wind energy is senseless, what seemed to be true when looking at all the failed projects. Contrary, in northern Germany, politicians on state or regional levels quickly realised the opportunity wind energy offered. Clean energy, jobs, a better infrastructure, to name a few. A dialogue between politics and the young industry was started. Consequently, lots of initiatives were taken on federal state level that made wind energy development benefit.

In the 1990s the situation improved enormously. Now, a much brighter and more influential political backing was given and an unprecedented increase of wind turbines was the result.

The northern German states served as an example for the rest of Germany. Though, wind energy development was not as fast as it was in the north, almost all states are willing to support wind energy and a considerable amount of power has already been installed.

18.4.4 Reliable Foreign Wind Turbines

Without doubt there were a couple of early German firms trying to develop reliable and economic wind turbines. However, without installation of the Danish and Dutch machines that were present in Germany since the early 1980s, the image of wind energy would probably have suffered a much longer time, than it did in the past. Advanced Danish wind turbines gave an immense benefit to German wind development. Numerous Danish and Dutch units were installed in the 1980s. In the 1990s, the Danish dominated the German market and installation figures. The technical quality of these machines has been very influential and supportive. They showed that wind worked year after year. Thus, it was the foreign machines that helped to improve the destroyed image wind energy had in the 1980s and, partially, in the 1990s. They helped in convincing people to trust this new technology. Still by 2012, Danish companies, or those of the Danish origin, have quite a strong presence in Germany.

18.4.5 A Working Network

At the very early stage in the German history, various wind power associations were founded. The first was dated 1974–1975, and was called "Verein für Windenergieforschung- und Anwendung" (Association for Wind Energy Research and Application). Eventually it was renamed "Deutsche Gesellschaft für Windenergie" (German Association for Wind Energy).

A considerable number of such associations followed in the 1980s and 1990s. These associations had to "fight" on many fronts: Typical areas were representation of the industry, authority affairs, pushing wind projects, improving the overall image of wind energy, securing political support and entering dialogues with opponents, to name some. In many cases such associations were responsible for a region or one federal state.

In 1996, the DGW and the "Interessenverband Windkraft Binnenland" (Interest Group for Inland Wind Energy) merged, and formed the well-known "Bundesverband WindEnergie" (Federal Wind Energy Association). The BWE's record for achieved goals in the wind energy sector is ever growing. Its network and influence have increased. Thus, the BWE as well as the rest of a vast working network have made a substantial contribution to progress in the German wind energy industry.

18.4.6   Strong Offshore Approach

Though Germany definitely did not pioneer in offshore for the last 20 years, the industry is heavily preparing for large-scale offshore applications. This tendency can easily be recognised when looking at the new generation of German multi-megawatt turbines that is being developed for the last years. Starting in 2004, REpower made the first step into the "offshore-class" with its 5M (5 MW) turbine, located at Brunsbüttel, northern Germany. Though installed onshore, the machine is aimed at offshore application. In the following years the 5M went into mass production and is now proving its suitability at a couple of offshore sites. A 6 MW successor was already realised.

Figure 18.23   Areva Multibrid M5000 at "Alpha Ventus" offshore wind farm, 2010 (left); REpower 5M located at "Alpha Ventus" offshore wind farm, 2010. Note the platform for helicopters (right) (Photo: Arne Jaeger).

Besides REpower, the German company Multibrid developed a compact 5 MW wind turbine. Activities for this design started in the late 1990s. Multibrid was taken over by French Areva. Between 2009 and 2010, six Areva M5000 and six REpower 5M were installed offshore, in 39 km distance from the German coast.

"Alpha Ventus" was the first German offshore wind farm and is being followed by a larger wind farm, comprising 80 machines rated at 5 MW. This second offshore wind farm, called Bard Offshore I, will be delivered by a third company, whose roots go back to 2003.

Bard Engineering specialised in offshore applications right from the beginning. The company consists of numerous subsidiaries all working in different areas dealing with offshore wind farms. Two 5 MW prototypes were set up near Emden, northern Germany, in 2007–2008. A first near-shore unit was completed in 2008, close to Hooksiel. Currently, a 400 MW wind farm is being installed near the Alpha Ventus site.

Figure 18.24   Bard VM 5 MW nearshore, near Wilhelmshaven, 2009 (left); Bard VM 5 MW prototype installed at Rysumer Nacken, Emden, 2008 (right) (Photo: Arne Jaeger).

References

1. *Windkraft-Journal* (1981–1990) Verlag Natürliche Energien, Eckernförde.

2. Handschuh, K., (1991) *Windkraft gestern und heute*, Ökobuch Verlag, Staufen.

3. *Wind-energy converter HSW 250* (1994) Final Report, Husumer Schiffswerft, EUR 14920, European Commission, Luxembourg.

4. Tacke, F., (2004) *Windenergie. Die Herausforderung. Gestern Heute Morgen*, VDMA Verlag, Frankfurt am Main.

5. Rave, K., and Richer, B., (2008) *Im Aufwind Schleswig-Holsteins Beitrag zur Entwicklung der Windenergie*, Wacholtz Verlag, Neumünster.

About the author

Arne Jaeger was born in 1989 in Duisburg, Germany. He became interested in wind turbines in 1993 during a stay in northern Germany. Since that time his devotion to this topic is unstoppable. Since the mid-1990s he has made uncountable trips to important wind power countries, such as Denmark, Germany, the Netherlands and even the United States. His main wind energy activities concentrate on both the history and high-quality photography of modern wind turbines. He visited several wind power fairs, for example, the industry's leading fair "Husum Wind `99" at the age of ten. Since the mid-1990s he has been continuously building up his own archive covering a wide and huge collection of books, magazines, in-depth literature and documents, all dealing with various fields of wind turbines and aspects related to them. He's currently (May 2013) finishing a degree (B.A.) in Media Management, Public Relations and Communications in Cologne, Germany.

Chapter 19

Direct Drive Wind Turbines

Friedrich Klinger

INNOWIND Forschungsgesellschaft mbH, Altenkesseler Straße 17/ D2, D-66115 Saarbrücken, Germany

f.klinger@wind-energy-research.de

19.1 Introduction

This chapter discusses the question of why it took thirty years for the realisation to set in that direct drive (DD) of a wind turbine is clearly the better concept, as well as acknowledge the pioneers and inventors who worked hard to design and implement this realisation.

Directly driven wind turbines already existed in Persia 3000 years ago. In the Persian–Afghan border region Sistan–Baluchestan wind turbines are still driving the millstones with vertical axes (Fig. 19.1).

Historical resources state vertical axes turbines around the year 644 AD and a description of that construction type is dated on 945 AD. Centuries later, reports about Chinese windmills were brought to Europe, as shown in Fig. 19.2. They were built of

Wind Power for the World: The Rise of Modern Wind Energy
Edited by Preben Maegaard, Anna Krenz and Wolfgang Palz
Copyright © 2013 Pan Stanford Publishing Pte. Ltd.
ISBN 978-981-4364-93-5 (Hardcover), 978-981-4364-94-2 (eBook)
www.panstanford.com

bamboo and textile sails and have been used to draw water from rice fields.

Figure 19.1 Windmill in Nashtifan, Sistan Province, Persia (Photo: Windmill, *Reconciliation of Man and Hard Nature*, Saeid Golestani, 2010).

Figure 19.2 Chinese windmill (Photo: Deutsches Museum, München, published by Erich Hau, *Windkraftanlagen*, 2003).

In the 12th century, the first windmills with a horizontal axis and a wooden gear were built in Europe. In Fig. 19.3, two

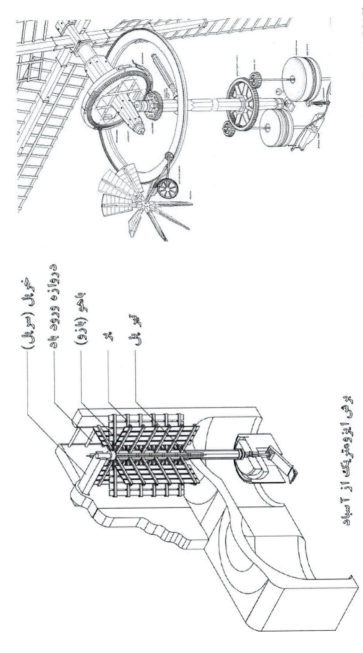

Figure 19.3 Comparison of historical windmills (Image: Windmill, *Recon-ciliation of Man and Hard Nature*, Saeid Golestani, 2010).

drawings by Iranian students compare the different concepts: on the left, the Persian windmill with a DD and a vertical axis, and on the right, the drive system of the European windmill with four wings at a horizontal axis and the transmission gear down to the millstones.

In form of post mills, tower windmills and Dutch windmills, horizontal systems have been used till the 19th century. They were more efficient than their Persian and Chinese predecessors, but much more complex in their construction. Usually the rotor had four wings and was mounted horizontally; a wooden gear wheel connected the upright shaft to further gear wheels that turned the millstones. They are archetypes for modern wind turbines. The much simpler Persian ancestral windmills, and hence the DD, were overseen for a long time.

19.2 The Drive Train Evolution

Since wind turbines used to produce electricity were introduced, a big variety of construction principles also appeared. It is true especially for the rotor, which rotates with two, three or multiple blades along a horizontal or vertical axis and converts kinetic energy from the air flow into mechanical energy. Therefore the power of resistance is used in the rotor. Increased conversion efficiencies close to 50% can be achieved with rotor blades that use the aero-dynamical buoyancy, as birds or airplanes do.

In a similar manner, different concepts for the conversion from mechanical to electrical energy have been established. All necessary components, including the generator, together nowadays are called drive train. The conventional, dispersed drive train is well established, and until 2010 can be found in 80% of all worldwide produced wind energy converters that are used for electricity production.

Thereby the enormous rotor moment in the size of, for example, 1 500 000 Nm at a 3 MW turbine is reduced by the factor 100 in a three-step gearbox, while the rotational speed is increased by that factor. That has the special advantage that a fast-running generator can be used, which is available in different varieties by most manufacturers of electrical machinery. The generator is small, light and has a very high electrical efficiency (Fig. 19.4).

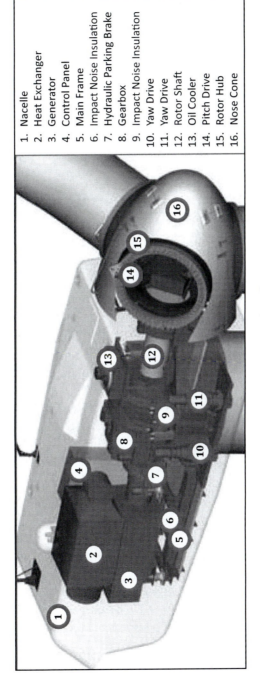

1. Nacelle
2. Heat Exchanger
3. Generator
4. Control Panel
5. Main Frame
6. Impact Noise Insulation
7. Hydraulic Parking Brake
8. Gearbox
9. Impact Noise Insulation
10. Yaw Drive
11. Yaw Drive
12. Rotor Shaft
13. Oil Cooler
14. Pitch Drive
15. Rotor Hub
16. Nose Cone

Figure 19.4 Drive train with a gearbox (Image: www.ge-energy.com/wind).

However, wind farm operators have had big problems with the drive trains ever since wind turbines came into use. The main problems are the long standstill periods through total standstills, expensive repairs and time-consuming maintenance work at the gearbox.

19.3   Direct Drive Revolution

DD is a drive train concept that was first successfully introduced to the market in 1995 by the German manufacturer Enercon, which replaced the gearbox and the fast-running generator with a low speed, circular multi-pole generator that was directly connected to the rotor (Fig. 19.5).

Figure 19.5   Gearless drive train by Enercon (Image: Enercon).

The so-called "gearless" wind turbine from Enercon, with a generator of 5 m to 6 m diameter was at first considered an exotic machine, but is highly established nowadays with a market share of 60% in Germany in 2011, although its price ranges over that of competitors.

Various types of DD concepts emerged since 1990; for example, the prototype of the Genesys-project, and the concepts by Vensys/Goldwind, Lagerwey, Zephyros, Mtorres, Leitwind, Hyundai, Impsa and Siemens. Some manufacturers used the proposals and experiences from the research team "Windenergie" of the Saarland University of Applied Sciences, which were developed under the lead of Professor Friedrich Klinger. The company Vensys is a start-up of this group.

An important indicator of the concept is the so-called "external rotorring" that applies permanent magnets to excitate a synchronous machine and therewith decrease the outside diameter of the generator (DDPM). In Fig. 19.6 a concept drawing shows the principle of a gearless wind turbine, as it was planned by Klinger in 1990.

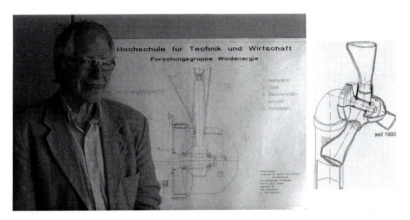

Figure 19.6   Professor Friedrich Klinger and the concept drawing of the DD from 1990 (Image: author).

With a sketch like this, Klinger and a group of students and researchers went on an excursion to north Germany in spring 1990 in order to visit the leading companies producing wind turbines like Tacke, Husumer shipyard and Enercon in Aurich (Fig. 19.7).

Figure 19.7 Students from Saarbrücken visiting Enercon in 1990 (Photo: author).

19.4   Prototypes and Beyond

The wind researcher team from Saarbrücken worked continuously since 1990 on studies for Siemens and Thyssen, comparing the two different drive train concepts. The department of hydropower generators at Siemens in Erlangen in cooperation with the Saarbrücken research group, initiated a development of a prototype of a 750 kW DD generator with inverter. Siemens offered the whole package for DM 690 000 (ca. EUR 345 000) to Danish wind turbine producers, but they kindly refused due to the high price.

When the Enercon E40 became popular already in 1995, a Frankfurt based banker Manfred Hecht put Klinger's research group in charge of designing a new gearless 600 kW system. By that time Hecht had built up a Fund (WONSEI, West-Ost-Nord-Süd-Ethische-Investition) with about DM 5 million. Together with Ciro Capricano, who just completed his study of technical physics at the University of Applied Sciences in Wiesbaden, and was working as the leading engineer in the wind research department, Hecht was following the development in Saarbrücken carefully. After nearly two years the prototype of the Genesys project was set up in the north of Saarland in January 1997 (Fig. 19.8).

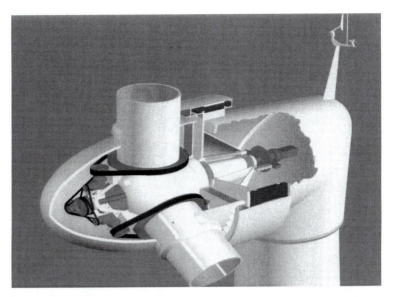

Figure 19.8   Genesys Prototype (Image: author).

Features of the prototype:

- Variable rotation speed by a inverter from SEGELEC and a control system from SMA Kassel;
- External rotor-synchronous generator with excitation system of permanent magnets;
- Rotor blades from the company Aeroconstruct (Keller);
- Rotor blade pitching by tooth belts;
- Blade pitching gear with a central control shaft and a spring-loaded disc brake, which brought the blades without energy storage but with the kinetic energy of the rotor into vane position, so that the rotor was able to stop safely.

The system was quite a sensation for the experts in the sector, not just because of the simplicity of the design, but also with a 46 m rotor it produced 20% more kWh per month compared to a Vestas V44 turbine in the same wind park. The client Hecht and his project manager Capricano planned to build a high number of turbines, but the banks refused loans for innovative developments and so the project ended in an economic disaster.

In 2000 the company Vensys emerged out of Klinger's research team together with different stakeholders such as Provento and the

Denker–Wulf Group. By that time the prototype of a 1.2 MW turbine, based on the Genesys system (DDPM), was nearly completed and was installed in 2003 in the north of the Saarland. Despite a fire in the nacelle caused by a condensator battery explosion, interest in the licence was still big that year. Wu Gang, president of Goldwind, Science and Technology, Urumqi, China, acquired the first licence and manufactured it. More than 8000 up-scaled 1.5 MW turbines came into successful operation until the end of 2012 (Fig. 19.9).

Figure 19.9 Vensys 1.5 MW, with rotor diameters of 62, 70 to 77 m (Image: Vensys).

In that way advantages of the DDPM concept were faster recognised by China than by the rest of the world. Other licences have been contratcted to Czech Republic (CkD), Brazil (Impsa wind), India (ReGen Powertech), Spain (Eozen) and in 2011 to Egypt (AOI).

There had been parallel developments of DDs in the Netherlands by Henk Lagerwey and Zephyros, by Manuel Torres in Spain and by

Leitner in Italy. The main differences in the DD machines can be found in the bearings. Basically the hub is connected to the blades in a rigid way as the rotor is to the generator, and is carried by two roller bearings. In Fig. 19.10, the bearing concept introduced by Enercon can be seen—the rotor hub is carried by a small locating bearing in the front and a bigger floating bearing in the back. Therefore the hub bolted to the rotor can be described as "flying", stored slightly in front of the two bearings.

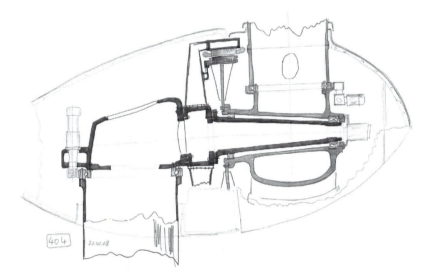

Figure 19.10 Enercon concept drawing (Image: author).

A further innovation to simplify the direct drive–bearing concept was introduced by Lagerwey (Fig. 19.11). A bearing that takes overhanging bending moments from the rotor in one unit consisting of three rows of cylindric rollers (Rothe Erde) or double row tapered rollers (SKF, FAG).

It was a 2.5 MW Vensys turbine that was first equipped with a Nautilus bearing from SKF in 2006, which was later used in most of the new developments as well as in offshore turbines by Siemens from 2009 and 2010. A totally new bearing concept was suggested by the wind researchers from Saarbrücken in 2006 and later used in a 2 MW project in China (Fig. 19.12).

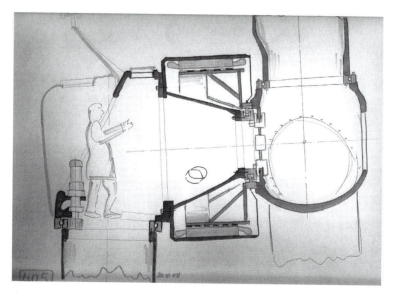

Figure 19.11 Lagerwey concept drawing (Image: author).

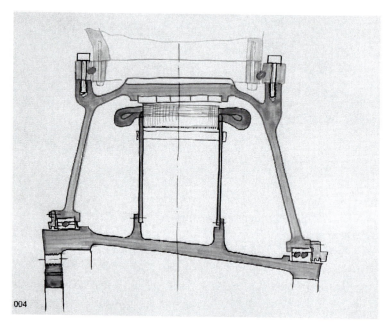

Figure 19.12 Bearing concept of a hub generator (Image: author).

In the concept of the hub generator, the casing of the direct driven permanent magnet generator is functioning as the hub itself. Therefore it could also be called double direct drive (DDD), as it cannot be more direct.

A robust 6 MW offshore design by Alstom, which was erected in 2012 as a prototype, is using a different bearing concept with the rotor hub and the generator runner mounted on separate bearings (Fig. 19.13). In order to prevent elastic deformation of the shaft that is transferred to the air gap of the generator, clutch elements are mounted in between (pure torque concept).

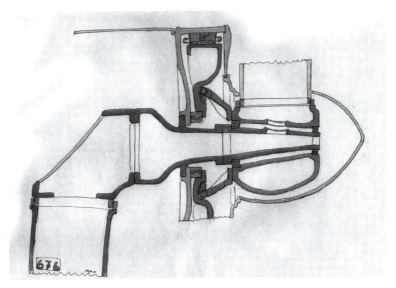

Figure 19.13   Alstom concept (Image: author).

19.5   Direct Drive Pioneers

The DD concept was already proposed by Hermann Honnef in 1930–1940. He was planning huge wind power plants with DD and three to five gigantic counter-rotating double rotors mounted on steel lattice towers (Fig. 19.14).

The power plant was supposed to be 500 m high and the rotors, of 160 m diameter, should produce 20 MW of electricity at a wind speed of 15 m/s. Moreover, Honnef was the first one to propose a direct driven offshore wind installation (Fig. 19.15).

Figure 19.14 Wind power plant by Hermann Honnef.

Figure 19.15 Honnef's offshore rotor (Photo: www.energyprofi.com).

Even if Honnef's 120 m diameter ring generators were most likely not realisable, he was still one of the DD pioneers. His work, however, came to an end with the beginning of the Second World War in 1939.

And yet another quiet pioneer should not be forgotten. Gerd Otto, a researcher based in the former German Democratic Republic, described in detail a DD wind turbine with permanent magnets in his scripts as well as in his patent application in 1986.

The concept of the Persian windmills, mentioned in the beginning, was not entirely forgotten. Wind turbines for elecricity production were models for vertical-axis turbines like the Darrieus- or H-rotor (Fig. 19.16). However, they were never successful.

Figure 19.16   (left) Darrieus Rotor, Heroldstatt, Germany (Photo: W. Wacker); (right) H-rotor Darrieus turbine (Photo: Stahlkocher).

In contrast to the Persian archetype, modern vertical turbines are not installed between walls in a slipstream, and can operate independent from the wind direction. That brings up a new problem: each rotor blade has to turn against the wind, which causes torque variation and hence reduces the life span of a turbine. Nevertheless, new concepts and development still come about, but could not really compete with horizontal-axis turbines.

When this text was written in November 2011, no DD offshore wind turbines were in operation. Moreover 80% of all other on- and offshore turbines to this date were running with a gearbox. The author is confident that it will be the opposite within a time of next 10 years and 80% of all wind turbines will be gearless. Big and reliable wind power plants will be needed in order to manage the transition to a renewable energy future.

About the author

Friedrich **Klinger** received diplomas in mechanical engineering from Technical College Saarbrücken and Technical University Karlsruhe, Germany, in the years 1956 and 1960, respectively and a Ph.D from Technical University Aachen, Germany, in 1977.

Prof Klinger started his professional carrier in 1965 in designing fatigue test equipment at Carl Schenk AG, Darmstadt, Germany, and from 1974 as head of development at Hannes Marker KG, Garmisch-Partenkirchen, Germany. In 1976, he became an R&D manager for testing machining at MTS Systems GmbH, Berlin, and later a member of the board of MTS Systems, Minneapolis, USA.

In 1982 he joined Saarland University of Applied Sciences as professor of mechanical engineering, when he also started developing direct drive prototypes Genesys and Vensys in a research group for wind energy, from where Vensys Energy AG started in 2000. Currently, as emeritus and CEO of INNOWIND Forschungsgesellschaft mbH, he is still into consulting and developing gearless wind turbines.

Chapter 20

How the Early 1980s Micro- and Power-Electronics Innovation in Germany Revolutionised Wind Energy Systems

Jürgen Sachau

Interdisciplinary Center for Reliability, Security and Trust,
University of Luxembourg, L-1359 Luxembourg

juergen.sachau@uni.lu

It was in 1980 when I first came in touch with a completely new field in systems and control engineering: using micro-electronics for control, integrated circuits, which at that time had recently developed into the first 4 bit and soon 8 bit, the so-called microprocessors. Up to then, digital signal processing and control used to be a field for expensive mainframe and mini-computers, controlling large installations for processing, manufacturing and production installations or military, air and space equipment.

20.1 University Pioneering from Theory to Practice

At that time, the electrical engineering institutes at Braunschweig's Technical University annually presented their laboratories for the

Wind Power for the World: The Rise of Modern Wind Energy
Edited by Preben Maegaard, Anna Krenz and Wolfgang Palz
Copyright © 2013 Pan Stanford Publishing Pte. Ltd.
ISBN 978-981-4364-93-5 (Hardcover), 978-981-4364-94-2 (eBook)
www.panstanford.com

students to make them familiar with their research and technology development. Germany's first technical university was founded as Collegio Carolino zu Braunschweig in 1745 at the time of Galilei, Kepler, von Guericke, Newton and Leibniz out of the need to join theory and practice—with the intention to realise mechanical constructions, such as perpetuum mobiles that could not work for theoretical reasons, and to develop manufacturing know-how for mechanic computing machines, which at that time failed due to lack of understanding about the accumulation of mechanical tolerances. And now, over 200 years later, putting theory to practice, it once more became a challenge: cybernetics and mathematical engineering were well advanced and the programming flexibility of the microprocessors would now allow to carry over-advanced methods to broad application in a wide range, unlocking creativity and challenging the engineers' spirit of invention.

In the university's laboratories, as students in the beginning of the 1980s, we were of course electrified to see first examples of what should become common in the next decades. At that time, benefits of technology in general were much under public discussion and we just had seen the Three Mile Island 1979 nuclear disaster. The energy technology professor fresh from nuclear business, reacting to the students union was open to discuss necessity and safety of nuclear power instead of other high voltage topics. Engineers became ready to take responsibility for the technology they developed, not few of us hoping to be able to contribute to society as engineers instead of creating surveillance, nuclear or war technology.

We had the opportunity for a round-trip on the transrapid prototype outside, where micro- and power electronics were enabling both magnetic driving, and lifting by advanced control in the instable equilibrium between gravitation force and permanent magnet field. According to the linear drive, for the lab a synchronous multi-pole ring-machine of 1 m diameter had been constructed. For investigating smooth movement, that is, transition to the next sections, not the whole number, but only two permanent magnets were needed. Fully equipped, this topology was perfectly suited for electromechanical generation at low rotational speed, that is, gearless wind converters. By means of contemporary micro-

and power electronics, the variable-frequency power—*via* an intermediate DC link—would become convertible into grid-conformant rigid 50 Hz. At that time, Aloys Wobben, later carrying on this technology founding Enercon, was a lab-engineer in the Institute for Electrical Machines, Traction and Drives founded in 1920 and led by Professor Weh, who as an emeritus is still today remaining active in design of electric machines for wind power.

Figure 20.1 Synchronous multi-pole machine with two E-shaped exciter magnets at the Technical University Braunschweig, Germany.

A further important institute involved in wind energy was Professor Werner Leonhard's Institute of Control Engineering. Profiting from the progress in the broad DFG programme for new electrical drives, micro- and power electronics could straightforwardly be developed for the 3 MW GROWIAN wind converter—a courageous step directly to a prototype of the double-fed induction machine, where the variable mechanical rotation, by means of micro-processor control was made self-adapting to wind fluctuations. Mechanical rotation could be superimposed by a flexible rotorfield, controlled *via* frequency and phase in the proper way for feeding both active power and reactive power suitably into the 50 Hz grid. This was part of an ambitious project of the German Ministry for Research and Technology (BMFT) in which I worked on my diploma thesis in microprocessor control, developing and prototyping a new nonlinear active damping method for synchronous generators, as for direct coupling of wind converters to the grid.

The GROWIAN prototype, the world's largest wind converter for a long time, was supported by major German companies under leadership of MAN and went successfully in operation.

Figure 20.2 GROWIAN (GROsse WIndenergie ANlage) wind turbine.

Interestingly, it was not the innovative electronics which caused problems, but the classical construction of the mechanical shaft, and thus it was easy to denunciate big wind power as premature and too expensive. Germany lost a leading position, needing two more decades to bring this technology in this power range to the market.

Figure 20.3 Blades and nacelle of GROWIAN before being lifted *via* its own tower.

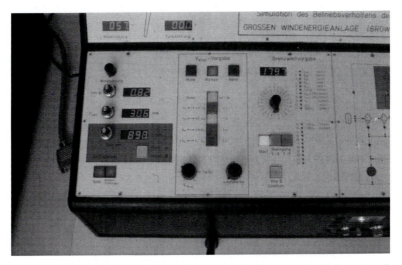

Figure 20.4 GROWIAN operational control simulator at the University of Kassel, Germany.

All this was enabled by programming flexibility of microprocessors, interacting with power-electronic switches and implementing advanced control methods in assembler code—programmed on a workstation with several tricks to save precious computing time, burnt into read-only memories with their small glass-window showing the chip, erasable by UV light. The memory packages were then mounted on processor-boards in 19″ racks of all lab-made cards, probably worth several thousands of euros in combined hardware and software, computing power today entirely found in single chips for less than a euro. Code had to be put into operation and tested with simple seven segment displays and numeric keypads, watching transients *via* analog output and oscilloscope. It was the time, when the first Apple computers had just appeared and the sensation was in the air, that IBM would bring something like a small "personal" computer on the market, not a playing-machine but something to be taken seriously.

20.2 Putting Technology to Work for Integrated Systems

With the wind-converter damping prototype running, I continued as one of the first PhD candidates starting at the recently founded university of Kassel with its free-minded founding president Ernst-Ulrich von Weizsäcker, going beyond established university structures, not having faculties, but something called interdisciplinary centres. Renewable Energies, Ecological Agriculture, Fringe Groups, this was the place to work on such exotic topics, at that time much under suspicion of the scientific establishment. Soon a follow-up project was drafted for the BMFT in cooperation with Professor Leonhard and Kleinkauf, one of his first PhDs and now at the University of Kassel, and we also participated in the first European demonstration of a stand-alone island supply with wind power, balanced by a battery unit for the Irish island of Cape Clear. The scope was generally enlarging towards wind-power systems for island grids. My Greek colleague Fotios Raptis, who soon became a friend, infected me with his enthusiasm about islands—Greece has more than 3000 of them.

Once more, microprocessor control together with power electronics turned out to be the key to high-quality power supply, basically coping with the need for rapid balance of wind fluctuations in order to avoid intolerable frequency deviations. In parallel to the demonstration projects in Ireland, Jordan, Brasil and Greece, my PhD work focused on the advanced microprocessor control, a new method of voltage vector control in a rotating reference frame. It allowed to reduce frequency deviations by 80% in these island configurations and was demonstrated in the 10 kVA plant in our machine laboratory.

Figure 20.5 Günther Cramer, founder of SMA, assisted by project engineer Roland Grebe, explains the Cape Clear power container to German Research Minister Riesenhuber and the Irish Prime Minister Haughey.

The wind/battery/diesel system was built by MAN, VARTA together with SMA's power conditioning and battery container. With its microprocessor-based energy management as a core component, it was finally shipped to the Irish island of Cape Clear. Together with representatives of the companies involved, the Irish utility and the local authorities, we held a two days seminar, when this European funded pioneer island wind-power demonstration system was inaugurated. Representatives of the THERMIE programme of the European Commission's Directorate General for Energy were there, among them my friend Komninos Diamantares, who should later become my colleague in Wolfgang Palz's renewable energy research programmes.

It was a cold, yet sunny day with little wind, mildened by the gulfstream which allowed even palms to grow here on this island, only accessible by ship. Between a handful of houses, nearly everybody went by foot, and now the Irish Minister flew in by helicopter with some press people for the open air inauguration ceremony directly at the power container on Cape Clear Island.

Figure 20.6 Workshop participants, among them Hubert Nacfaire and Komninos Diamantares from the EC, Professor Kleinkauf and the author from University Kassel, Germany, and Günther Cramer, founder of SMA.

Figure 20.7 Inaugurating the wind/diesel/battery system on the Island of Cape Clear. From the left: G. Cramer; Dr Riesenhuber, Minister of Research; Irish Prime Minister Haughey; O'Ceadagain from Cape Clear Cooperative Company and Dr. Glynn from the Irish Ministry of Research (Courtesy of SMA press archive).

Light, sound, drink cooling, all powered by the demonstration plant, the procedure became somewhat delayed but was finished in time before the batteries were discharged, just in time to avoid start-up of the noisy backup diesel. A certain pioneer atmosphere was underlined by a footwalk to the cosy pub in this remote European location at the rough Irish Sea, which had damaged previous undersea cables. Was it possible that this was the first impression of what was going to become the future of Europe's energy system?

20.3   European Modular Systems Technology

Within Kassel's newly founded Institut für Solare Energiever-sorgungstechnik (ISET), I built up the systems technology research and technology development, devoted to the technical integration of the different renewable energy technologies into entire energy systems. Before the feed-in law introduced in Germany in 1991, that completely changed everything, islands and remote locations appeared to be the most reachable markets for wind and sun energy and that was what we focused on. Fortunately, Wolfgang Palz from Brussels had stimulated the foundation of a number of European Economic Interest Groupings in the renewable energies field. At their core, I participated in the 1991 Cork meeting founding EUREC Agency with the goal of joining European forces for research and development of renewable energy technologies, in between representing some 50 European research centres.

It was not only meeting the founding researchers of the initial handful of European players in the field and having marvellous dinners, it was tough work as well, as these meetings were the occasions to set up the networked European projects. Often, well timed as the last opportunity to get all the signatures and forwarding the polished proposals to Brussel's before the Commission's strict deadline. On the occasion of our Petten meeting, as one of the first EUREC Agency projects, the MEGA-Hybrid initiative was born. The core idea was addressing the problem of integration, that is, to develop a technology that would allow to progress from the status of engineering single installations forward to a technology of systems. Thus, megawatts of plants combining use of wind, solar and biomass with storage and backup should become economically competitive. And thus my systems' engineering department at the Hessian Institute ISET found its way into Wolfgang Palz's renewable energy research and technology development (RTD) programmes. We initiated and coordinated the modular, expandable, and generally adaptable (MEGA) systems technology development. The first project was implemented together with Peter Helm from WIP, Munich, Platon Baltas from CRES, Athens and Manuel Cendagorta from ITER, Tenerife.

The MEGA-Hybrid core technology established grid-compatible single AC-bus and single communication–bus coupling of all components, enabled through state-of-the-art micro- and power electronics. We proposed and with the PhD collaborator Alfred Engler developed the first self-synchronising microprocessor control for power-electronics inverters, also starting to use field-buses which soon became ethernet-based and network-compatible. Components of different structures and sizes thus became integrateable in all different system contexts. MEGA Hybrid's AC-modular technology paved the way for a tree of subsequent European RTD and demonstration projects and their national counterparts, often of much larger size.

Starting from wind and PV for island grids, within a decade it became the standard in Europe, also leading to today's AC-coupled parallel PV grid-power conditioning. Consequently, we developed the power-electronics prototype for the Kassel based large-scale production of string-inverters for photovoltaics. Naturally matching the modular nature of the generators, they turned out to be the key success factor for large-scale integration of PV in the built environment. Electrical planning and safe and secure installation could now be given into the hands of local craftsmen, no more requiring special engineering.

Under the guidance of Wolfgang Palz, the European Renewable Energy RTD activities were rapidly growing and, delegated by our Minister, I came to Brussels to support him in the 5th Framework Programme. Integration of renewable energies and thus systems technology became more and more important, as wind- and water-power but also bioenergy and photovoltaics were rapidly becoming competitive. Within the European projects, Kassel's systems engineering was further managed by Philip Strauss who had already joined the MEGA-Hybrid project as a PhD collaborator, and my systems engineering activities became a core-part of Germany's Fraunhofer Gesellschaft for applied research in its Institute for Wind Energy Systems, IWES.

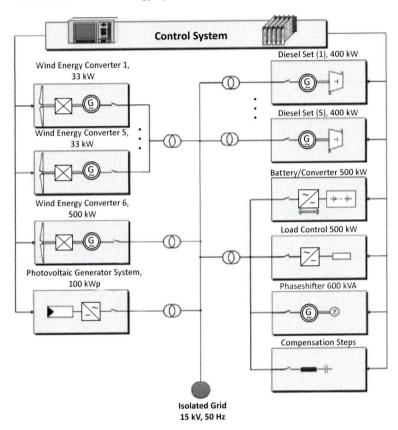

Figure 20.8 Modular Systems Technology on Kythnos Island, as being transferred to more Greek islands like Sifnos, Serifos and Astipalea islands. International Conference Renewable Energies for Islands, Chania/Kreta June 2001.

In the European Research Directorate, in the following five fascinating years of RTD management and strategy development with its incomparably stimulating, open minded and multi-faceted atmosphere, meeting and working with most of the major players in the field of Renewable Energies, soon the need for integration became more and more evident. Accordingly, the Advanced

Electricity Storage Programme, which I proposed and drafted for Europe's Joint Research Centre, was adopted already in 1997, with micro- and power electronics as key enabler for integration and coordination of the diversity of components. With the new challenges in engineering research, I could not resist to come back to control engineering reality, returning as professor to Kassel, soon being offered to build up the energy systems and control engineering activities for Luxembourg's new university and the Interdisciplinary Centre for Security, Reliability and Trust.

20.4 Defending Progress

In between, Germany's wind converter manufacturing had considerably grown, facing a major market, and back from Europe's Directorate General for Research, I was supporting their association to protect their innovations, especially with regard to the intellectual properties rights. These turned out to be an important competition issue, non-European companies massively trying to get into the European market. Initiated and coordinated by REpower, in the core of the discussions were the advanced control methods joining micro- and power electronics with their, in between, considerable cost degressions. Now, more than two decades later, the GROWIAN power conditioning and control technology from Braunschweig was dominating with its capabilities of smooth integration and even support of the electricity grid. For control of the flexible rotorfield in the proper way to feed in active and reactive power into the 50 Hz grid, a diversity of variants had emerged.

Counteracting a patent offensive of a large company from outside Europe, it became necessary to clarify the more formal nature of their numerous applications and solidify the previous state of art towards the European Patent Office. It took some profound literature and patent research and a number of difficult discussions with developers, component manufactures and patent attorneys. Otherwise the markets could have been dominated by large companies getting patents granted for a technology that had long been known or for which the original patents in micro- and power electronics fields had already expired.

Figure 20.9 Windpark Grömitz (Courtesy of REpower Systems AG, Photo: Jan Oelker).

20.5   On the Path towards 100% Renewable Energies

With my new affiliation in the middle of Europe, Luxembourg's Interdisciplinary Centre immediately after its foundation became member of the Distributed Energy Resources Laboratory Network of Europe (DERLab). Still and again, integration of the new sustainable energy technologies is remaining the major challenge. Based on flexible grid-coupling through power electronics, it is the micro-electronics and data-processing technology that has to match the emerging distributed medium and low-voltage active-grid structures on the way to full solar supply, with more and more European communities being committed to follow this path as soon as possible.

On the ground of secured communication structures, methods for maintained reliability, extended protection and best usage of power lines are developed. Robust distributed optimisation of generation, storage and consumption dynamics are the measures to reduce vulnerability and optimise technoeconomics. The transition to such active-grids is expected to reach major coverage with technology and services within one to two decades in Luxembourg and Europe. Luxembourg with its both urban and rural structures and its elaborate and reliable information and communication technology (ICT) infrastructure being an ideal pioneer area, SnT, together with Luxembourg's utility CREOS and the support of Luxembourg's Fonds National de Recherche is committed to the long-term active-grid strategic framework.

With a data-processing layer adjoined to the existing electricity grid infrastructures as precondition for a "smarter", that is, active-grid, progress is crucially depending on new distributed system and control technology solutions. Especially the grid operators like CREOS are facing new challenges. With their leading responsibility in integration, they need to maintain reliability, safety and security in a much more complex situation with an increasing number of distributed feed-in and storage units to be integrated, with wind power in a substantial share.

Figure 20.10 Luxembourg's pump-storage at Vianden.

In the concert of Europe's structural variety, the conversion process of the current electricity distribution grids to the future active-grid structures will find a diversity of complementary solutions in competition of central and decentral technologies for generation and storage. Progress in versatile, robust and flexible decentral solutions crucially depends on implementing the adequate methods for measurement, control and coordination, employing flexibility of micro-electronics, well integrated with power electronics.

The way to sustainable energy supply with its major challenges to:

- accommodate energy production of all sizes;
- increase reliability and security of supply;
- optimise regional grid use and balancing;
- customise power quality for different needs;
- empower customers for active market participation; and
- enable new players, products and services

is fundamentally changing the role of the distribution service operators. National and European policies continue to be committed to progressive usage of wind and other renewable energies, especially for distributed electricity generation. Complemented by measures for energy recovery and efficient use of energy and with major potentials in distributed combined heat/power production, the way is also paved for broad introduction of electric transport substituting its major consumption of fossil fuels.

Following a long and stony way from the first enthusiastic attempts, when nearly anybody dared to imagine a major share of renewables in our powermix, it seems nearly a miracle that within three decades renewables became so significant. In wind energy systems as well as for all other renewable energies, progress in micro- and power electronics has been and will be the core transformation driver towards participative, social and sustainable society. In the meantime, the challenge has completely turned around: to accommodate all renewable energies up to 100% coverage in our infrastructures in Europe and worldwide. Contributing an estimated increase of annual GDP by several percent, the dream of sustainable energy supply with 100% renewable energies is in reach within three to four decades, or might speed even double once more?

About the author

 **Jürgen Sachau** did his electrical engineering in computer and control systems at Germany's first technical university in Braunschweig with Prof Werner Leonhard. He then elaborated his study through his PhD in digital control of independent power grids.

He joined the University of Kassel in 1984 with Prof Werner Kleinkauf and become project leader for decentral energy systems. From 1989 onwards, he started building up the Department of Systems Engineering of the German Energy Research Institute (ISET)—today Fraunhofer-IWES—and their strategic programme. He has participated in the foundation of the European Agency of sustainable energy research institutes and companies, such as EUREC-Agency and has thus contributed to a number of EU conference boards and scientific committees on renewable energies, as well as to a number of committees of IEA and IEC. He has also participated in EC project evaluation and assessment. He is editor of the journal *European Transactions on Electrical Power* and has served as editor-in-chief of the *International Journal of Solar Energy* during their print era for over one decade. He is also the founding editor of the *International Journal of Sustainable Energy* and has published several hundred articles.

From 1995, he guided three clusters of over 50 non-nuclear energy projects in four EC-programmes at the European General Directorate for Research in Brussels, after which he took over the responsibility for supply quality management of the EU-funded solar power demonstration programs and the advanced storage strategy at the EU's joint research centre (JRC), Ispra. He was also professor at the Energy Institute of the University of Kassel, where he taught systems and control engineering from 1992 to 2003, after which he founded the Systems and Control Engineering Department of the University of Luxemburg. Currently, he is scientific director in Luxembourg's Interdisciplinary Centre for Security, Reliability and Trust, where his major responsibility is security and reliability of future electricity grids fully powered by distributed generation.

Chapter 21

How the Electricity Feed-In Law (Stromeinspeisungsgesetz) Came to Be Passed by the German Parliament, Enabling Renewable Energies to Establish Their Position in the Market

Ulrich Jochimsen

Netzwerk Dezentrale EnergieNutzung e.V., Haus der Natur, Lindenstr. 34, D-14467 Potsdam, Germany

jochimsen@netzwerk-den.de

The 1990 German Feed-In Law, allowing the sale into the grid of electricity generated from renewable energy sources, came into force on 1 January 1991. This piece of legislation emerged from within the framework of a political constellation unparalleled in the history of the country's electricity production and supply industry. That this had been at all possible was first and foremost due to the efforts of the Labour MP Dr Hermann Scheer (SPD) who, having recognised the unique opportunity at hand, resolutely brought together and united the key players involved—a group of MPs from all parties—thus harnessing the forces across the great political divide.

Prior to this, Germany had witnessed many dramatic and decisive events taking place at home and abroad: the two oil

This chapter was translated into English by Philomena Beital.

Wind Power for the World: The Rise of Modern Wind Energy
Edited by Preben Maegaard, Anna Krenz and Wolfgang Palz
Copyright © 2013 Pan Stanford Publishing Pte. Ltd.
ISBN 978-981-4364-93-5 (Hardcover), 978-981-4364-94-2 (eBook)
www.panstanford.com

crises of 1973 and 1979, which had shocked the industrial nations and highlighted Germany's need to become less dependent on energy imports; the anti-nuclear movement's demonstrations in the 1970s and 1980s, which ultimately led to the founding of the Green Party; the German government's 1980 commission of inquiry on the use of nuclear energy; the aftermath of the 1986 nuclear accident in Chernobyl and the Labour Party's (SPD) resolution that same year to phase out nuclear energy within the space of a decade; the collapse of the German Democratic Republic and the subsequent integration of that state into the Federal Republic of Germany.

The outstanding contribution made by the extremely well-organised South German hydroelectric power industry in cooperation with the region's forest and sawmill owners—and above all their parliamentary representative at the time, the Conservative MP Dr Engelsberger (CSU)—deserve special mention at this point.

However, had it not been for Hermann Scheer, one of the most important figures in the field of renewable energy policies, this multi-party parliamentary initiative would never have materialised, let alone been so successful—circumventing, as it did, the influential party whips in the process! It was he who had founded the solar energy association EUROSOLAR (The European Association for Renewable Energy) in 1988, in defiance of the unambiguous objectives of the powerful German Mining and Energy Workers' Union lobby within his own party. EUROSOLAR was one of the first to actively push for the 100% replacement of fossil and nuclear energy by renewable energy sources.

We must bear in mind that, given the existence of the unrelenting three-line whip (obligation to take the party line), any politician who dared to "take a chance on more democracy" —to quote Willy Brandt's famous expression from the year 1969—would find himself accused of high treason. Yet this "treason" is exactly what Hermann Scheer had organised. **The "Inspector General for Water and Energy", a constitutional relic dating back to 29 July 1941, but nevertheless still valid to this day—that is, the "Führer" principle[1]—now found itself jeopardised to a high degree by the more contemporary and diametrically opposed principle of democracy.** This was a strategically skilful move at a

---

[1]Energie und Führerprinzip, http://www.ulrich-jochimsen.de/EnergieFuehrer-prinzip.html (German language only—for brief information in English, see footnote 3).

time when the managing directors and top lawyers of the electrical utilities in the West were busy trying to seize possession of (i.e., privatise) the former East German electricity sector.

In 1990, with the legislative period drawing to a close and pressure mounting, the German government cooperated with the big energy companies in an attempt to destabilise the initiative behind the Feed-In Law.

At our first meeting at the Ministry of Economic Affairs in Bonn in May 1990, we—the wind energy, solar energy and bio energy "cranks"—found ourselves at the long negotiating table face to face with the united forces of the energy supply industry. Behind them stretched a huge photograph of an open cast lignite mine, which occupied the entire wall. Seated on the left were officials of the Ministry of Economic Affairs and on the right members of parliament and representatives of the hydroelectric power sector.

At that memorable meeting we were informed that negotiations about the future of the Feed-In Law would only be continued if Manfred Lüttke, the representative for the hydroelectric power industry in the federal state of Baden-Württemberg, packed his bags and left. Immediately, Erich Haye, Dr Ivo Dahne and I have made it quite clear that, if this was the case, we were going to get up and go too. Amidst the euphoria that accompanied German reunification, with its high ideal of "living at peace and in freedom" at last, neither the energy supply industry nor the German government—who, following a parliamentary resolution, were meant to present a legislative proposal before parliament— could have allowed themselves to push things too far. This would only have caused antagonism ahead of the celebrations with "our brothers and sisters in the East". Moreover, the energy supply industry in the West was completely tied up with its endeavours to appropriate the East German energy supply industry, which involved the scandalous "GDR power supply agreements", the wording of which has still not been disclosed to this day. In any case, the representatives of the energy supply industry felt confident that, sooner or later, they would be able to rid them-selves of the Feed-In Law.

Due to the developments taking place concurrently, only one alternative remained open to the electrical utilities: They would either have to eradicate the Feed-In Law or else take over the entire East German electricity sector. After all, there

was a danger that the co-generation plant (combined heat and power) virus might spread westwards into the Federal Republic.

Via the Network for the Decentralised Use of Energy, which we had set up in November 1990 through the People's Chamber, the highest organ of state power in the former East Germany, we succeeded in instigating judicial review proceedings at the Federal Constitutional Court against the "GDR power supply agreements", which had been drawn up by the electrical utilities in the West with the intention of taking over the entire network. These companies would not have been willing to run the risk of initiating parallel proceedings against the Feed-In Law, for fear that both lawsuits might be dealt with at the same time—in which case it would have become only too apparent that the fundamental structure of the German power industry dates back to the year 1941, when the NS-Regime was at its most brutal.

Our success: The impending lawsuit forced the electrical utilities to hand back part of their assets to some local authorities. On 27 October 1992—in several respects an important date in the history of the Constitutional Court—it came to oral proceedings. Not at the Constitutional Court in Karlsruhe, however, but at the railway repair works in Stendal instead. Not only was the venue itself out of the ordinary: the court surprised us all by introducing a settlement proposal[2]—an anomaly in its history. By so doing, it managed to prevent the necessary reappraisal of the laws governing the German energy sector, created during the Third Reich. The court's president, Roman Herzog—who was later granted the office of Federal President in recognition of this settlement proposal—saw to it that the proceedings "ran dry" by allowing the case to continue for too long. And so the goal was achieved: One by one, the 164 East German plaintiffs withdrew their charges, after receiving "lucrative" offers from the West German energy supply industry, in keeping with the rule of mechanics whereby machinery will only function if it is well oiled—in other words, after allowing their own palms to be sufficiently greased.

Rupert Scholz, chief negotiator for the German parliament in the constitutional commission made up of both the Bundesrat

---

[2]Jörg Henning, Transformations probleme nach der Wende am Beispiel der Stadtwerke Halle GmbH, Forschungsberichte des Instituts für Soziologie, Universität Halle, ISSN 0945-7011.

(upper house, representing the federal states) and the Bundestag (lower house), acted in a similar manner: He did his utmost to stop the German constitution of 23 May 1949 from being transformed into one for the whole of reunified Germany. Not at all in keeping with article 146 which states that "this basic law shall cease to apply on the day on which a constitution freely adopted by the German people takes effect".

Many people were probably asking themselves: Where on earth would we be then? Should it be revealed at this early stage that Theodor Maunz, who taught both Rupert Scholz and Roman Herzog, had written commentaries for Gerhard Frey's *Deutsche Nationalzeitung* (formerly *German National and Military Newspaper*) right up until his death in 1993 and that the German Reich still lives on in legal terms? Should all wonderful takeover and bankrupting opportunities in the acceding territory (i.e., the five new federal states) be lost on account of a new and transparent constitution? Should a growth in power and revenues be sacrificed at the altar of such fantastical notions as "sustainability" and "democracy"?

21.1 The German Renewable Energy Association (Bundesverband Erneuerbare Energie) Is Founded to Safeguard the Feed-In Law

The Renewable Energy Association was founded in 1991 as an umbrella organisation for small and medium-sized operating companies, in order to safeguard and strengthen the Feed-In Law. The management and top lawyers in the energy supply industry were unable to extend their influence over the rapidly developing renewable energy sector protected by this law, as long as they had not successfully warded off the legal proceedings against the "GDR power supply agreements" and taken possession of the energy production and supply network of the former GDR. At the beginning of September 1994, the last obstacles had been overcome and the Feed-In Law came under immediate attack. Only after some considerable effort and by standing shoulder-to-shoulder and closing the ranks right down through society was it possible to curb the onslaught. However, the delaying tactics employed by the large electrical utilities cost small and medium-sized operating enterprises a fortune, with some unable to survive financially as a result. This is how these large companies promote, protect and continue to breed new centralistic structures within the energy sector.

Since attack can sometimes be the best form of defence and with a new political constellation brought about by the change of government in 1998, the German Renewable Energies Law (Erneuerbare-Energien-Gesetz)—in full the Act on Granting Priority to Renewable Energy Sources—promoting large-scale generation of electricity from various renewable energy sources, was passed and implemented in 2000, as successor litigation to the Feed-in Law. Once again it was essentially Dr Hermann Scheer MP who, now a member of the ruling party, promoted the campaign und protected the law right up to his death in 2010, while those "responsible" within the government had to be dragged to the ball, as was the case with the successful lawsuit at the European Court of Justice, concerning the approval of the "subsidies" granted under the Feed-In Law.

Up until the end of the 1990s, the front line in the battle had been clearly defined: on the one side, decentralised renewable energy and on the other, centralistic fossil and nuclear energy sources. Since then, however, the political demands of the population have forced those in favour of centralised structures to become active within the decentralised renewable energies sector—even if only to provide themselves with a devious alibi. In the eyes of the centralised energy supply industry, which had grown so mighty in the spirit of the Kaiser, Siemens and Deutsche Bank, the current political state of affairs cannot possibly last much longer. Meanwhile, the industry acts according to the American slogan: "If you can't beat them, join them!" When the Feed-In Law was replaced by the Renewable Energies Law, the energy sector itself succeeded in becoming a beneficiary of the "subsidies" available for the supply of renewable energies.

Those wishing to see and contribute to a just and sustainable ecological and social development in Germany, would be advised to sever ties with the unscrupulous profiteering which is being carried out behind the camouflage of deceitful labels such as "sustainable" and "renewable". Otherwise their own credibility might one day be shattered and they could find themselves completely worn out and exposed to the ridicule of the established fossil and atomic energy sector. During their term of office, Chancellor Gerhard Schröder's red-green coalition government strengthened the dictatorship of the electricity supply industry significantly, under the guise of the so-called "liberalisation" created by Chancellor Kohl. And all this took place under the protection of Schröder's accomplice, Economic Affairs Minister Werner Müller (independent in political terms perhaps, but certainly not with regard to corporate interests). The energy supply sector in 2011 is demanding subsidy guarantees for its coal-fired power stations and monopoly guarantees for its new generation of large power plants and high-voltage transmission lines.

21.2 The Ongoing Destruction of Our Public Services and the Very Basis of Our Existence

Shortly before the legislative period 1998–2002 came to an end—and while all the protestors were on holiday—the German government shamelessly ruined their political record with regard to energy policy. Chancellor Schröder and his Economic Affairs Minister Werner Müller (a former E.ON employee and eligible for a company pension) flouted the decision reached by the Monopolies Commission and the Federal Cartel Office and gave the go-ahead for the takeover of the Ruhrgas AG Company by E.ON. The explicit aim of the merger was to limit competition in the German gas market. E.ON was to rise up to become a global player, financing its international conquests by means of guaranteed revenues, in other words by charging "unjustified and excessively increased" prices to its customers in Germany. At the same time, the concept of the multi-utility company as the ideal provider of services was being praised to the skies—"from the power station to the wall socket, from the borehole to gas cooker and from the source to the water tap". Already in 2002, gas prices in Germany were far higher than those of other EU nations.

Aristotle, who perceived that form is a more decisive factor than substance, would also have recognised that concentrating our attention on renewable energy sources alone will do nothing to ease the burden on our environment without the necessary structural changes within the energy sector. Those in favour of a centralised system know all too well that people with no knowledge of the past have no future. After all, there used to be numerous decentralised electricity producers both in Germany and Japan, but these were brutally eliminated and expropriated and are now more or less forgotten. And the centralists are sure to gain control of the present situation too. If, in the decentralist camp, we have no leaders with a knowledge of the history of our electricity production and supply networks, then the ripe fruits of all our labours are simply going to fall onto the laps of the proponents of centralism. Hence their strategy first of all to undermine the long-established hydroelectric power sector because after that it will be easy for them to gain control of the wind energy sector with its lack of historical awareness.

Those in favour of centralism cannot forget for one moment that it was the decentralised hydroelectric power industry under the leadership of Manfred Lüttke, vice-president of the German Renewable Energy Association, who bestowed on them the incredibly successful and—for their intents and purposes—highly dangerous Feed-In Law of 7 December 1990, written down clearly on a single sheet of standard-size paper! In 2002, Chancellor Schröder's old government was literally washed back into power by a very narrow majority as a result of both the severe flooding caused when the River Elbe ran at an all-time high and the threat of war in Iraq. The advocates of centralism among the realists in the Green Party subsequently informed the Renewable Energy Association that a new generation was now in power and that the time had come for the older generation to remove themselves quietly, important though their role may have been in the past. The association should now open itself up to the financial market, to large size wind farming, monocultures and offshore wind power.

Having served on the board at the Renewable Energy Association from its founding on 14 December 1991 until April 2007, I feel obliged to point out the reasons why it was founded in the first place and the aims it originally pursued. I continue to support those aims and promote them as a directive for the association's policies in the future. From 1962 to 1968, I worked at the Institute for

Experimental Nuclear Physics at the Nuclear Research Centre of the Institute of Technology in Karlsruhe. I can still picture quite clearly the terrible responsibility—and the helplessness—of those who, having designed the atomic bomb as an instrument with which to contain National Socialism, were later forced to witness the devastation and suffering caused by their well-meant invention after the bombing of Hiroshima and Nagasaki. This example illustrates what the chemical industry and energy sector—with their close ties to the state—are capable of if they are not kept under control.

The energy supply industry is ruthlessly transforming the concept of renewable energy sources, as a strategy to preserve the provision of our essential public services, into the exact opposite— the business of death and destruction. The word "sustainable" serves only as a wrapping to conceal the true nature of the products they put on the market and the destruction of our natural surroundings caused by their unethical business practices. Their ultimate aim is to make the public subservient to the oligopolies.

In the federal state of Brandenburg—an hour's drive from Berlin—whole areas, which were once green pastures and thriving landscapes, are turning into desert. Lusatia was formerly a lignite-mining region and so-called government "revitalisers" are now describing how savannah areas are beginning to appear there. And this in Lusatia, of all places, whose name derives from the Sorbian word *luzicy* meaning "swamps" or "water-hole". Only a century ago, this was one of the most water-rich and fertile areas of Central Europe! This progressive destruction is ultimately the work of eight decades of German policy in the energy and chemical sectors. Seen in moral and cultural terms, this is quite clearly wrongdoing of immeasurable proportions, protected by the dominant sciences jurisprudence and economics, now bereft of any real content. This destruction is an expression of what happens when the ruling doctrine of the day is constantly repeated in parrot-fashion. It was people with no awareness of the past who strengthened the old power and economical structures in the course of the last 65 years. Instead of analysing and reappraising these structures, they transfigured and camouflaged them. With their legal expertise, they defended the Kaiser's Mining Act of 1871 and Hitler's central constitutional institution "Inspector

General for Water and Energy".[3] To this very day, these forces have been able to prevent genuine liberalisation with regard to energy policy, as well as the establishment of an effective supervisory and regulatory authority.

For whole generations to come, the landscapes ravaged by unrestrained energy policies will stand out as conspicuous monuments to the ruthless exploitation and desire for further conquests of the RWEs, E.ONs, EnBWs, Vattenfalls and Co. These supra-regional multi-utility companies are supported legally and politically by the billions of euros of tax-exempt accrued liabilities available to the atomic power industry. These multi-utility companies can use this capital income unrestrictedly, for example, to buy out rivals or for breaking into new markets. At the same time, decentralised renewable power supplies are being depicted in a defamatory manner as a means of exploiting the public at large.

---

[3]The Energy Industry Act of 1935 restructured the German energy sector and laid down the framework conditions for a cheap and secure energy supply, in order to prevent any detrimental effects on the national economy which competition might have. Local authority districts were controlled by the Ministry of the Interior, and private companies by the Ministry of Economic Affairs. From 1938 to 1941, the German cabinet did not come together at all. At the height of Hitler's power and only one month after he launched his campaign to invade Russia, the Office of Inspector General for Water and Energy was created and the government ministers hitherto responsible for the energy sector read in the newspapers that they had been replaced, first by Dr Fritz Todt and later Albert Speer. The energy supply industry in post-war West Germany was structured according to the 1935 law, which provided for monopolies in power generation, transmission, distribution and supply. The Inspector General, however, was retained in the form of an inconspicuous footnote. This chapter has not yet been reappraised and so this relic lives on. After the war there were 11 Verbundgesellschaften (high-voltage network associations) in West Germany. Prior to market "liberalisation" in 1998, about 1 000 electrical utilities existed, 8 of which were involved in large-scale power generation and high-voltage transmission and about 80 in regional distribution. After "liberalisation", however, only 4 large companies existed in Germany.

About the author

Ulrich Jochimsen was born in Niebüll/ Schleswig on 28 June 1935. He grew up in a free atmosphere, taking part in Boy Scout activities for five years. An amateur radio enthusiast, he received his licence (DJ1PZ) in 1953. After serving a three-year apprenticeship as an electrician, he became the youngest radio officer in the merchant navy in 1955 and then sailed around the world for another three years. From 1957 to 1962, he studied electrical engineering in Bingen on the Rhine and spent a year as an exchange student (*via* the German Academic Exchange Service) at the Ryerson Institute of Technology in Toronto, Canada (1959–1960). He worked as an engineer at the Institute for Experimental Nuclear Physics at the Institute of Technology in Karlsruhe from 1962 to 1968. In 1966, he started his own company VIDEO-DIGITAL-TECHNIK, specialising in television studio technology. In 1973, he developed the pocket radio-telephone, that is, mobile phone, and founded the Institute for Telecommunications Technology and Systems Research that same year. He was the only representative of the state of Hesse in the federal commission on technical communications in 1974–1975, where he presented his concepts of the BLACKBOX (a wall socket for the individualised use of the telephone in the private home, separating the user's appliance from the monopolised telecommunications network, an idea very much ahead of its time) and the mobile phone (using the higher frequency bands for individualised communication). In 1976, he invented the ENERGIEBOX, a mini co-generation plant designed for use in individual homes for the decentralised production of energy and heat—as an alternative to building further nuclear power stations. He carried out a study on the ENERGIEBOX for the minister-president of Hesse from 1977 to 1978. Since 1978, he has been involved in the struggle for the use of decentralised renewable energies. On 26 April 2006—the 20th anniversary of the Chernobyl disaster—Ulrich Jochimsen was awarded the Bundesverdienstkreuz (Order of the Federal Republic of Germany, similar to the British OBE). Also in 2006, he received the German EUROSOLAR prize for his ENERGIEBOX.

Chapter 22

Wind, Women, Art, Acceptance

Brigitte Schmidt

Art Wind Farm, Lübow, Solarzentrum Mecklenburg-Vorpommern, Dorf Mecklenburg, Haus Nr. 11, D-23966 Wietow, Germany

solar.simv@t-online.de

"Sun worshipper"—that was how a journalist from the "Schweriner Volkszeitung"[1] once called me. I was born in Plessa, in the heart of the brown coal region of Eastern Germany, not far from the legendary Horno.[2] At the age of 4, I began to absorb 'the knowledge of the world', which my father had kept over the war years, and I wanted to become a scientist. At the age of 10, I installed an electrical wiring in my parents' flat; at 11, I illegally passed my moped license; and at 13, I learned typing without looking. My father Franz Kießlich—an artist and teacher—and I, armed with sketch-pads, wandered through the "moonscape" of the desolate territory of the once open-cast brown coal mines. My mother, a sales assistant and someone who knew the telephone books by heart, fought daily against the coal dust, being unable to dry the washing

---

[1]Article in *Schweriner Volkszeitung* (SVZ), 25.09.2006.
[2]Horno was one of the villages demolished and resettled by Swedish concern Vattenfall AB in 2005 due to lignite mining investments.

Wind Power for the World: The Rise of Modern Wind Energy
Edited by Preben Maegaard, Anna Krenz and Wolfgang Palz
Copyright © 2013 Pan Stanford Publishing Pte. Ltd.
ISBN 978-981-4364-93-5 (Hardcover), 978-981-4364-94-2 (eBook)
www.panstanford.com

in the fresh air. But it was a losing battle. My Grandfather burned to death in a coal dust explosion in the legendary briquette factory of Plessa, the place with the first conveyor bridge in Europe, on the verge of the 20th century.

22.1   How Did It Start with the 'Wind'?

I was sitting on a plane in the late autumn of 1996, and was sleepily paging through the in-flight magazine and was suddenly widely awake. Was it my face and that of my project partner from Hamburg Halstenbek, Rosemarie Rübsamen, in the magazine? The title of the article was "No Fear of Megawatts and Millions—an East-West Story of Success". Our story was now in numerous articles and our efforts were honoured with the EUROSOLAR Prize in 1996.

Figure 22.1   Brigitte Schmidt and Rosemarie Rübsamen with Dr Hermann Scheer, EUROSOLAR Prize ceremony, 1996.

It all began in 1991, with EUROSOLAR and meeting with Dr Hermann Scheer. I began to rethink energy, after 20 years of experience in teaching and research at the University of Wismar, at the University Eduardo Mondlane in Maputo, Mozambique, and studies at the Technical University in Kiev.

"Produce electricity yourself! Use wind power!" Tempting ideas. Ulrich Jochimsen and I drove to the "scenes of crime"—to the electricity rebels in Schönau and to Denmark to the world famous Folkecenter for Renewable Energy, to Preben Maegaard. It was being produced; electricity from wind and that was a reality in 1992. Why should we not also produce wind power in Mecklenburg-Vorpommern, in the crippled, voiceless or dumbfounded East Germany after the reunification?

Now always after giving lectures, presentations and tours of the Lübow Wind Farm, I ask my visitors, "what are the three basic requirements for a successful wind energy project?" Location, financing and grid supply—is the answer. The correct implementation of the basic conditions or parameters decides over the feasibility of a wind project. That was also how we wanted to start 19 years ago.

22.2 Planning Phase

It was not a problem to find a windy location in Mecklenburg-Vorpommern in 1993. However, we wanted it to be in our vicinity, in the community of Lübow. The community was supposed to be the 'landlord' of the ground on which the wind energy farm was to be erected. The community should be co-owner of the wind farm. The community was supposed to benefit from the tax advantages of the wind farm, a private-owned wind farm.

After discussions with the mayor, who wanted to know if the wind power generation was meant for electricity suppliers, he was immediately on board when we answered his questions. He suggested Mühlenberg near Lübow as a good location. "Well, let's go now and have a look at this place," was my immediate reply. In the bright moonlight, Mühlenberg lay hauntingly in front of the cemetery grounds, not far from the oldest church of Mecklenburg in Lübow from the 11th century. We enjoyed the magic of the moment, however, we decided to continue with all further investigations and explorations in the daytime.

I tracked down 13 possible locations for the wind farm, but one could not call that real planning. Then, an important 'wind woman' appeared, Thea Hefti from Switzerland. She was from an insurance company and was deeply convinced that planning without the necessary money for licenses was simply impossible. She provided a substantial amount for the licenses and referred me to Rosemarie Rübsamen, who was a successful wind farm planner and an anti-nuclear pioneer from Hamburg. That was the third woman to join our team.

Figure 22.2   Rosemarie Rübsamen, Brigitte Schmidt and Thea Hefti—the Windwomen.

The question about the organisational structure then arose. A book by one of our team members, Rosemarie Rübsamen's *Energiegemeinschaften* (Energy Communities) helped very much in exploring the advantages and disadvantages of company structures. We decided to found a limited company—the URS mbH, a company for the usage of renewable energy. 'U' stood for Urban, back then the managing director of the first East German photovoltaic factory in Wismar. 'R' stood for Rübsamen and 'S' for Schmidt. The location of the headquarters of the company was clear; it had to be in Mecklenburg-Vorpommern, in our village. There were too many examples of companies in East Germany being the "extended workbenches" of large production companies from the old federal states.

The Women's Wind Farm in Lübow is a wind farm in which the whole system of the realisation of the project, that is, of systematic planning, of construction, and system of the operation—was and still is in the hands of women. Involvement or investment in this project was always allowed for all townspeople. It is a citizens' wind farm.

Systematic planning—what an overblown concept, especially without even a telephone connection in the house. In East Germany before the reunification, telephone lines were like bananas—scarce goods. After having returned the borrowed and really heavy wireless telephone to Ulrich Jochimsen, I was left with the neighbour's telephone. Standing in the corridor, I used the phone to gather information on the characteristics of wind turbines, preliminary construction enquiries, legal regulations for the construction in Mecklenburg-Vorpommern, planning conditions for wind energy facilities in protected areas, etc. I can vividly remember the late-night discussions concerning the selection of the wind turbine company to deliver our equipment.

"We take Vestas", that was the majority decision of all those in our team who took part in the discussion. The wind power equipment of the Danish company Vestas was already well established on the market in those days. Asynchronous machines with a gearbox—robust, already proven and tested in Denmark. Also Nordex with their tubular towers were discussed. Enercon deviated from the traditional concept of asynchronous machines. They did not need a gearbox, did not need tons of steel lifted up to the nacelle. Enercon uses a ring generator—a synchronous machine that generates AC power. The control system includes an inverter which at any wind speed rectifies the fluctuating voltage and frequency. The DC power is then inverted in thyristor inverters at the bottom of the tower and transformed into AC power with normal grid voltage and frequency.

The first wind turbine manufacturer of the world, Enercon, was in those days able to offer a power grid certificate. The technical principle was not so new for me, since I knew this from my long-standing research at the Technical University in Wismar, from common three-phase shaft generators on ships and the principle of high-tension DC power transmission (HGÜ) which I was lucky enough to experience in Mozambique at the Cabora Bassa Dam.

Being a specialist in reliability, I felt uneasy when I thought about the over 180 possible failures that could occur during operation of a wind turbine, and since every single minute the turbine is not operating, there is a financial loss. I asked myself: how should it all be maintained?

The offer from Enercon was a partner concept. Enercon insured, so-to-say, the wind and guaranteed continuous production, if the wind was blowing. Immediately the next controversy in the team flared up. Even though the general opinion was that no re-insurer would insure this risk of Enercon, my decision was certain that if we did not use this benefit for the Lübow Wind Farm, which was essential for survival, then I would leave the team.

The preliminary building application for the project proved to be suitable for approval. After nearly three years of planning, a letter from the authorities fluttered into the office. Impatient as I was, I ripped it open, glanced at the first few lines and caught my breath. "(...) a couple of storks", and my heart dropped because I knew that meant we could not build the turbine here.

Figure 22.3 The stork and the wind turbine.

I threw the letter to the floor and broke into tears. We should have paid an ornithologist over the last year to count the birds on the site. Our efforts were all for nothing, the permission refused because of the presence of the birds. Then Ditmar Schmidt walked in. He had been my husband since 1971, a colleague, a partner through thick and thin. He picked the letter up from the floor, read it and gave it back to me. What the letter actually read was: "Since only a couple of storks (...), the building permission is granted."

22.3 Citizens' Participation, but How?

Financial investment of the citizens in renewable energy projects, in citizens' projects, is the obvious and well-known principle that fits with the decentralisation of renewable energies.

There is another important aspect—feelings. And we strongly believed that feelings should be expressed. The best instrument for the expression of feelings is art. The Mexican artist, David Alfaro Siqueiros, pointed this out in the best possible way: "Over the course of the last four centuries, virtually all principles, after which one has strived, have disappeared from art. We have to realise anew that a piece of art is created by humans for humans as an instrument for the transfer of feelings and expression of movements. That is why we have to develop our creative elements and adapt to our projects or tasks."

The decision was made; we would try to undertake a connection between technology and art. The aims of the first artistic wind farm were: Increase the acceptance of the farm through involvement of local artists and children; integrate educational elements through the conveyance of messages and feelings; distinguish the technical object—the tower—by using colour, since the white tower is the most disturbing factor of the so-called timber line which goes up to a height of 15 m; create a demonstration and pilot project with study options; convey art as a part of the solar movement; increase the advertising impact.

The selection of three very different local artists began very carefully at a point in time when it was not yet clear if we would get the building permission.

Franz Kießlich, born 1917, was a lithographer, a painter for the advertising industry, an art teacher in Triwalk. On erection of the wind farm, he was 79 years old.[3]

---

[3]The NDR local television channel reported on our artistic wind farm and about him.

Figure 22.4 Brigitte Schmidt with one of the artists, her father Franz Kießlich,[4] 1999 (Photo: Jan Oelker).

Hans W. Scheibner, born 1944, was an artist from the Leipzig school and stage designer from Maßlow and he was integrated in the preparation phase.

Figure 22.5 The artist Hans W. Scheibner.

---

[4]Franz Kießlich is one of the artists featured in the book "*Windgesichter— Aufbruch der Windenergie in Deutschland*", ed. Jan Oelker, p. 291, published by C-Macs Publishingservice, November 2005.

Willi Günter, with an exceptional talent for colours, lived in Trisbet near Schwerin in those days and spontaneously joined our team.

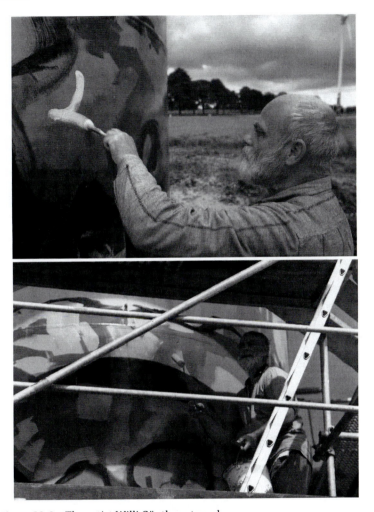

Figure 22.6   The artist Willi Günther at work.

In the meantime, we got building permission for four E40 Enercon wind turbines, with a power output of 500 kW. So, we were missing the fourth artist. We did not want a local artist, due to the pilot nature of this art project. Once again EUROSOLAR provided me with support and assistance. Jürgen Claus, born 1935, professor at the Academy of Media Arts in Cologne, resident in Baelen, Belgium, enriched the team also as a theorist of the "solar art." He made a short film about the Art Wind Farm, Lübow.

Figure 22.7 The artist Jürgen Claus at work on the "Preservation of the Oceans".

The single overall statement of the art project was provided through the appeal to retain the beauty of our nature in the broadest sense. The key to procurement was the single choice of motifs by the artists. The "Sun over the Sea" or "Preservation of the Oceans" were referring to the book *Planet Sea* by Professor Jürgen Claus, published in 1972.

Jürgen Claus himself wrote about his completed work on the Lübow Art Wind Farm: "The 'Sea tower' which we designed should remind us for a long time that the wind farm is only a few kilometres from the sea. The sea is a laboratory, an origin not only for climatic facts or data, but also for artists, a trove of wonderful forms and shapes worth retaining. A few of which gleamed from the 'Sea tower', for example, a radiolarian at the entrance, a cephalopod in the middle, or a Neptune Volute on the right side of the tower. For the observer, the short texts on the tower should close the gap on expanse of open sea: depth, rhythm, mirror, breaking of the waves (...)"

The execution of the project "Preservation of the Oceans" was carried out by Uwe Morzinek from Hamburg, Gorch-Christoffer Tieth from Schwerin, and the company HTW Wismar.

Figure 22.8   "Preservation of the Oceans" by Jürgen Claus: (a) seahorse; (b) cephalopod; (c) radiolarian.

The motif "The Tree of Life", chosen by Franz Kießlich, was a merging of two cultures—a visual transformation of willow, the typical northern German tree, with the typical African woodcut art. An African sculpture "Tree of Life" is a carved figurative group from a piece of sandalwood or blackwood. The willow knots, distorted hands, at the top end of the artwork signified STOP, STOP with the crazy waste of energy. A special feature of this design is the colouring in four segments following the four seasons: winter, spring, summer and autumn. Local children decided on the placement according to the compass.

Figure 22.9   The motif "The Tree of Life" by Franz Kießlich.

An extraordinary encounter in the Environmental Authority in Grevesmühlen brought us the colours back, and this is what happened: Public art also has to be authorised, as does everything in Germany. I took the four art drafts out of my bag and spread them out on the authority's desk. Everything according to the representation of tree trunks was held in a brown tone. The civil servant asked quite rightly, why everything was so brown. Could one not design it colourfully according to the four seasons? This suggestion took away our fear that the authorisation might have been turned down, and the artists completed their artworks in a blaze of colours.

"The Creation" by Hans W. Scheibner, expressed agricultural reality. Tractors driving downwards and crashing was a symbolic description of the agricultural situation after the reunification.

Figure 22.10   The motif "The Creation" by Hans W. Scheibner.

The motif "The Mecklenburg Countryside" by Willi Günter, was a sea of colours with lots of small details waiting to be discovered, as we see it in Mecklenburg from spring to winter.

To further enhance the level of acceptance in agreement with the artists, we organised a children's painting event, which entailed that children from the Pestalozzi Secondary School in Wismar,[5] from the Grammar School in Dorf Mecklenburg[6] and from the Secondary School in Lübow.[7] The children designed a meadow-ring around the base of the tower about a metre high. It should look like a real meadow with beetles, flowers, butterflies and blades of grass.

Figure 22.11   "The Mecklenburg Countryside" by Willi Günter.

Figure 22.12   Children painting the wind turbine (left); Finished children's painting (right).

Also, the "Save the Seas" tower had a "ring" of stones, shells and crabs designed by the children. When the "executors" Morzinek

---

[5]Judith Burba, Anja Bierwolf, Ramona Bierkandt, Astrid Nagel, Rommy Schäfer, Dörthe Simon, Susanne Ritter, Susanne Körner, Fanny Hamann, Bianka Kupfer, and teacher Barbara Ternes

[6]Carlo Wunrau, Mathias Paech, Robin Haselbach, Dörte Boxberger, Michaela Zinke, Kerstin Meurisch, Wiebke Häberei, Sissy Witt, Richard Weger, Kristin Eggers, Karla Haase, Janina Manzke, Henrike Barkmann, Stephanie Kupsch, and teacher Hans Kreher

[7]Manja Ernst, Stine Stefan, Nicole Kurth, Mandy Podszus, Christian Schäfer, Wiebke Voß, and teacher Ilse Dänhardt

and Tieth arrived with their spray guns and compressors to apply the blue base colour, they sprayed over the children's work. As a result of this inadvertent mistake, the tower was attacked by "graffiti artists" every year. We learned a lesson about community acceptance from that experience, and my task is to invite these "artists" to the wind farm and organise a day off from school, so that they can sandpaper away their "artworks". Frank Kießlich touches up the free spaces from year to year.

The realisation of the art project was only possible after intensive research. Three alternatives or variations were examined to solve the problem of how the artists would get their towers painted. One way was for each artist to paint while sitting in a basket attached to a crane. Another option was for the artists to paint in the production hall of the tower, although the production then had to be adjusted to the work pace of the artist. The third alternative was the possibility of using normal scaffolding around the tower.

All three choices were more or less unaffordable, so another possibility was additionally checked—self-adhesive foils on the towers. Luckily, fortunate circumstances came to our aid. A construction company from Wismar wanted to subscribe shares in the wind farm and heard about the art project. Their enthusiasm was so big, that we received a special offer for the scaffolding for the four towers over a period of four weeks.

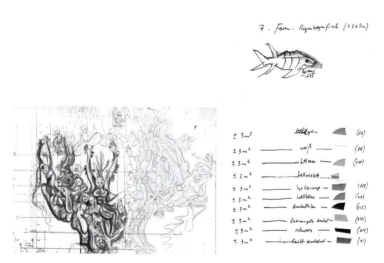

Figure 22.13   Examples of design drawings with matrix for calculation of the amount of paint.

We were also looking for potential solutions for the preservation of the adhesion of the paint on the dried primary coat of the towers. The solution was that Enercon would have to deliver the towers without the normal green coat of paint. It took a lot of time to solve the difficult task of deciding on the quantity of paint for each motif and also the special colour mixing in the paint production company. For this reason a to-scale grid was put over an artistic paper print, after which all the surfaces of each colour were calculated and multiplied by the scale. The artists worked with 42 different colours, with which they painted 600 m² of tower surfaces and used 153 litres of paint.

The beginning of the painting process was quite interesting. The towers stood a week inside the scaffolding, but none of the artists appeared. The artists had between 3 months and 2.5 years time to identify themselves with the project. Nearly two weeks had passed after the towers had been put in scaffolding and there was still nobody around. I drove to the tower with my father, Franz Kießlich, took a brush in my hand and started with the first strokes. He tore the brush out of my hand and perplexedly began painting for nearly three weeks. One by one, the other artists began to appear at the wind farm.

Figure 22.14 Wind turbine in the scaffolding.

It was possible to stick to the planned inauguration date of the wind farm, 22 June 1996. Nearly 400 visitors came to the wind farm, accompanied by jazz music, and witnessed the "christening".

Figure 22.15   Inauguration of the Art Wind Farm, Lübow.

Each wind turbine was christened with a female name from four different generations. The wind turbine "Save the Seas" got the name Sophie (Sophie Schmidt), my five-year-old granddaughter, representing the generation which will implement the energy turnaround. She was still so young, that she could just barely hold the bottle of sparkling wine with which the dedication of the name was carried out.

"Mecklenburg Countryside" got the name Lotte (Lotte Bösch), the oldest citizen of Lübow, who brought a baking tray of cake every day to the building site during the construction phase. She was so fascinated by the new technology and the accomplishment of constructors that she could not sleep in the night before the inauguration, so she got up, fetched her traditional costume from Mecklenburg, and wrote a poem.

Figure 22.16   Sophie Schmidt (left); Lotti Bösch (right).

"The Creation" got the name Heidrun (Heidrun Rudolph), an employee of the Department of Trade and Industry of Mecklenburg-Vorpommern, representing the generation that constructed the wind farm. Under her guidance we received the last financial support for wind turbines in Mecklenburg-Vorpommern.

The wind turbine with the topic "The Tree of Life" got the name Rosa (Rosa Fischer), the grandmother of Rosemarie Rübsamen. At the inauguration, Wolfgang Lüdtke, the mayor, received the first solar car in Eastern Germany from the wind farm organisation. It was a green "Meck-Mobile", produced by the company Solar Nord in Wismar, with a photovoltaic roof and without doors for the summer. Dr Ditmar Schmidt was the development engineer.

22.4   No Fear of Megawatts and Millions

We needed a financial concept for the DM 5.5 million (EUR 2.75 million) of investment costs of the Lübow Wind Farm. The responsibility for this was in the hands of Rosemarie Rübsamen. The annual income was calculated on an independent wind survey. An investor brochure was put together with the help of external tax expertise. Doubt arose with the external forecasted output of 1 million kWh/year per wind turbine, so that our experienced planner Rosemarie Rübsamen had to put together another wind survey, after which the output had to be corrected downwards to only 800 000 kWh/year per wind turbine. After a preliminary inspection and always looking to be on the safe side, a further safety margin was calculated and the output corrected downwards again. This farsighted and cautious approach later proved lifesaving for the survival of the Lübow Wind Farm.

And this is what happened: In 1996, it was experienced all over Europe that here was 10% less wind. Additionally to our own capital, we had taken out two loans to help financing the Lübow Wind Farm, a DtA and an ERP loan through KfW, a the German reconstruction bank, with a repayment phase of 12 years. Through the lower output as a result of the reduction in wind, we experienced financial constraints in paying back the loan and so we could have not stick to the 6% rate of return. To avoid an overdraft of 10% or more, the limited partners themselves invested new funding in the wind energy company over a limited period of time at a 6% interest rate. In this way, 16 years later after the construction, we are still owners of the wind farm. Many other companies, who did not plan so carefully, have lost their wind farms.

22.5 Friends from All over Europe

Our aim was to get capital resources through deposits from the citizens of the region. The citizens of Lübow and the region were very sceptical after the reunification, so we reduced the share limit down to DM 1 000 (EUR 500). But that was also not sufficient. So we recruited investors among my European friends and Rosemarie Rübsamen among her German friends. The result was the first European financed wind farm with investors from Austria, Switzerland, Sweden, and East and West Germany.

The financial participation of our friends could not be valued highly enough but the foreign financial participation also could not be claimed on tax. German investors in those days were able to use investments in wind energy by writing them off on tax.

We had thought of a special concept for the community. We paid the 20-year leasehold on the municipal ground in advance and hence gave the municipality the possibility to financially participate in the wind farm as a silent investor. Today, after the repayment of the loans for the Lübow Wind Farm, a five-figure amount of tax flows into the local budget.

So far not only interest in the wind farm was shown by young visitors from Germany and from Europe but also from Mexico, Togo, Ghana, Mozambique, China, Russia, Bangladesh, Thailand, Israel, etc.

Figure 22.17   Groups of young people visiting Art Wind Farm, Lübow.

Figure 22.18   Women of every age climb up wind turbines.

I can especially remember a delegation of professors from the United States, who had found the art wind farm on the internet and then decided to visit us. Whoever "climbs" the wind farm, receives a certificate. One professor confirmed to me that a large part of "green investments" worldwide were made by women. Especially women climb the wind turbines.

22.6 The Justice of the German Electricity Industry

In December 1995, the day of the ground-breaking ceremony was approaching. Late in the evening a fax with two attachments arrived in my office. "Dear Ms Schmidt, help me!" was the first sentence. What had happened?

The operator of a wind power plant near the town of Güstrow in Mecklenburg-Vorpommern had received a letter from energy supplier WEMAG Schwerin with approximately the following content: "The German electricity feed-in law is in our opinion unconstitutional. We have filed a suit. Should we win then you have to pay the complete money back to us which we have paid you for the electricity (...)" There were two signatures, one of which was Berthold Fege's. I have remembered his name since later, after the WEMAG had dismissed him, he became a member of the organisation Solar Initiative Mecklenburg-Vorpommern e.V. which I founded in 1997. A letter with the same content was sent by the WEMAG to the bank financing the operator of the wind energy park.

I immediately forwarded the documents to Dr Hermann Scheer, President of the European Renewable Energy Association, EUROSOLAR, and member of the German parliament, who then in a letter to the WEMAG made it clear that Germany is not a Banana Republic, and that the electricity feed-in law was approved in parliament.

The WEMAG apologised; the filed suit was decided on 10 years later against the German electricity industry. This, however, did not stop the electricity industry filing a further suit against the next renewable energy law (EEG). This incident lead to a huge campaign of uncertainty stretching from Mecklenburg-Vorpommern to

Bavaria. These cases were documented by EUROSOLAR. As a matter of fact in those days the estimated 300 MW of wind power in Germany was not reached.

Figure 22.19   An article in the *Neues Deutschland* newspaper, with headline "Wind Energy in Trouble. A campaign of giant energy suppliers against the renewable energy law has consequences—less contracts", 7 May 1996.

What has this got to do with the Women's Wind Farm in Lübow? Well, I had to put our investors into the picture immediately, before the ground-breaking ceremony. What would happen if the electricity industries filed suits were successful? Would we go ahead with the construction of the Lübow Wind Farm? A few male investors left us. The female investors were unanimous in saying "Now more than ever!" So we went on with the construction!

Three of the four locations got a pillar foundation because of partial permafrost in the soil. Respectively, two reinforced concrete piles in each of the four foundations had to be rammed 11 m to 13 m into the ground. It was already dark, winter, cold, minus 19 degrees and a massive pile driver with a hardwood head and a weight of 40 tons hailed down on the end of the first reinforced concrete piles. The ground under me began to vibrate and the hardwood head instantaneously caught fire. The storm pulled my fur hat and I could not really see the dramatic fire spectacle. This repeated itself so long until the calculated depth was reached.

The next morning the sun was shining, the snow glowing on the field. And the extruding remains of the reinforced concrete piles were being shortened with a pneumatic hammer by a worker from the Danish supplier.

22.7 Operational Availability

The output of the single wind turbine E 40, levels off at between 780 000 kWh/year and 820 000 kWh/year. Thanks to the negotiated partner concept right from the beginning, the Lübow Wind Farm has an availability of 96.5% to 100%. After nine years of operation, the rotor blades were all exchanged.

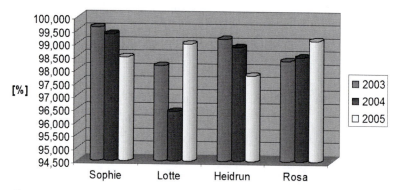

Figure 22.20 Operational availability of the four wind turbines at the Art Wind Farm, Lübow.

Figure 22.21 Changing blades.

22.8   What's Next?

Founded in 2001, the 100% renewable energy region Lübow-Krassow, with eight municipalities, profited for a long time solely from the output of the Lübow Wind Farm. Today seven further

Figure 22.22   Art Wind Farm, Lübow, in May.

municipalities have been added to the region and the plan for 100% renewable energy is 180% overfilled. The next wind farm is being planned.

About the author

Brigitte Schmidt, born in 1948, completed her degrees as an electrician from professional schools of Lauchhammer and Bitterfeld, Germany, in the years 1965 and 1967, respectively. She obtained her Dipl.-Ing. (electrical drive and automation) degree from Technical University of Kiev, Ukraine, in 1973 and Dr.-Ing. (reliability of asynchron-motors) degree from Technical University, Wismar, Germany.

Since 1992, she is concurrently chair of Engineer Association of Mecklenburg-Vorpommern VIW e.V. for renewable energies and EUROSOLAR Association Mecklenburg-Vorpommern. She is also a board member of EUROSOLAR Association Germany, Bonn, since 1992. She is the managing director of Wind Power Plant Lübow Ltd. since 1993; Solar Association "Solar Initiative Mecklenburg-Vorpommern" since 1997; and Solar Transport Ltd. since 1997.

Dr Schmidt is a chartered engineer and government consultant for renewable energies. She is an international renewable energy and regional development consultant, advising on renewable energy policies, sustainable development and industry strategies for governments and small and medium enterprises. She is also a very experienced trainer, project manager and university professor. Dr Schmidt is specialised in training in renewable energies for politicians, engineers, public, schools, students, professors from eastern European countries; strategic planning, work programmes and project management; cross-border programmes. Her experience in industrial sector includes planning and management of wind power plants.

Not only is Dr Schmidt an expert in planning and management of European projects and training of university staff and other organisations in the field of renewable energy (solar energy, wind energy, biomass, solar architecture, energy-efficiency and solar reconstruction), she excels in the production of presentation and training materials for renewable energies.

Chapter 23

California: Wind Farms Retrospective

Arne Jaeger

Angerbenden 23, 40489 Düsseldorf, Germany

ajwind@web.de

23.1 Introduction

In the aspiring global wind power industry today there is hardly a book that does not deal with, or at least mention, one particular event in wind power history: The great Californian Wind Rush. It may sound a bit like the old wild west times and the thrilling gold rush that everyone wanted to enjoy—and it was.

Backed by a huge package of incentives, the first large wind farm development began in California in the early 1980s. What followed was an unprecedented rush for wind turbines. This rush changed some pieces of landscapes from abandoned and seemingly worthless places to wildly popular points-of-interests companies fought for. Within a couple of years lonely rolling hills and flat desert terrains were dotted with thousands of different units quickly whirling away. Wind turbines with two blades, three

Wind Power for the World: The Rise of Modern Wind Energy
Edited by Preben Maegaard, Anna Krenz and Wolfgang Palz
Copyright © 2013 Pan Stanford Publishing Pte. Ltd.
ISBN 978-981-4364-93-5 (Hardcover), 978-981-4364-94-2 (eBook)
www.panstanford.com

or four spinning clockwise or counterclockwise or others resembling an egg beater created a totally new spot for tourists and for some—a new kind of income.

What happened there in the Golden State was a new and unforgettable experience for everyone, in both respects. A rise is often followed by a collapse, and so it happened in California. Since there was no extension for tax credits, one of several incentives, there was no rush for the future. By 1986 a harsh crisis overshadowed the young wind industry and many companies failed to survive that crisis.

However, some companies managed to continue. And despite a bad crisis, wind power has emerged to become a substantial energy resource today. The industry is still growing worldwide, and obviously people have learned from the Californian lessons. Thus, it is worth to have a closer look at that phenomenon.

This article is supposed to give an impression of what is generally referred to as the "Californian wind farms". A retrospective of the 1980s is dealt with in detail followed by an overview of the situation 25 years later.

The experiences and the know-how gained at the wind power industry's largest test field had a deep impact on the whole development that followed until today. Thus, when talking about wind power, it is essential to know something about California.

23.2 Historical Development

23.2.1 The 1970s US Wind Activities and Background of the Late 1970s for the Californian Case

In the 1970s the foundation for a far-reaching wind development was set. The 1973/74 oil crisis led to a global shock, and alternative ways of producing energy were researched. Denmark was one of the leading countries to initiate its own wind energy development. The private sector in Denmark was the motor for making the country a synonym of modern wind energy utilisation. Parallel to Denmark, the United States was the second country in the 1970s to start wind energy research on a large scale. In 1974, the Federal Wind Energy Program kicked off. This was the largest research program for wind energy in the whole decade. The governmental support played a major role. The overall aim of the research program was to quickly develop reliable wind turbines of various sizes ready for mass production and application in the United States. Support was given to development of small and intermediate size wind turbines for remote application and large wind turbines for use in grid-connected wind farms.

In 1978, another important decision was made in the US wind development. The young industry realised that the utilities would play an important role in future wind power development. Thus, political regulations had to be created to harmonise the utilities with the new energy source. The Public Utility Regulatory Policies Act (PURPA) was passed and aimed at the creation of a bridge between smaller power production units (wind turbines) and the big utilities. The "Small Power Producers", as they were called, should be given interconnection with the utilities' power lines, be paid fairly for feeding the grid with their produced power. Furthermore a basis was given for an overall improved cooperation between the utilities and the new group of small producers.

Apart from governmental measurements, the state of California set its own goal aiming at a 1% share of renewable energy in 1987, and a 10% share by the year 2000. In the late 1970s an initiative was taken by governor Henry Mello, who brought this ambitious goal on a political level by passing an act named after him. The act led the state to introduce an additional set of incentives making California the most attractive state in the United States for the wind industry. In addition to the PURPA benefits and a package of federal tax credits, California offered liberal state tax credits, accelerated state depreciation of the wind turbines, and offered low interest loans and small business loans. These individual incentives were available since 1978 and 1981, respectively.

What made California even more attractive for investors and developers was the fact that the California Energy Commission (CEC) played an active and influential role in wind energy development. While other states had similar or even more lucrative incentives, the CEC researched and produced reports on potential California wind resource areas and how to develop them. Thus, investors and operators (i.e., small power producers) were given enormous financial and strategic and organisational incentives.

Knowing where to put up wind turbines gave California a highly valuable benefit and clearly made a difference to other US states who did less or no research on their wind resources [1, 2]. Thus it was more risky and difficult to develop wind farms outside California despite all the tax credits and incentives the other states offered.

23.2.2 1981–1985: The First Five Years

The very first wind farms were placed in three areas: The Altamont Pass area in the north, the Tehachapi Pass north of Los Angeles and the Boulevard Desert close to San Diego. The San Gorgonio area, near the famous city of Palm Springs east of Los Angeles became the third largest area but its development began three years later. All of these areas were identified by the CEC as having high potential compared to many other sites in the state that were either left out for various reasons or developed much later.

Wind turbines were first set up in 1981. In Altamont it was Fayette[1] and US Windpower (USWP)[2]; in Tehachapi, Storm-Master[3] wind turbines were favoured and in the Boulevard Desert, the four-bladed Butler[4] turbines made up the wind farm. The range of these early wind turbines was from 40–56 kW.

In the first year, wind farms were generally quite small in the beginning—totalling around 150 units in the whole state. While USWP both acted as producer and operator of wind turbines and sold the electricity to the utilities, others, like Windpower Systems, only delivered their machines to their customers (such as Zond Systems in Tehachapi).

The American concept featuring a lightweight construction, passive yawing and high rotation speeds was very popular at

---

[1]Fayette was a company from Pennsylvania established in 1978. Fayette quickly became one of the biggest manufacturers and operators by 1985 with some 1600 units deployed in Altamont.

[2]USWP was the largest US manufacturer/operator with ca. 4200 units in California by 1992.

[3]Windpower Systems was a company based in San Diego. Its StormMaster turbine was known for its slender blades but often failed catastrophically due to their extreme light-weight construction.

[4]Butler was a company that produced a wind turbine designed by Terry Merkham, an industry pioneer of the 1970s.

that early time. The wind turbines used for "wind farming" were almost all derived from 1970s federal research activities. These designs were generally based on private engagement. Their development was furthered on a federal basis in order to obtain reliable and economically viable machines for a future US market. Some of the Californian wind turbines were tested at Rocky Flats, a test centre in Colorado. Often a Rocky Flats test certificate made a wind turbine more attractive and trustworthy on the market.

The first years of wind farm development were characterised by a lack of regulation. It was totally sufficient if a wind turbine— no matter which type—was just put up. No one asked for reliability, any references or if the machine was actually working. This inevitably led to fraud. For example, it happened that tube towers with a wooden board bolted at the top were put up on a field. A photo taken from the far distance led investors to think it was a professional wind farm. It was anything but a wind farm.

Figure 23.1   Panoramic view of Danish Nordtank 65/13 turbines with silver-grey stepped tubular towers, typical for early Danish wind turbines, Tehachapi, 2010 (Photo: Arne Jaeger).

In 1982, wind farms expanded slightly, and more American manufacturers entered the market, such as Carter and ESI. Contrary to the American market dominance there was a punctual but remarkable development at both Altamont and Tehachapi. Among the hundreds of American machines two Danish Vestas turbines were put up in 1982 in Altamont. Additionally, five should be Belgian WindMaster 75 kW machines were placed nearby and started operation the same year. And Oak Creek Energy Systems, a

Tehachapi developer, chose six Danish Danregn Bonus 65/13 machines among the American units already operating at its site. The Bonus and Vestas units were 65 kW machines equipped with a 3-bladed 15 m rotor operating upwind. They represented the typical Danish design philosophy of heavy weight construction, slow rotation speed and active yawing.

Figure 23.2 Early American Century CT-6 000 60 kW with double tail vane at Tehachapi, 1985 (Photo: Preben Maegaard).

Backed by early negotiations of Danish company representatives (amongst others: Wind Matic) since 1982, more and more Danish machines were chosen by American developers in the years to come. The three companies mentioned before were followed by Micon, Wind Matic and Nordtank, all of whom were major Danish wind turbine manufacturers. Since 1983 they delivered their machines in thousands to California. Parallel to Denmark, Bouma and Polenko from the Netherlands, MAN from Germany and Howden from Scotland are the European companies of higher importance during the rush years. The federal and state incentives intended to develop an American wind industry more and more became a blessing for the Europeans. In 1984, Americans still made up large parts of the market and had thousands of units sold. But more foreign manufacturers entered the market. European wind turbines became quickly established in California due to their different but more reliable technical approach, a fact that particularly applied to the Danish. Two terms were popular at that time: "American Design" for all US manufactures (see previous paragraph) and "Danish Design" referring to economical and reliable Danish wind turbines.

The Danish influence in wind farm development led to a dominance of 3-bladed wind machines. But still many different concepts with two, three or four blades operating up- or downwind appeared. Anything having blades, a generator and a tower was "considered" a wind turbine. Various "mixes" of blades, towers, nacelle designs, rotation directions, colours and other aspects of visual appearance were typical for California's wind rush and created a colourful and diverse mass of spinning rotors dotting the landscape.

The 1970s US research also gave birth to unknown designs such as the vertical-axis wind turbine (VAWT) that took a great upward trend in the California wind rush. There were two major companies, FloWind[5] and Vawtpower,[6] who developed the vertical Darrieus turbine further on the basis of intensive 1970s research. The two firms offered machines ranging from 100 kW to 250 kW. With their egg beater and onion-like appearance, FloWind turbines gave the Altamont and Tehachapi Pass a new look while VawtPowers caught the attention of people in the San Gorgonio area. Unlike conventional wind turbines, the roughly 500 VAWTs deployed in the three resource areas had two aluminium blades instead of fibreglass blades. Manufacturing such blades was based on a special extrusion process, first applied by Sandia Laboratories in the mid-1970s. VAWTs were only built from 1983 until 1985. Technical inferiority to horizontal-axis turbines finally made them disappear from the market. The biggest problem with VAWTs was that they had to be started with a motor before

---

[5]FloWind was founded in 1982. After the rush years, FloWind faced financial problems and remained a small light in the industry. In the early 1990s FloWind developed an advanced three-bladed 300 kW VAWT but disappeared in 1996.
[6]VawtPower sold 40 units for a wind farm at San Gorgonio.

electricity was produced. They were not able to start themselves. Various design changes were attempted to improve their operation, such as the use of glass fibre blades. But none of them succeeded.

Figure 23.3 American vertical-axis Flowind F-17 turbines rated at 170 kW at Altamont Pass, 1985 (Photo: Preben Maegaard) (left); A Transpower turbine installed in San Gorgonio Pass area, Palm Springs, California. Note sail-wing blades, the wheels and the drive train located on the ground in the foreground (Photo: Flemming Hagensen) (right).

Probably the world's most unconventional approach to use the power of wind was the one of Transpower and West Coast Wind Power. These machines certainly worked with the power of wind but not like any other wind turbine did. Instead of a 360° revolution the Transpower machines had two "blades" tied to a rope that was supported by two wheels. The rope, pushed by the wind *via* the blade, was revolving around a wheel that drove a generator. The blade consisted of somewhat of a sail wing and led one to think of a classical sailboat. Transpower built a 200 kW and West Coast Wind Power a 150 kW model. For sure, these "wind turbines" were underrepresented in the wind rush and only a handful were built. They could be found in Tehachapi, San Gorgonio and Salinas Valley.

The dramatic development that took place in the Golden State might become more obvious when looking at numbers. While there were "just" 150 machines installed in 1981, the state had roughly 1150 new turbines online in 1982 and some 2400 turbines were added in 1983. The largest addition was in 1985 with a plus of 4 989 new wind turbines. The new wind farms spread like wildfire. Though California was not considered as a large test field for dozens of different designs, it was in practice. Despite all the research that was done before the wind rush, wind turbines still did not have the technological maturity they should have. In the mid-1980s no one was completely sure which design or which type would succeed in the end. The whole world of wind power was gathered in California. Thus, the rush always included some kind of "Darwin attitude".

Figure 23.4   Sixty-five kilowatt Lolland wind turbines manufactured by Danish Windpower Productions installed at Tehachapi. In the 1980s it had the reputation of being the "Rolls Royce of Danish wind turbines". It is based on the late 1970s "Blacksmiths" design developed by NIVE. In this issue an important key person was Preben Maegaard of the Nordic Folkecenter for Renewable Energy (Photo: Arne Jaeger).

With rising numbers of installed machines came a growing demand for larger units. Larger machines promised more cost reduction and higher tax revenues. In the early years the average size was around 50 kW, but in 1985 it reached 100 kW. Especially for Americans rapid up-scales were common. Manufacturers believed that high average wind speeds could increase output if larger generators were used in machines with smaller rotors. At first there seemed to be no obstacle to use a 75 kW generator for a wind turbine rated at 56 kW like Fayette did. After three years of operation experience, Fayette made the step to 400 kW. ESI made a jump from 65 kW to 200 kW, which was a massive step at that time. Hurrying led to a loss of proper wind turbine development. Quality of production, adequate siting and site development suffered. Soon, operators faced tremendous problems with hundreds of inoperative machines. For some, too rapid up-scaling, failing machines and uncertainty about the future of the tax credits turned the wind rush into a nightmare. It was the other side of the coin.

Figure 23.5 Danish Micon 65 kW operating without nacelle cover, Altamont Pass, 2010 (Photo: Arne Jaeger).

In Denmark, however, the 1980s saw a completely different philosophy. On a private level, manufacturers started their development with tiny machines around 15 kW and continuously up-scaled them in small steps. Design was always adapted to size, and technical know-how was gained during all steps. A strong network was built, and know-how was shared. The success story had its roots in the mid-1970s. In the end the Danish philosophy proved much more reliable and economic [3]. An easy maintenance and better safety measurements made the Danish models highly popular. Without rapid up-scaling, technical and economical risks were kept low. This advance led to a massive Danish presence in Californian wind farms.

The young industry existed under enormous pressure: The tax credits were destined to end on 31 December 1985. Hence, time was quickly running out—be it Danish companies that produced seven days a week and year-round or American developers hurrying to get their giant blocks of machines installed before the magic deadline. In the mid-1980s the wind business turned into an extremely risky adventure for everyone. Danish suppliers sent their units to the United States without knowing if the tax credits were extended. American developers ordered machines on the assumption that they were extended. The months before 31 December saw many Danish machines staying packed in US harbours and the desert fields. Nobody knew what would be the answer of the White House concerning the tax credits extension. Hundreds of machines never left containers and were shipped back to Europe.

Main information:

- Continuous and strong rise of wind farms from 1981 to 1984. Figures decreased slightly in 1985 and 1986.
- American wind turbines were first used and ruled until roughly 1984.
- Meanwhile Danish (European) machines dominate more and more.
- Americans used light-weight constructions while Europeans preferred heavy-weight designs.
- Too-fast growth of wind turbines, especially on the American market.
- European designs definitely surpassed American designs.
- California's wind farms consisted of a very diverse mix of concepts.
- Three main resource areas: Altamont Pass, San Gorgonio Pass, Tehachapi Pass. There were a total of seven different areas.

Figure 23.6 American ESI turbines being raised at San Gorgonio Pass, 1985 (Photo: Preben Maegaard).

Figure 23.7 Vestas machines forming the famous "Wind Wall" at Tehachapi (Photo: Arne Jaeger).

23.2.3 1986–1990: Crash, Market Shrinking and Repairs

With the expiration of the extended tax credits on 31 December 1985 came a harsh and overwhelming crash not only for the US wind industry, but the whole wind industry. For a long time everyone hoped for another tax credit extension, but was left in doubt for too long whether this would become reality or not. Wind energy found little sympathy in the White House around 1985/86. One reason was a low oil price at that time.

Nearly all American manufactures went bankrupt since they could not balance the sudden gap between investing in new developments and the lack of new orders. Of the dozens of firms that were involved in the wind rush, only two survived. These companies were the biggest and most successful before the crash—namely USWP and Fayette. They acted as both manufacturer and operator, giving them enough financial strength to survive.

But the United States was not the only one hit. In Denmark, almost all companies faced enormous financial problems due to cancellation or lack of orders and too high expectations about the future of the wind industry, especially the US market.

The disappearance of the tax credits led to a market and industry shrinking since 1986. In order to survive, it came that manufacturers, operators or foreign companies of the energy industry merged, shared operation management or took over wind farms or wind turbine production. The many operators who went bankrupt often left huge areas of developed or undeveloped "wind land", as suitable areas were called. In the American wind sector, a handful of operators and manufacturers still existed by the end of the 1980s. The industry was consolidated and slowly came back to life.

Around the mid-1980s, Danish companies had taken over the US wind turbine market. Due to less Danish bankruptcies and more suffering American companies, the Danes enjoyed slightly more opportunities in the second half of the 1980s. For example, DanWin realised some projects between 1987 and 1989. A few more Danish firms that came up around 1985 were involved in this "post-rush-phase" but their projects remained at the planning stage.

The second half the 1980s was principally a time of re-development at the engineering level. In the rush years, manu-facturers often had paid less attention to development than to production. Hence, development of a wind turbine was several steps behind production. This simply meant that wind turbines were produced that did not match the idea of what is called "technically mature". Selling them was preferred—a fact that applied to both American and some European turbines.

After 1985, manufacturers and operators started programs to rebuild, modify, restore or repair their turbines. This was to bring failed wind farms back to operation, make revenues and improve the image of wind energy that suffered substantially after the failure of thousands of wind turbines across California. "Retrofitting" became a famous term describing the exchange of wind turbine components for overall improvement. Components exchanged were mostly blades but also gearboxes, generators, nacelle covers and other parts.

A good example of a highly serious technical problem was the shutdown of a cumulative 1 000 Micon turbines in the San Gorgonio Pass between 1987 and 1988. Due to blade failures several wind farm blocks, each consisting of hundreds of Micon machines, were brought to a halt—one after another. The Micon turbines, rated at 108 kW, used Danish made AeroStar blades that either showed cracks in the blade root or improperly working tip brakes [2, 4]. For millions of US dollars, the blade tips were fixed, faulty components were exchanged and the industry had learned another hard lesson. Such technical improvements remained a common practice in the 1990s up until today.

Figure 23.8 Danish Micon 108 kW, Tehachapi, 2010 (Photo: Arne Jaeger).

Another new challenge for American and European manufacturers were Asian manufacturers like Mitsubishi or Sumitomo. Mitsubishi started its own development in 1980 without any participation in the rush years. A cheap and reliable 250 kW model was used in Tehachapi and Hawaii in 1987. A first set of 20 machines was followed by another 660 between 1989 and 1991.

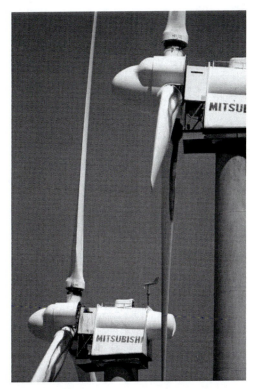

Figure 23.9   Japanese Mitsubishi MWT-250 running at Tehachapi/Mojave, 2010. Note oil leakage at the blade root (Photo: Arne Jaeger).

There were more Asian firms trying to enter the US market. A dozen of highly advanced two-bladed Delta machines made in Japan were put up in the Altamont Pass. Another company, called Sumitomo, built a 200 kW light-weight prototype at Tehachapi. But neither Delta nor Sumitomo managed to realise larger installations and thus vanished the years later.

Main information:

- Extreme crash after 1985 tax credits expiration
- American market strongly diminished—European market troubled

- Industry shrinking since 1986. Many bankruptcies, takeovers, cooperations, etc.
- Lots of repair, modification, retrofitting since the mid-1980s
- New Asian influence, especially Mitsubishi, Delta, Sumitomo

Figure 23.10   Japanese Mitsubishi MWT-250. Approximately 660 machines are running at Tehachapi/Mojave. They were put up between 1987 and 1991 (Photo: Arne Jaeger).

Figure 23.11   DanWin 160 kW at Tehachapi, 2010. This machine won an industry design award in 1988 (Photo: Arne Jaeger).

Table 23.1   Top 10 wind turbine manufacturers 1981–1991

American	Size (kW)	Period	Number installed[a]
USWP	50, 100	1981–1991	4025
Fayette	56, 75, 95, 250	1981–1986	ca. 1500
Energy Science Inc.	50, 65, 80, 200	1982–1985	726
Earth Energy Systems	17.5, 20	1984–1985	630
Enertech	25, 40	1983–1986	536
FloWind	170, 250	1983–1985	530
Carter	25, 250	1982–1986	331
Century	60, 100	1982–1983	316
WindPower Systems	40	1981–1984	311
WindTech	75	1983–1984	261
European			
Vestas (DK)	65, 90, 200, 225	1982–1991	2357
Micon (DK)	30, 60, 65, 108, 250	1983–1988	1630
Nordtank (DK)	65, 75, 150	1983–1987	1207
Bonus (DK)	65, 100, 120, 450	1982–1990	ca. 1000
Wind Matic (DK)	65, 95	1983–1985	345
MAN (D)	40	1984–1986	323
DanWin (DK)	120, 160	1985–1987	233
HMZ (B)	50, 200, 250, 300	1982–1985	184
Howden (GB)	60, 330, 750	1984–1985	96
Danish Wind Power Production (DK)	65, 110	1985–1986	71
Asian			
Mitsubishi (JPN)	250	1987–1991	660
Total[b]			**17 243**

Source: Refs. 5–10.

[a]Cumulative figures based on reports from the California Energy Commission, the Electric Power Research Institute and company brochures. Figures are approximate and do not automatically match the real number of installed units.

[b]The total varies steadily from year to year. Total number of installation given in this table is approximate and not specific to the year 1991.

23.3   The Retrospective in Detail

23.3.1   What Can Be Seen Today

When visiting California's wind farms in 2010, no matter which resource area, visitors would quickly recognise which concepts and turbine designs, respectively, survived and still operate today. The many turbines whirling away today on the rolling hills and plains deserts are mainly, without surprise, of Danish origin. The Danish dominance of the 1980s has survived until today. When driving the roads leading through the resource areas, all steps of Danish development of three decades can be spotted. Danish machines have also found their way to newer sites like Pacheco Pass,[7] Solano County and several minor places in California.

American representation can be limited to two companies whose turbines still cover the landscape: USWP (Kenetech) and Enertech. All remaining designs did not survive the 1980s or disappeared throughout the 1990s or 2000s. Of the other European designs, the German MAN Aeroman and Renck-Tacke[8] units in Tehachapi and the Scottish Howden (inoperative) at Altamont can still be spotted.

Figure 23.12   American Enertech E-44 40 kW turbines operating at Altamont Pass, 2010. Note the blade tip brakes (Photo: Arne Jaeger).

[7]Among other Danish models, Wincon delivered 87 units to a site in Pacheco Pass in 1987.

[8]Renck-Tacke was the predecessor of the later famous German Tacke.

USWP machines strongly dominate the Altamont area with the 56–100 kW model, which was installed more often in the state than any other turbine type. In Solano County, a site in the north of California, USWPs make up a remarkable portion of the wind farms developed there. In 1993, USWP changed its name to Kenetech. Some hundreds of its KVS-33, 300–400 kW turbines were installed in Altamont and Solano but also sold to other developers at San Gorgonio.[9]

Figure 23.13   American USWP 56–100 (100 kW), 2010. Several thousands were raised at Altamont between 1983 and 1992. It is strongly dominating this area. Note lacking nacelle covers (Photo: Arne Jaeger).

Enertech machines remained operational throughout the whole 1990s and 2000s in the Altamont Pass, as did the first sets of Wind Matics, employing the early Riisager-designed[10] rotors. They are still overlooking Highway 850 among voluminous Dutch Polenko machines.

---

[9] Some hundred KVS-33s were installed in other US states (Minnesota, Texas) as well as in Europe (Spain, the Netherlands). Two sites in northern Germany were originally intended to be developed by Kenetech but were left for German firms after the demise of Kenetech in 1996. Technological spin-offs appeared in the Ukraine (1993–2001) and in Spain (Abengoa Wind Power 1991–1993).

[10] Riisager was an influential Danish pioneer who, in the 1970s, created a rotor design that was given his name.

Figure 23.14 Another example of lacking nacelle covers on American USWP 56-100 turbines. (Photo: Arne Jaeger).

Figure 23.15 American Enertech E-44 40 kW turbines operating at Altamont Pass, 2010 (Photo: Arne Jaeger).

From the early Bonus and Nordtanks 65 kW from the rush period, to advanced medium-sized Vestas machines around 200 kW of the late 1980s, the Tehachapi Pass still today gives a strong impression of the turbulent and dynamic development the wind industry has gone through. Drivers using Interstate 10 starting in Los Angeles will be surrounded by some thousands of Micons, Vestas, Nordtanks and other 1980s turbines when heading for the city of Palm Springs in the San Gorgonio area.

But the development of wind turbines was always of fast nature and did not cease with the mid-1980s crisis. The advent of mid-size wind turbines in California was just a matter of time. Already in 1988/89 a first set of 500 kW units (Floda 500) was built at San Gorgonio. Consequently, representatives of the mid-1990s state-of-the-art machines in the 400–600 kW range, megawatt and multi-megawatt designs of the early 2000s and the more recent years can be spotted in newer developments across whole California.

For example, 20 Dutch Nedwind 500 kW turbines were put up at San Gorgonio in 1994. Kenetech's KVS-33, a 360 kW model, is well represented there since the mid-1990s. Zond's Z-48, a 750 kW model, Vestas 660 kW and Mitsubishi's MWT-600 have quite a dominance in the San Gorgonio area since the late 1990s and early 2000s.

In Tehachapi, Vestas put up dozens of its ubiquitous V-27 225 kW between 1990–1993, while German Tacke installed four of its TW-600 in 1994. They were followed by some 500 kW Vestas at a nearby site the following year.

The long awaited and late introduction of megawatt machines in California came with the realisation of four Nordex N-54 1 MW in 1999 at the Energy Unlimited site in San Gorgonio. The Tehachapi "Oasis" wind farm in 2002 comprising 60 Mitsubishi units rated at 1 MW was the first larger project in the state using Megawatt turbines.

Large-scale Megawatt machines are strongly represented at Solano County with hundreds of Vestas and GE turbines all rated at 1.5 MW and above. These wind farms were built between 2003 and 2009. And even completely undeveloped sites are now taken into consideration. The Hatchet Ridge wind farm in northern California is using 44 Siemens 2.3 MW turbines while in Kumeyaay, east of San Diego, 25 Gamesa units equipped with 2 MW generators are operating. At Mojave, a site close to the Tehachapi Pass, the Los Angeles Department of Water and Power has built the massive Pine Tree wind farm with some 90 GE 1.5s (1.5 MW) becoming the department's first wind energy project. The most ambitious project in the state in 2011 was being developed by TerraGen and Oak Creek Energy systems. With its final 720 multi-megawatt turbines ranging from 1.5–3 MW built by GE and Vestas the Alta Wind Energy Center[11] will deliver a substantial contribution to the ever rising energy need of the state and will back California's top position in the US ranking. The first 140 machines were already completed by mid-2011.

The steady technical improvements mentioned in chapter 2.2 result in a clearly visible operation philosophy. The elements most often lacking on American and European turbines are the nacelle covers and spinners. It is a standard situation if, for example, 7 out of 10 USWP turbines at Altamont Pass lack nacelle covers. This is to simplify maintenance. A maintenance team consisting of just a hand full of engineers is often responsible for hundreds of machines. In order to keep all turbines operative, nacelle covers are often taken down and left on the ground or are disposed at nearby sites. The missing nacelle covers make the work easier and quicker for the maintenance team. Despite the missing nacelle covers, the wind is strong enough to run the wind turbines and does not affect the aerodynamics. Few people passing the resource areas notice the wind farms in detail, and generally no one is annoyed about this fact.

Another example of simplification of maintenance is noticeable on many Danish turbines. In the 1980s, almost all of the Danish manufacturers equipped their machines with the same components. On some wind farms, interchange of components is practiced— most commonly blade interchange. Thus, it happens that Bonus turbines[12] are equipped with blades intended for Nordtank[13] machines and vice versa. Such a practice quickly becomes visible when spotting the Patterson Pass wind farm comprising of 336

---

[11]Located at Tehachapi/Mojave.
[12]Blades had green tips.
[13]Blades had red tips.

Bonus and Nordtank machines located in the Altamont Pass. Such practice is only possible because both turbine types used the same blade manufactured by Alternegy[14] until the mid-1980s.

Leaving wind turbines unprotected and exposed to weather and climate has never been considered a problem. The hot but dry climate of California allows machines to run this way and components can be stored on the ground with few or no protection measurements. There is no aggressive humidity causing any form of destruction to the components.

Another highly discussed issue is clearly visible in the Altamont Pass. For decades the fast spinning American machines often caused heavy troubles between wildlife protectors and their organisations, wind farm operators, residents and other institutions. The machines killed thousands of birds annually. Thus, the Altamont Kenetech modified some strings of machines with one, two or three black or black-striped blades. This was to create visual imbalance and easier recognition for the thousands of birds passing the Altamont area every year. Though lots of further measurements were taken by Kenetech and other companies, the wildlife problem was never solved entirely and still interest conflicts continue today. Inoperative turbines have become rare. Unlike two decades ago, turbines only stand idle for maintenance reason but not due to failure of technical concepts.

23.3.2 Why California's Wind Farms Are Unique

Today there are thousands of wind farms in the world and just few countries still without a wind farm or at least a single turbine. And even before the Californian wind rush, wind farms existed in some countries (United States, Denmark). They were, however, of comparably small size and consisted of few units.

There are several positive and negative characteristics only to be attributed to the Californian development that strongly separate it from all other developments in the world:

- California's wind farms were the first. They have impressively given proof that it is possible to use wind power on a large-scale. Wind energy is able to satisfy the needs of hundreds of

---

[14]Alternegy was the largest Danish blade manufacturer in the 1980s. After the crash, it was surpassed by famous LM.

thousands homes. Back in the 1980s, the public and the rising wind industry including its critics were given a very strong success story—despite the failure of many units. For many years the wind industry was given a benefit.

- Until today there was no wind development in the world in which so many different wind turbine manufacturers were involved. More than 30 various manufacturers from eight different countries of origin are counted in just one large wind development in a single US state.

- There has never been another wind development bringing together such a high amount of different wind turbine concepts. The US lightweight and vertical-axis concepts, the heavyweight Danish designs, various concepts from other manufacturers are "mixed together". Thus, a highly valuable amount of know-how and experience was gained and a wild and colourful visual impression was created.

- In the Californian wind development, the highest number of installed units was realised. In 1985 alone 4989 units were put up. It made them the largest wind farms worldwide for decades.

- California's wind farms are the most retrofitted, repaired and modified in the world.

References

1. Vosburgh, Paul N. (1983) *Commercial Applications of Wind Power*, Van Nostrand Reinhold Company.

2. Righter, Robert W. (1996) *Wind Energy in America: a history*, University of Oklahoma Press, Norman.

3. Stoddard, Woody (1986) "The California Experience" in: *DANWEA '86, Conference Proceedings*, The Association of Danish Windmill Manufacturers, Centec Business Consultants, Copenhagen.

4. "Wind Power Monthly News" Magazine, Vol. 1, 1985-Vol. 6, 1990.

5. *Wind Power Parks: 1983 Survey*, Strategies Unlimited, Mountain View, CA, Interim Report August 1984, EPRI AP-3578.

6. *Wind Power Stations: 1984 Survey*, Strategies Unlimited, Mountain View, CA, Final Report June 1985, EPRI AP-3963.

7. *Vindmølleparker i Californien*, Situationsrapport Juni 1983, Teknologirådets Styregruppe for Vedvarende Energi.

8. Results from the *Wind Project Performance Reporting System*, 1985 Annual Report, California Energy Commission, August 1986, P500-86-013.

9. Results from the *Wind Project Performance Reporting System*, 1989 Annual Report, California Energy Commission, October 1990, P500-80-010.

10. Various company brochures.

About the author

See Chapter 18, *Overview of German Wind Industry Roots*, by Arne Jaeger.

Chapter 24

Emergence of Wind Energy:
The University of Massachusetts

James F. Manwell

Department of Mechanical and Industrial Engineering, University of Massachusetts,
Amherst, Massachusetts 01002, USA

manwell@ecs.umass.edu

Although wind energy had a long history in the United States, by the 1960s it appeared that wind energy was no longer relevant. At that time fossil fuels were cheap and plentiful and nuclear power was "too cheap to meter". Unnoticed by many, however, the seeds for wind energy's renaissance were already germinating. As of 1956, M. K. Hubbert was already predicting that oil's days were numbered. With the publication of Rachel Carson's book *Silent Spring* in 1962 many people became aware of the environmental consequences of industrial development. By 1970, the first Earth Day reflected the beginning of the new awareness. As it turned out the University of Massachusetts in Amherst (UMass) provided one the fields where some of these seeds took root.

Wind Power for the World: The Rise of Modern Wind Energy
Edited by Preben Maegaard, Anna Krenz and Wolfgang Palz
Copyright © 2013 Pan Stanford Publishing Pte. Ltd.
ISBN 978-981-4364-93-5 (Hardcover), 978-981-4364-94-2 (eBook)
www.panstanford.com

24.1 The Renewable Energy Vision of William Heronemus

For many years UMass was an agricultural school and the transformation to a university did not begin until 1947. The 1960s witnessed a significant expansion, with construction of many new buildings and the hiring of many more professors. For the future of renewable energy, one of the most significant new hires (1967) was Capt. William Heronemus, who had just retired from the US Navy. His charge was to oversee the establishment of an Ocean Engineering Program. He did in fact do that, but of most significance was that he brought with him a vision of an energy future which was radically different than that which was most people were expecting. In 1971 he wrote the following:

"In the immediate future, we can expect the 'energy gap' to result in a series of crises as peak loads are not met. The East Coast will be dependent on foreign sources for most of its oil and gas. The environment will continue to deteriorate in spite of ever-increasing severity of controls. Air pollution, oil spills and thermal pollution are likely to be worse, not better in 1985. In the face of the continuing dilemma: power *vs.* pollution, a third alternative [to nuclear and fossil energy] must be sought. It may be found in the many and varied non-polluting energy sources known to exist in the United States or its offshore aggregate. These energy sources, tied together in a national network, could satisfy a significant fraction of our total power needs in the year 2000. That favourable outcome could result from a serious research and development effort started now and a design and construction effort started in 1985."[1]

William Heronemus's essential vision was no less than to completely replace conventional forms of energy, fossil as well as nuclear, with renewable energy sources. The sources he had in mind were solar, wind, and ocean thermal differences. Together with other faculty members, Professor Heronemus established the Energy Alternatives Program at UMass. This included participants from mechanical, civil, industrial and electrical engineering. The

[1]Research Applied to National Needs (RANN) Directorate (1971) *A National Network of Pollution-Free Energy Sources*, Proposal submitted to the National Science Foundation, Washington DC.

program was dedicated to working out the details of this paradigm and to educate the engineers of the future who would help make this vision a reality.

One of the key technologies in this future was to be offshore wind energy. This was a remarkable idea, especially in so far as there were barely any working wind turbines on land at that time. Because there were so few working turbines, Heronemus proceeded to learn as much as he could about the most relevant experience and vision of others. These included Albert Betz, Hermann Honnef, and Ulrich Hütter of Germany, Poul la Cour and Johannes Juul of Denmark, E. W. Golding of the United Kingdom, and Percy Thomas and P. C. Putnam of the United States. To their earlier concepts he added support structures for multiple rotors, floating turbines, fleets of such turbines, and a hydrogen storage system to firm the power. An example of one of his conceptual designs of a floating offshore wind turbine is shown in Fig. 24.1.

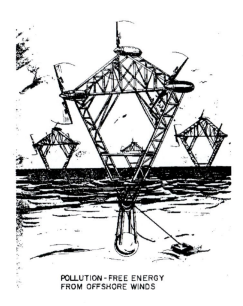

POLLUTION - FREE ENERGY
FROM OFFSHORE WINDS

Figure 24.1 Heronemus' conceptual design of floating offshore wind turbine, UMass, 1971.

These concepts were ahead of their time, and even today there are no offshore wind turbines in the United States. Many of the ideas were fundamentally sound, however, and are still relevant today.

24.2 Significance of the First UMass Wind Turbine, WF-1

Because of the paucity of actually existing wind energy converters, Heronemus proceeded to develop a relatively small wind turbine, which would be useful in and of itself and which would also serve as a platform to investigate the technology for larger turbines of the future. This turbine was known as WF-1 (short for Wind Furnace-1) and was conceived of as the core of a residential wind heating system.

The WF-1, designed and constructed at UMass between 1973 and 1976, ended up being historically one of the most significant wind turbines from that era. Although small by modern standards, at the time of its completion in 1976 it was the largest operating wind turbine in the United States. In many ways, the WF-1 has heavily influenced the entire modern global wind industry: the first generation of American wind-energy engineers was trained by working on its design and operation, and many of the WF-1's innovations appear in modern turbines. Today's "mature" wind turbine design bears what was once new on the WF-1: three fibreglass blades, near-optimal blade shape, blade pitch regulation, variable speed operation, and computer control.

The WF-1 could be considered the first "modern" US wind turbine, and in many ways it was more advanced than many later models of the late 1970s, 1980s and even 1990s. By Robert Righter's reckoning,[2] the WF-1 marked the beginning of the modern wind-electric era. Earlier wind-electric generators included small generators such as the Jacobs Windcharger, as well as larger models (some purely experimental), such as the Brush turbine, the Russian Balaclava (1930s), the Smith-Putnam (1930s), the Danish Gedser (1950s), and the German Hütter turbines (1960s).

The WF-1 was the forerunner of the turbines built by US Windpower of Burlington, MA. US Windpower went on to become the largest and most successful (for a while) wind turbine manufacturer in the United States. US Windpower eventually became Kenetech Windpower. Many of Kenetech's assets were acquired by Zond Systems, in turn purchased by Enron Wind, and then finally purchased by General Electric, which is now the major wind turbine manufacturer in the United States.

---

[2]*Wind Energy in America*, University of Oklahoma Press, 1996.

The WF-1 was paired up with a building known as the Solar Habitat, located on the UMass Amherst campus (Fig. 24.2). This facility was designed to demonstrate that a number of energy efficient building practices, such as solar hot water heaters, could be incorporated into a popular housing style, the ranch house. The WF-1 successfully demonstrated that wind heating could work, but the concept of a "wind furnace" never caught on. Relative to grid quality electricity, space heating was not economical, and subsequent technological advances and regulatory changes favoured the direct production of electricity from wind power.

Figure 24.2   The WF-1 and the Solar Habitat at UMass, 1976.

Many of the students who worked on WF-1 went on to work for major wind turbine manufacturers in the United States when the industry was just beginning. At least three students who worked on the WF-1 took jobs with US Windpower very early on. Later on others joined Northern Power Systems, Hamilton Standard, Kenetech, Zond, Carter, Fayette, Enron Wind, to name just a few. Two of the original WF-1 students were principals of ESI, a one-time major manufacturer of wind turbines. Many veterans of the WF-1 project still work in the wind energy industry, such as at GE Wind, the DOE's National Renewable Energy Laboratory, Northern Power Systems (in VT), and at Second Wind (in Somerville, MA).

24.3   Another Path to Renewable Energy

My own path to renewable energy was a different one than Professor Heronemus, but the paths both led to UMass. I was born in New Bedford, MA, once the home port of the American whaling industry. When I was still quite young my parents moved to Ohio, where my father became the superintendent of a reform school for juvenile delinquents. I lived on the grounds of that school until I was 18. The boys at the school were all from Cleveland. Nearly all of them came from poor families that had migrated to northern Ohio to seek work in its once large industrial sector. The contrast between the reform school and the surrounding town was always inescapable.

Growing up in the 1950s and 1960s included a *mélange* of activities from the mundane to the eye-opening: wilderness training with the Boy Scouts; a fascination with math, science, languages and history; long canoe trips on the lakes and rivers of Ontario and Maine; a road trip to Alaska, two weeks in Germany; a year as an exchange student in an English boarding school; a two-month hitch-hiking adventure from London to Istanbul, and a return to the United States by ship to enter Amherst College just as the Vietnam War was escalating. My college experience (which coincidentally was in the same town as UMass) and its aftermath were no less eventful and by the mid 1970s I was pondering what to do next.

The emerging issues of international realignment, nuclear power, environmental destruction and oil embargoes all led me in the same direction: a new vision was needed: equitable, inter-national, rational, and sustainable (i.e., based on solar energy). It was at that point that I met Professor Heronemus, and it became apparent that one way to help realise that vision was for me to begin studying engineering. As a result, I joined the Department of Mechanical Engineering at UMass as a graduate student in 1976 and studied there until I completed my Ph.D. in 1981.

During these years wind energy in the United Stated went through a succession of booms and busts, depending on the attitudes and inclinations of administrations in Washington and the governors in a few states. Wind energy technology began to go through its own evolution. Some of the turbines from those days were patterned after earlier designs; others were based on wishful thinking. Regardless of their provenance, however, it must be said that many of the turbines did not work very well or for very long.

24.4 Engineering and Education for a Renewable Energy Future

In order to make the vision of a renewable energy future a reality it became quite clear that the turbines needed to be more reliable and the way they could be integrated into a complete energy system needed to be better thought out. This all required serious engineering at the highest level. Turbines must be able to extract energy from the wind, convert it to a usable form, and deliver the energy to its point of use, all with structures that are light, resilient, and inexpensive, and within the confines of an electrical system which is required to supply firm power on demand. The fluctuating nature of wind, the structural response of the turbines, and the need for very long lives, imposed design requirements that were very challenging. These requirements were not well understood in the 1970s. It also became apparent that to make a wind energy system function properly requires expertise in many areas, including meteorology, aerodynamics, structural mechanics, materials, control, mathematical modelling, machine design, electrical systems and cost engineering, to name a few.

What setting is better for contemplating such problems than a university? Once I finished my Ph.D., it seemed to me that the most interesting and useful thing I could do would be to continue and expand the renewable energy education and research program at UMass that Professor Heronemus had started, so that is what I set out to do. This whole endeavour has had many twists and turns over the last 30 years, but it has resulted in opportunities for many students to learn about wind energy and to undertake research that had real significance.

The UMass wind energy program has been active in wide range of research areas, including hybrid power system design, wind resource analysis, wind turbine operation and design, and offshore wind energy, environmental impact and energy policy. In the paragraphs below, I will summarise our work in two of these areas: hybrid power systems and wind resource assessment.

Hybrid energy systems have been defined as "combinations of two or more energy conversion devices (e.g., electricity generators or storage devices), or two or more fuels for the same device, that when integrated, overcome limitations that may be inherent in either." In our work at UMass, we have focused particularly on those hybrid systems, which included, at least, wind turbines and diesel generators; these are usually referred to as "wind/diesel systems".

Hybrid power systems are intrinsically interesting because (1) they are potentially quite useful to many people in the world with limited access to conventional electricity and (2) they are technically quite challenging. They are also interesting because they serve as microcosms for some of the same issues that are just beginning to be encountered as large amounts of renewable generation begins to be added to much larger central power networks.

Work at UMass has involved a number of topics in this area. The key question was how to provide a very large fraction of the electricity in an isolated electrical network from a fluctuating resource, such as is the natural wind. We considered fluctuations over all time scales ranging from seconds to seasons. The studies we undertook included systems engineering as well as time domain and frequency domain simulations. To study the electrical aspects of such systems, we designed and built a complete hybrid power system hardware test bed. This test bed included a wind turbine simulator (using a purpose-built DC motor–AC generator with computer control), diesel generators, electrical load simulator, rotary converter (for voltage support), optional battery storage and specialised controllable devices (for power control and/or frequency control). A schematic of the UMass simulator is shown in Fig. 24.3.

The hybrid system studies lead to the development of the Kinetic Battery Model, which has since become used throughout the world in a wide range of battery applications. Further development and expansion of this model is still continuing at the present time.

One of the major accomplishments of our hybrid power system work was the development of the Hybrid1 computer code and its successor, Hybrid2, which employs both time series and statistical techniques. Hybrid2 became the standard for detailed engineering analysis of hybrid power systems and it is still widely used throughout the world.

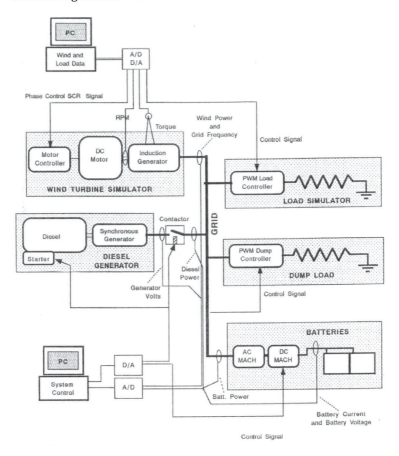

THE UMASS WIND/DIESEL SIMULATOR

Figure 24.3 UMass Wind/Diesel System Simulator, 1994.

Our work with hybrid systems has also been extended to include load management *via* distributed space heating, desalination of sea water, and the use of hydrogen as a storage medium. The desalination work has recently been further extended to a detailed study of a distributed generation community scale desalination project.

Some of our earlier work with hybrid power systems was reflected in the book *Wind Diesel Systems: A Guide to the Technology and its Implementation* (Cambridge University Press, 1994). This book was one of the key outputs of an International Energy Agency working group in which we were privileged to participate.

Wind resource analysis in the wind energy field has two very broad purposes. The first has to do with characterising the conditions which wind turbine itself must be designed to accommodate. The second has to do with the performance of the turbine (in terms of energy generation). Topics of interest in the first area have to do with turbulence (both temporal and spatial) and extreme events. Topics in the second area have to do variations in the wind that result in short-term electrical fluctuations and more slowly fluctuating mean electrical power.

In the area of wind resource analysis, our work has made use of what has by now become traditional data collection with anemometry (for both mean wind speed and turbulence investigations), through the use of SODAR (SOnic Detection And Ranging), and most recently LIDAR (LIght Detection And Ranging). For the processing of wind data, our early work included some pioneering use of time series analysis, modelling with long term statistics, spectral analysis, and data synthesis techniques (using Markov processes).

Recently we undertook in studies of the portion of atmospheric boundary layer in which the largest wind turbines of today operate (40 m to 150 m above the ground). These studies utilised instrumentation installed at multiple levels on very tall towers, as well as SODAR and LIDAR. One of our projects involved monitoring the wind offshore in Nantucket Sound; some of that work is illustrated in Fig. 24.4.

One of our most interesting projects was the creation of a wind turbine test facility at the top of mountain ridge not far from the University. The heart of the facility was an ESI turbine, which was originally installed in California. We brought the turbine back to Massachusetts and completely refurbished it before re-erecting it on the ridge. The turbine, which is shown in Fig. 24.5, was challenging in many ways—it had a 2-bladed, downwind, teetered rotor, for example. The blades were stall regulated, the nacelle was free to yaw as the wind direction changed, and the control system was initially relatively primitive. This facility provided an excellent opportunity for our students to study the behaviour of a real turbine in a complex environment and to design, implement and test a variety of modifications.

Figure 24.4   UMass and US Coast Guard installing offshore monitoring equipment, c. 1999.

Figure 24.5   UMass wind turbine test facility, c. 1996.

Finally, we have developed a significant wind energy educational program at UMass. For many years, we have offered a course on the fundamentals of wind energy. This course is directed at senior level or graduate students in Mechanical Engineering and it covers the most important topics in wind energy engineering. A second course is Wind Turbine Design. It builds upon the first one, and provides in-depth examination of how turbines should be built. Our experience with these two courses provided the background for our textbook *Wind Energy Explained: Theory, Design and Application* (Wiley, 1st ed., 2001; 2nd ed., 2009). Finally, a recently developed course on offshore wind energy provides the engineering background for this newly developing field.

About the author

James F. Manwell attended Western Reserve Academy in Hudson, Ohio, and then Brentwood School in Brentwood, Essex, England. He subsequently graduated from Amherst College with a B.A. in biophysics and then from the University of Massachusetts with an M.S. in electrical and computer engineering and a Ph.D. in mechanical engineering. He is presently professor of mechanical engineering at the University of Massachusetts and the director of the university's Wind Energy Center.

Prof Manwell has been working in field of wind energy for over 30 years, both within the United States and internationally. His research interests have focused on assessment of the wind resource and wind turbines' external design conditions, hybrid power system design, energy storage and offshore wind energy. He has authored the textbook *Wind Energy Explained: Theory, Design and Application*. He was involved with the International Energy Agency's WECS R&D activity, Annex VIII, which dealt with autonomous wind systems, and in conjunction with that activity, co-authored the book *Wind–Diesel Systems*. Subsequently, he became the US representative to the International Electrotechnical Commission's program to develop design standards for offshore wind turbines (IEC 61400-3), served on International Science Panel on Renewable Energies, and helped bring a large wind turbine blade test facility to Massachusetts. He is presently a member of the IEC group (TC 88 61400-MT3) which is developing a second edition of the offshore wind turbine design standard.

Chapter 25

An American Personal Perspective

Steven B. Smiley and Susan J. Kopka

P. O. Box 155, Omena, Michigan 49674, USA

smiley27@earthlink.net, cnslrtroy@earthlink.net

25.1 Introduction

The decade of the 1970s transformed the environmental and energy landscape of America, building the foundation for the modern emergence of wind power in the United States. As college graduate students in environmental/natural resources and economic studies Susan Kopka and Steve Smiley were transformed by the events of these times—particularly the environmental and energy crises. We experienced the immediate doubling of gasoline petrol prices, high inflation and interest rates, and long lines at fuel stations—with projections of worse to come. The popular books of the time were *The Limits to Growth*[1] and *The Population Bomb*.[2]

---

[1]Meadows, D. H., Meadows, D. L., Randers, J., Behrens III, W. W., (1972) *The Limits to Growth*, Universe Books.
[2]Ehrlich, P. R., (1968) *The Population Bomb*, Sierra Club/Ballantine Books.

Wind Power for the World: The Rise of Modern Wind Energy
Edited by Preben Maegaard, Anna Krenz and Wolfgang Palz
Copyright © 2013 Pan Stanford Publishing Pte. Ltd.
ISBN 978-981-4364-93-5 (Hardcover), 978-981-4364-94-2 (eBook)
www.panstanford.com

Amory Lovins wrote the article "Energy Strategy: The Road Not Taken" and the US President Jimmy Carter, elected in 1976, initiated massive energy efficiency and renewable energy policies and programs, providing grants to almost anyone who might have a potential solution to our energy problems.

The authors, Smiley and Kopka, began their professional careers in this era of promise, a continent apart and unaware of each other's work. As the principal investigator working on a grant from the US Department of Energy, Kopka and her research associates inventoried, catalogued and documented the history of passive energy heating and cooling in 19th century vernacular architecture and landscapes in the United States, within a program called "Small and Appropriate Technologies", which was granted with USD 300 000, and was written to the Purdue University School of Horticulture, where Susan was working as assistant professor. At the same time Smiley had also applied for small grants under the same grant program, granting up to USD 10 000 to anyone who had a reasonably good research or energy application idea. Smiley was awarded a grant to conduct a windmill technology exchange with the Nordic Folkecenter for Renewable Energy (FC), led by Preben Maegaard. Kopka and Smiley would meet more than 10 years later and combine their professional experiences and aspirations, and marry in 1992.

25.2 From Denmark to Alaska

Smiley first met Preben Maegaard at the Third Alaskan Alternative Energy Conference in Anchorage, Alaska. In 1979 they planned a technology transfer of wind power know-how from a Danish group of pioneering technicians to a group from his hometown of Homer, Alaska.

In October of 1980 six Danes came to Alaska. Over a two-week period the team built an 11 kW Danish-designed windmill from scratch, constructing everything but the gearbox and electric generator. Alaskans John Rogers, Tim Murnane, Frank Kohout, and Otto Kilcher provided machine tools, welding, electrical and carpentry skills. The windmill blades were built in Tim Murnane's boat building shop using mahogany—the only good wood available in Homer wood shops. Jacob Bugge, John Carlson, Bendy

Poulsen, Hans Pederson and Birger Kühn applied their skills while Dale Vandewalker kept the group well fed during the two-week project.

Figure 25.1   Installing the 11 kW wind turbine in Homer, Alaska, 1980.

After extensive analysis and fierce design debates, finally a 3-bladed 11 kW windmill with 7.6 m rotor diameter was built with a twisted blade made of mahogany. Tip air brakes were built combining the machine shop skills of John Carlson and Otto Kilcher and the carpentry skills of Tim Murnane and Birger Kühn.

Figure 25.2   The 11 kW wind turbine, Homer, Alaska, 1980.

This technology transfer involved the construction of a windmill based on the Danish concept, which became the backbone of the majority of the world's windmills at the time—3-bladed, upwind, grid-connected stall regulated asynchronous induction generator machine with fixed pitch, active yaw, and safety tip air brakes. This design concept was used in the manufacturing of thousands of windmills installed in Denmark and California during the 1980s.

Figure 25.3 Steve Smiley and Preben Maegaard in Thy, Denmark, 1980s.

25.3 Political Winds

In 1980, Ronald Reagan was elected as President while our Danish–Alaska technical exchange was under way in Homer, Alaska. Smiley remembers driving to the voting booth with Preben Maegaard in the car when the radio announced that Ronald Reagan had defeated Jimmy Carter for the US presidency. Alaska is five hours behind the east coast time of the United States, which means that the election results were projected before Smiley had a chance to vote. The Iranian–US hostage crisis weakened Jimmy Carter, and to this day Middle East oil politics indirectly impacts wind energy, both positively and negatively. The Alaska State Legislature also changed, with Republicans taking power for the first time in many years.

Based on the Danish–Alaskan technical exchange, Steve Smiley and Frank Kohout, the Homer electrical technician, were awarded USD 30 000 to continue the windmill work and research including a trip to the new US government wind test centre in Boulder, Colorado. Within two months of the elections, the national and state funding they received was taken away. All of Carter's energy initiatives and programs were immediately dismantled (including the solar panels at the White House!) as soon as President Reagan took power. This was the first of many big disappointments faced due to the changing political winds and it seemed to prove the understanding that "all energy choices and prices are based on politics and policy."

The massive growth and development in California wind power at that time was a result of Federal and State tax incentive policies which provided combined 50% tax credits to windmill owners. These tax credits allowed strong planning support from the California government and the "Standard Offer Number Four", or SO4, wind power purchases of USD 0.10/kWh or more. Americans provided the largest wind power experiment the world had ever seen. For better or worse, the American wind industry was built by accountants, bankers and lawyers, passing the tax benefits to those that could best use them.

Numerous windmill designs were tested, including the following: sails on circular tracks, vertical-axis Darrieus designs, up-wind and downwind horizontal-axis machines, two, three and four blades, free yaw, active yaw, mechanical and electric yaw. California provided a real world venue for testing wind turbine concepts. The Danish design concept proved the most reliable and there was a time in the 1980s when roughly 90% of windmills manufactured in Denmark were shipped to California.

While California provided a great proving ground, it also planted the seeds for a financial and economic policy model, which hindered steady wind development in years to come. The fifty percent tax credits and fair-priced SO4 power purchase agreements fuelled the emergence of wind power. However, when these policies and prices disappeared at the hands of politicians such as Ronald

Reagan in the late 1980s, the world wind turbine manufacturing industry went into decline.

Understand: All energy prices & choices are based on politics and policy!!

Figure 25.4 Ronald Reagan's wind power policy.

25.4 Lessons from California

In 1985, Preben Maegaard from the Folkecenter asked Smiley to help organise and lead a tour of 12 Danish wind turbine manufacturer representatives to visit the California wind industry and to attend the American Wind Energy Conference in San Francisco. They travelled in two large vans from San Diego, continuing on to Palm Springs, Tehachapi, Pacheco Pass, up to the Altamont and San Francisco, visiting every wind farm/project developer possible. The group included leaders from many of the early Danish windmill companies, who played important roles in furthering the technology; including Wincon, Wind World, Vind-Syssel, Danwin and Lolland.

The most interesting aspect of the tour was the intense investigations of the windmill junk yards. At first it was odd to spend so much time looking at broken windmills and parts, but then they began to appreciate that seeing how things broke was as important, or even more important, than seeing what worked. Nowhere in the world could one find so many broken experiments.

Figure 25.5   Karl Erik Olsen, Preben Maegaard, and Flemming Østergaard visiting Californian wind farms, 1985.

Figure 25.6   Visiting Californian wind farms: Preben Maegaard and Steve Smiley (left); inspecting American wind turbines (right).

Figure 25.7   Visited windmill junk—object of interest.

Figure 25.8 Intensive study of the wind turbine technology.

This tour of California occurred just after President Reagan's second term of office and when most American renewable energy policies were disappearing. In 1986, Vind-Syssel Wind Energy Systems, one of the early Danish manufacturers led by Flemming Østergaard, with Smiley's assistance attempted to enter the United States wind energy market. Vind-Syssel, advancing Folkecenter's designs and working with LM Glasfiber, pioneered the hydraulic activated air brakes on a 150 kW wind turbine. With supportive government policies ending, including tax incentives and good SO4 power purchase contracts, it was tough times in the world wind industry. California wind project developers told manufacturers they would buy windmills if they "delivered the windmills, brought the financing, took all the risks and gave them the profits." On one of the marketing trips to California, when asked a technical question about the Vind-Syssel wind turbines, Flemming Østergaard joked, "That's the first time anyone has asked me a technical question, [usually] all we ever talk about is financing!" During this same period 90% of the Danish wind turbine manufactures went bankrupt or were reorganised.

Figure 25.9 Steve Smiley and the LM blade with air brakes for the Vind-Syssel wind turbine.

25.5 Small Wind Power

Without the support of electric utilities for wind power there was a strong focus on "off-the-grid" small wind power. Even though small home systems were expensive for the energy they produced, avoiding long connecting power lines in out-of-the-way places in the vast American landscape created a significant demand for small wind turbines. Small wind power pioneers such as Mike Bergey of Bergey Windpower, and Elliott Bayly of Whisper in Duluth, Minnesota led the development and modern re-emergence of small wind in America. While these small wind turbines with 600 W to 10 000 W peak output were not the most durable, they were affordable and worked well in moderate wind areas.

Over a twenty-five-year period Kopka and Smiley installed and operated six different windmills, mostly of the "Whisper" series, from 500 W to 3 000 W. All were battery-charging systems, integrated with solar electric photovoltaics. Elliott Bayly of Whisper pointed out that 95% of his small windmills sales included solar electric charging, and the Whisper controls always included a solar PV connection. Most systems were supplemented with small gasoline or liquid propane generators to charge batteries during low wind, low sun periods, and with the evolution of improved direct to alternating current DC/AC inverters, home power systems went from 12 VDC and 24 VDC up to the standard US 120 VAC, 60 Hz home wiring. For many years Smiley operated a duel home electric system with both 24 VDC and 120 VAC. Battery storage systems working at 24 VDC allowed smaller DC electric wiring requiring half the wire amperage capacity of the 12 VDC systems. One hundred amperes at 12 VDC did not delivery much energy—1 200 W.

25.6 Living off the Grid

Susan Kopka left Purdue University in 1985 to build a homestead in northern Michigan as a practical demonstration of bioregionalism, or as the poet Gary Snyder put it, to "find home and dig in". Kopka was one of several women in Leelanau County living off the grid when she met Smiley in 1991. Sharing a common philosophy of providing a living demonstration of their renewable energy principles, they continued to build a model integrated renewable energy home. Their wind-powered home grew from a small, off-the-grid 30 m^2 (320 ft^2) cabin in the woods with a 500 W windmill, to a modern 220 m^2 (2 400 ft^2) integrated renewable energy home and studio garage. The sustainable energy sources include 3 kW wind power, 1.1 kW solar PV, 3 kW solar thermal, 30 kW biomass heating, solar and biomass thermal storage, 34 kWh electric battery storage, wood cook stove, passive solar, and six sugar maple trees shading the home for summer cooling.

The home was completely off the grid for the first fifteen years, and then connected to the electric grid as a back-up when needed. Running on an average of 300 kWh per month with all modern appliances, including efficient lighting, computers, stereo, microwave oven, toaster, refrigerator, freezer, washer and dryer, the home is an example of how there are no limitations with renewable energy.

Figure 25.10   Steve Smiley (left); Kopka and Smiley's home, 2012 (right).

During this period *Home Power* magazine from Oregon provided the best advice on how to make your own power. Remote Oregon and California marijuana growers were extremely interested in off-grid electric supply providing a big market for the development and advancement of modern small wind and solar PV systems.

25.7   Romanian Grid

With years of practical experience in home power, the Folkecenter invited Smiley to deliver small windmills for testing and to present his practical experience. This collaboration also resulted in the FC technical exchange with a Romanian utility research group lead by Cristian Tantareanu. This technology transfer led to the installation of a home power wind and solar electric system in an off-the-grid farm near Surducel, not far from the western Romanian city of Oradea. The installation was completed with horse drawn transport, hand tools and no concrete. The project was a unique combination of low tech and high tech electric systems. Romania provides an excellent example of the problems with modern electric grids, where centralised power distribution cannot be economically justified when required to run great distances to small rural electric users. This demonstrates how distributed renewable energy systems, especially wind and sun, can out-compete centralised power grids, with the potential to deliver electricity to billions of citizens throughout the world.

Figure 25.11   Home wind and solar electric system in an off-grid farm near Surducel, Romania.

25.8   Fighting with Windmills

The political and policy environment for wind power in America was at its low point in the late 1980s and early 1990s. Advocating for wind power in Michigan, while witnessing the successes demonstrated in Denmark, was like living in a separate reality. It was a separate reality because electric utilities in the United States had no desire to support wind power. If one asked an electric utility in Michigan if he or she could connect a windmill, he or she was told it would not work. After that, a letter would come from the electrical engineering department of Michigan

State University, whose academic programs are funded by electric utilities, telling one to forget about wind power. Smiley fought Michigan State University for five years to get them to stop discouraging wind power as an energy-generating option. In recent years they have been supportive.

Figure 25.12 V44 600 Vestas wind turbine at Traverse City. The owner and power purchaser Light & Power offers "Green Rates" for residential and business customers who purchase electricity generated from the wind turbine. The windmill is one of the largest wind turbine generators in Michigan, and produces around 800 000 kWh of clean energy every year (NREL/ Traverse City Light & Power, Photo by Don Rutt).

Working as a wind power consultant with Traverse City Light & Power (TCL&P) Steve Smiley developed the "Green Rate Wind Project," the first North American demonstration of a wind power green pricing program whereby the utility customers volunteered to pay a small premium for accounting for all of their electric consumption with wind power. The 1996 project consisted of the largest commercial wind turbine installed in North America at that time, a Vestas V-44 600. This wind turbine was the only machine sold by Vestas that year in America. With the development of this first electric-utility-scale commercial wind project in Michigan, a state with ten million population, the community municipal electric utility director received a letter from the Michigan State University Electrical Engineering Department recommending against the project. The letter told the utility director, that "only Democrats would consider such a wind project." Fortunately, the nearly 100% Republican utility board of directors and City Commission ignored the professors political opinions.

25.9   It Is (also) about Money

Over 250 residential and commercial customers of the 8 000 customer municipal utility volunteered to pay an additional USD 7.50 per month on average to add this new wind power to the generation mix, at no additional cost to any other customers. Steve Smiley adapted these green pricing policies, which had been advocated by David Moskewitz and Ed Holt from the State of Maine, to fit the practical situation at TCL&P. While "green pricing" is not the best policy model to advance renewable energy, it was one of the few practical policy tools on the table at the time that would be adopted by electric utilities. Sophisticated TCL&P customers told Smiley "why should we pay more for good things, it's the bad things that should be penalised." The best clean energy policies are now understood to be feed-in tariffs and black pricing.

While the TCL&P wind power green pricing program was not expanded in Michigan, it was adopted for solar electric programs and widely copied in North America, led by work in Colorado with Randy Udall and Rudd Mayer, and the City of Austin Texas. Austin copied the TCL&P model and installed roughly 250 MW of wind power, while Colorado and other western states installed over 1 000 MW of wind power with this policy model—roughly USD 2.5 billion of installations.

The American wind industry grew in the mid-1990s based on the tax policies of accelerated depreciation schedules established in 1986 (5 year/double declining balance, MACRS) and the 10 year production tax credit of USD 0.015/kWh established as part of the Energy Policy Act of 1992 under president George H. W. Bush, Sr. Such policies, however, primarily favoured those that have "passive income" taxes to offset, those in the top income brackets. The result of these policies meant that only rich people or capital corporations with taxes to avoid could benefit from wind turbine ownership. In order to maximise profits, these policies allowed wind projects to be sold to new owners after five to seven years after most of the tax advantages and profits are accounted for.

It is the opinion of the authors that this financial model led wind manufacturers, especially with General Electric buying into the wind market, to cut corners and costs by reducing the durability of the gearboxes and other components—as nothing else much mattered to investors after seven years of operation. The results of these policies and their version of planned obsolescence came to fruition when large series of gearbox failures pushed major wind turbine manufacturers to the brink of financial failure in the following decade.

These tax-based policies, which come and go with the changing political winds, are not healthy for the wind industry in the long run. A strong wind turbine manufacturing base cannot be built with such uncertainty. Durable renewable energy generation systems will not be built with short-term profit maximisation as the primary economic motivation. Renewable energy policies such as the German feed-in tariffs which guarantee a 20-year fair power price to make a market like agriculture crops, with guaranteed access to the electric grid, are what are required to build a solid wind turbine industry. With fair price policies such as the feed-in tariffs, wind power markets and ownership are open to all citizens, not just for the 1% of the population who can afford it.

25.10 The Matter of Scale

The middle part of the first decade of 2000 saw rapid growth in the installation of the multi-megawatt wind turbines. In 2007 it was a seller's market—the prices were high for commercial wind turbines with delivery times of over two years. The wind industry raced for larger rotors on taller towers, with 100 m rotors on 100 m towers becoming more common for 2 MW and 3 MW generators. Meanwhile, a hole in the market was developing with the increasing absence of mid-scale wind turbines. With supply constrained and the absence of mid-scale wind turbines, Steve Smiley and Preben Maegaard (XMIRE), worked with Tupac Canosa Diaz as project manager, to initiate a technical exchange to build a Michigan based mid-scale wind turbine.

Figure 25.13 Steve Smiley, Jane Kruse, and Preben Maegaard.

Steve founded Heron Wind Manufacturing to develop the wind turbine. The H-777 wind turbine is an advanced evolutionary step in the Danish FC integrated gearbox drive train adding a variable speed, variable pitch, sweep twist adaptive rotor (STAR®) blade system to produce what is expected to be the most cost-competitive mid-scale wind turbine in the American market.

As transmission line capacity limits the development of large multi-megawatt wind turbine projects, distributed wind power, integrated with other renewable energy sources and energy storage may become the next trend in the United States. As Steve Smiley says, "the wind is everywhere, and so should be the windmills". Hermann Scheer's and Preben Maegaard's plan for a 100% renewable energy world may become a reality sooner than we hope.

About the authors

Steven B. Smiley received his Bachelor of Science in education with a major in economics and minor in geography and American history from Central Michigan University. He received his Master of Arts in geography and economics in 1975 from Central Michigan University and conducted post-graduate studies at the University of Calgary, Alberta, Canada. His principal research was titled "A Framework for Rural Community Land-Use Decision-Making".

Steven and his wife of twenty years, Susan Kopka have lived "off the grid" and "on the grid" integrating solar (electric, thermal and passive), wind and biomass energy at their home in northern Leelanau County, Michigan.

Steven formulated and managed the development of the first community green power wind project in North America, the Traverse City Light & Power, 600 kW wind turbine, "Green Rate Wind Project" that became a model for many communities including Austin, Texas and Colorado. He has been involved in all aspects of permitting and operating small and large windmills.

Steven led the development and installation of the first three utility scale wind turbines in Michigan, in Traverse City and Mackinaw City, Michigan, totalling 2 400 kW of total electric capacity.

In 2007 he established Heron Wind Manufacturing LLC, a Michigan-based new commercial wind turbine manufacturing unit with a target market of community wind power projects. His firm Smiley Energy Services LLC conducts renewable energy economic feasibility studies, renewable energy technology transfer, planning and design and project development services.

Susan J. Kopka studied landscape architecture at Michigan State and Ohio State Universities, graduating from OSU with a Bachelors of Landscape Architecture (BSLA, 1976) and a Master of Science (1978). While studying at OSU, Susan was a graduate teaching assistant in landscape architecture and worked for the Upper Arlington, Ohio Public Parks Department. From 1978 to 1980, Susan was an associate with the multidisciplinary planning firm Planning Resources Inc. In 1980, Susan became assistant professor of landscape architecture and principle investigator of a $300,000 Department of Energy "Small and Appropriate Technologies" grant at Purdue University in West Lafayette, Indiana.

Susan moved back to her native Michigan in 1985 and settled into a now 30 year live-in demonstration project of living "off the grid", in the Traverse city area. Susan and Steve married in 1992. Susan has taught art at Northwestern Michigan College (NMC) for 21 years and, after receiving Masters in Clinical Social Work (2004, LMSW), Susan is now the sole proprietor of a private clinical practice Common Counsel (www.commoncounsel.net).

Chapter 26

Residential Wind by Way of Illustration

Igor Avkshtol

Department of Advanced Product Engineering, Beacon Power LLC, Massachusetts, USA

aukshtol@beaconpower.com

> *Wind is song of whom and of what?*
>
> (Velimir Khlebnikov, "Wind Is Song")

26.1 The Story

The subject of this story is a small wind turbine, thereafter, Turbine. A short history of its design, introduction to the market, and support is the plot. It is of residential scale; that means it is intended for a single home, rather energy efficient, or remote cabin. The goal is to provide the reader with some usable information. If you think about buying a turbine for your backyard, it may hint on the right questions to the dealer. If you design your own turbine, it should help you avoid some pitfalls. If you own a wind turbine, it may help you understand it better, show the ways to improve, rationalise, avoid unnecessary expenses, and maximise return on investment.

Wind Power for the World: The Rise of Modern Wind Energy
Edited by Preben Maegaard, Anna Krenz and Wolfgang Palz
Copyright © 2013 Pan Stanford Publishing Pte. Ltd.
ISBN 978-981-4364-93-5 (Hardcover), 978-981-4364-94-2 (eBook)
www.panstanford.com

Figure 26.1   The Turbine.

There is much more in the wind than the ability to rotate things hanging above the ground. Poets, the conscience of mankind, identify the wind with freedom and adore it—impeccable credentials.

The real names of people and companies associated with Turbine are not disclosed here. Though apologies are in order, seeking the permit would turn up irrational: in the time of writing the publication was not certain, the time for writing was limited, and accommodating individual wishes would be damaging for the integrity of the story. Besides, though technical stuff is discussed here, worlds apart from the detective genre, some mystery adds to the intrigue. The people and story are real; the names are "secret de Polichinelle"—just google it.

26.2   The Company

Some words about the company, thereafter Company, which developed Turbine are due at the beginning. In this story it will play a humble role of background—that doesn't do its justice. A university professor who taught aerodynamic founded it. It hatched from student projects. It carried some academic flavour. Introducing Turbine, Company made some mistakes, but never reduced to foul play. While Turbine being under development,

Company went through change of generations in leadership; the process was additionally complicated by archetypical Father and Son conflict. Several people at the top of the company didn't have children, they called themselves "dead genes club"—hardly a lot of extra dedication to work came from that fact, but some might. The work on Turbine was combined with other projects—not a trivial task for a small company. In nutshell, the design environ-ment was not particularly favourable, but it was certainly good enough for success.

During the debate about healthcare reform, the Vice President of Company, by that time retired, sent a mass email, apologising to the recipients for that single case of violation of his own rule against such actions. He told his friends, acquaintances, and former co-workers about his dismay of the way the debate was developing. He raised his voice against deceitful methods that became SOP for politicians, lobbyists, and predatory business interests. He took to heart "quality of our lives and the nature of this thing we call America..."

Geographically the root system of the design team was widespread: Armenia, Belarus, Germany, Latvia, and Oklahoma.

26.3   Engineers and Designers

Leading Engineer was an Oklahoma native with Native American ancestry. He graduated from Oklahoma University with such impressive grades that he was accepted to a General Motors design team right from the university bench. He lived in Detroit for several years designing air conditioning systems for passenger cars and testing them. He was on hostile terms with the grey sky of Michigan and some GM policies—that brought him to Company. He coordinated all the work and did most of Turbine mechanical design. His hobby was audio amplifiers; that allowed him to contribute a lot to the Turbine electronic controller. His work was indispensable in discovering some weaknesses of Turbine and improving it after the initial introduction to market.

Alternator Designer was born in Armenia to a musical family. To become a self-made man he took engineering at local university, volunteered to the army after graduation, and was injured in a border conflict. He moved to Germany following his parents. There he obtained a Master Degree in Electrical Machines. He was contracted for a project and left Company before Turbine manufacturing started. In Germany he was accepted by Bosch and successfully continued his career in electrical machine design.

Electronic Designer obtained his degrees from universities in Latvia and Belarus. After gigantic Soviet industry collapsed, he worked on grid-tie inverters for small wind turbines. He got involved in a power electronics project in the Folkecenter for Renewable Energy, Denmark. He struggled to finish it, as his soldering skills lagged behind in comparison with the theoretical knowledge. However the difficulties, the spirit of green comradeship cultivated there as well as vision for the future, great personalities in the leadership, and natural beauty of the place impressed him immensely. After moving to the United States he accepted a job offer by Company and felt like his dreams came true.

The newly elected Company president was looking for fresh ideas. Turbine seemed to be the right one: it had to replace an old bird that had been in production for a decade. New materials and devices, rear earth permanent magnets, inexpensive fibreglass, semiconductors (MOSFETs), and digital signal processors, were bagging for implementation. Peak power tracking and high efficiency DC-to-DC converters promised numerous advantages. Once there was a quiet period between stages of other project he set the team to work on Turbine.

Relying on a heavy Russian reference book and advanced German software, Alternator Designer worked fast on the permanent magnet alternator, while listening on his computer to the music composed by his genius Grandpa. Important technical decisions were validated during sometimes-heated discussions between Alternator Designer, Leading Engineer, and Electronic Designer.

26.4   The Turbine

The Company had vast experience with the battery charging application. So it was decided to start with it. It was meant to go for grid-tie, battery-less version in the next stage. At the time of the story efficient MOSFET transistors and Schottky diodes were available only for comparatively low voltage. So the turbine was designed for 24VDC.

The alternator produces a three-phase voltage of variable frequency and amplitude. In most cases it is rectified and, depending on application, converted to a different level of DC (battery charging) or to a constant voltage and constant frequency AC (grid-tie or off-grid AC application). Most small wind turbine manufacturers use a set of three brushes and three slip rings to pass the power over from the alternator on the top of tower to the ground equipment. For Turbine it was decided to install a rectifier on the tower and use a set of two brushes and two slip rings for power transfer.

Figure 26.2   The Turbine in the American West.

For its higher efficiency the electrical textbooks usually praise three-phase AC over DC or single-phase AC. It is indeed better if compared at the same voltage. However, the output voltage of three-phase uncontrolled bridge rectifier is higher than the input one—for the same current density DC requires slightly less copper of the cables; for the same amount of copper it produces less losses. The difference between DC and three-phase AC in cable losses is about 27%. Company has never regretted using two-wire DC from the tower top down. But there was a period of time when installation of rectifier on the tower top was questioned—until its lightning protection was enhanced.

The most time-consuming part of Turbine design was the electronic controller. From the perspectives of never-ending fight for shorter time to market it made sense to move its design to a specialised company. The later was experienced in controllers for photovoltaic applications. However, some negative signs of outsourcing started showing up soon. Firstly, the design time stated in the contract was considerably exceeded. Secondary, the company tried to utilise control algorithm common for photovoltaic (PV) application in the Turbine controller—"marriage of black toad to white rose" could never work. Obviously, the electronic design company underestimated the difficulties of designing for wind.

Years later, Electronic Designer happened to be in New England; there was a renewable energy fair in a city park. Electronic Designer dusted his Company cap feeling like he got a date with Turbine and headed to the park. However, the fair was small, not a single wind turbine. There were several PV stands, though. Electronic Designer engaged one salesperson in discussion of hybrid systems, but the person was very anti-wind, claiming the difficulties of site selection and reliability issues killed wind as a PV competitor or collaborator. Electronic Designer was surprised to see the renewable field splitting on fighting sects. When he shared his observation with Company president, the later confirmed it: "the solar industry has never been our friend; it's a shame, but that's just how it is."

Company confidence in its design ability was additionally boosted by an unfortunate event. A tornado struck where a Company turbine was installed. It remained standing, all blades and body intact, despite the surrounding totally devastated. The economics for Turbine looked very favourable, too. An advertising campaign was launched in specialised publications to "fire" potential customers in anticipation of manufacturing; "tornado tough" was the marketing pitch.

When the design company delivered the Turbine controller with corrected control algorithm, it turned up to still have several problems. Some of them looked like design sloppiness: parts were getting hot, LED lights were blinking when they were not supposed to, and safety margins were tiny. Electronic Designer managed to correct the most obvious of them, but some were not easy to rectify.

By that time, Company had a long list of customers and even accepted an initial payment from those striving for faster delivery. Shipping to customers had been postponed several times, mostly due to start-up difficulties at company branch in China and the delay with designing of controller. At that point of time Company had to bite the bullet and choose between damage to its reputation because of poor quality of controller or because of another delay. Angry customer calls and need for cash flow skewed the balance towards the former. Knowing that Turbine was still in raw state, Company took a brave decision: Turbine warranty was set to five years.

> *... But the first storm is hopeless and gone*
> *And from now on the journey is spring.*

(Velimir Khlebnikov, "On This Day of Blue Bears ...")

Figure 26.3 Guy wire failure.

And warranty claims did not fail to come up in large numbers. The very beginning of manufacturing is always associated with correcting defects, hopefully small. All usual stuff was showing up: stainless screws were too brittle, blades needed bigger washes, brush springs lacked stiffness, and cast iron tower mount was subject to excessive fatigue. But the controller was the "weakest link". Some decisions by its designers came from applying PV experience to wind controller.

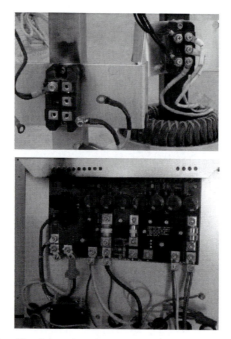

Figure 26.4 Rectifier (above) and controller (below) struck by lightning.

PV panels are easier to predict, their maximal power and voltage can be calculated in accordance with a dedicated section of NEC. Turbine behaviour depends on many factors. It is impossible to predict everything on the design stage. Sufficient margins are a must; reliability is of the utmost importance. For example, using MOSFET transistors instead of diodes provides for a better efficiency at low power—a very important factor. However, transistors require comparatively complex drivers and they are more vulnerable to overload and noise—questionable qualities for wind application.

Voltage coming from the wind turbine depends on the alternator RPM and, consequently, wind velocity. If the later exceeds certain value, Turbine furling mechanism kicks off moving blades away from wind. Furling is simple and reliable, gravity based, but as slow as any mechanical device. Density of air, turbulence of wind, verticality of tower, precipitation, and temperature affect it. The conditions for a particularly high voltage from turbine are not observed every day. But, even if the damage happened once in a season, multiplied by the number of customers and the length of warranty period the quantity of controllers in need of repair would not be sustainable.

While continuing shipping Turbine with controller of known deficiencies and repairing under warranty any damage to it as well as improving Turbine mechanics where needs arose, Company started designing a new controller. This time it relied on in-house expertise. It was not a modification of the original controller, as incremental changes wouldn't do. Interaction with Turbine customers and analyses of their problems brought in new ideas, some of them were not short of leap for the wind industry. Company had good understanding of customer needs, turbine physics, ability of modern electronics—firm base for a solid design of new controller.

The power circuit was considerably simplified; Schottky diodes took place of MOSFET transistors where that was practical. The control algorithm became more sophisticated and included several modes, some unique at the time of introduction.

The following description of Turbine operational modes should give the reader an idea of what the modern residential scale turbine is capable of. Here they are explained on Turbine example, that is, with battery-charging application in mind. For grid-tie operation in the description below the DC link in AC-DC-AC converter has to take place of the battery voltage, active front-end—the diode rectifier and shorting transistor, and grid—the battery; there is no need for slow mode in the grid tie.

26.5   Normal Operation—Peak Power Tracking

The controller loads turbine such a way that the rotational speed of blades is optimal for the given wind velocity from the perspectives of maximal power taken from the wind and eventually converted to the electrical energy consumed by loads or/and stored by batteries. The controller sets the rotational speed of turbine by creating the voltage on turbine output. That comes as a result of turbine physics. The controller creates a voltage on its input by very fast connecting turbine output to the battery and then shorting the turbine (boost) or disconnecting the turbine (buck). This connect–short or connect–disconnect cycle is repeated many thousand times per second, so only average values become meaningful. For the boost the voltage on the input of controller becomes below the battery voltage, as part of the cycle the turbine is short-circuited, that is, the voltage is zero then. For the buck it becomes higher than the battery voltage, as part of the cycle the turbine is allowed to accelerate with the battery being disconnected.

The emf (= open circuit voltage) of alternator of given construction depends on rotational speed. The alternator and the controller are parts of the same circuit. If alternator emf is higher than the voltage on controller input, it creates current in the circuit (i.e., mechanical torque on alternator shaft) that slows down the alternator and blades. If alternator emf is lower, it can't create any current and torque—the alternator and blades accelerate. Eventually there is equilibrium in the circuit: the rotational speed of turbine is stable for the given wind velocity and input controller voltage.

On the stage of initial testing the characteristics of turbine are taken; for each value of wind velocity the rotational speed of blades providing for maximal power from the wind is specified. As the turbine is tested at different wind velocities, the peak power curve images; it is a combination of numerous points—maximal power at certain rotational speed for the given wind velocity, the later as a parameter. It is not a secret for an educated reader that it is a cubic curve.

For practical use, the curve is recalculated the following way. Rotational speed is converted into voltage on the output of turbine, same as input voltage of controller. Power divided by rotational speed is torque. It is converted to current based on alternator characteristics. The result is a modified peak power-tracking curve: voltage as a function of current. It is stored in controller memory.

The controller has a current sensor and it sets its input voltage as a function of measured turbine DC current. Unlike PV industry, where universal controllers are common (they look for the power peak), the wind controller is intended for a particular turbine with specified characteristics. Searching for peak power usually leads to unnecessary oscillations of turbine speed.

26.6 Slow Mode

Off-grid customers know the painful truth: there is no correlation between their need for electricity and power production of PV panel or wind turbine. Most of the time the customer craves for energy, and wind controller does its best by peak power tracking. But there are inevitably intervals of time when the wind is plentiful while need for energy is low and the battery is already fully charged. In such a case Turbine controller initially engages a dump load. It does that gradually while maintaining optimal battery voltage (most other controllers just apply full available dump load at once, give it some time and disconnect it). If dump load is not sufficient or totally absent, the controller turns to something opposite to peak power tracking—it makes the blades operate at such speed that the amount of power taken from the given wind is small. For that it has yet another curve in its memory, again voltage as a function of current, which provides a low flow of power to the battery to keep it fully charged, but prevent over-charge and damage. An additional benefit is that the blades become slow and produce low audible noise. Company pioneered this mode in Turbine controller. It has become a common feature of turbine controllers throughout the industry since then.

26.7   Stop Mode

If you go to a driving school, the first thing they teach you is braking. That seems counterintuitive; the name "driving" implies different skills. Though the purpose of wind turbine is producing power, it is frequently important to stop it. Slow mode not always reaches its objective of preventing battery overcharge. In that case, the controller stops the turbine. There is need to do that for safety reasons or to prevent mechanical damage from very strong winds as well. Savvy turbine owners sometimes install a shorting switch to stop the turbine. But the controller "knows" better than simple shorting, it is equipped to do that intelligently. The turbine is stopped when the energy accumulated in its rotating parts is taken away. The more power is taken, the faster the turbine stopped. Sounds very similar to peak power tracking, doesn't it? There is one significant difference: the goal is to take as much energy as possible, not to produce maximal sustainable output of electrical power for the given wind. Unlike peak power tracking, losses are also useful in decelerating the turbine. If the turbine is just shorted, the only real place where energy is dissipated is alternator winding. Its resistance is not that high—power is limited. When the controller stops the turbine, it creates losses in the winding too, but it also directs power to the battery and/or dump load—the deceleration becomes faster and more reliable. Eventually the controller short-circuits the turbine through transistors on its board (shorting is a part of boosting as described above; transistors are already there, in combination with controller intelligence using them for stopping is only logical).

26.8 Blown Fuse Mode

If the battery is selected properly and it is in a good shape, its internal impedance is low, its voltage is comparatively stable. The battery is connected to the output of controller; it uses it to limit its input voltage coming from the turbine. But what if the battery fuse is blown or battery is accidentally disconnected (not all turbine owners are electricians, sometimes they don't realise the full consequences of their actions)? The controller can short-circuit the turbine, that is, make the voltage zero. Problem with that is the need of power for controller to operate. With battery disconnected for whatever reason and turbine shorted, the controller can use the energy stored in filter capacitors on its board for short time, but they have to be recharged at least from time to time for continuous operation. The mode is realised the following way. As soon as the controller detects that the battery is lost, it shorts the turbine. At the same time it senses the voltage of filter capacitors. When it drops below preset level, the controller "releases" the turbine for a split second to give capacitors a pulse of current, checks the voltage again and repeats that until filter caps are recharged to upper preset level. Then the cycle is repeated, starting with shorting. Thanks to the automatic blown-fuse mode controller remains operational and protects itself from excessive voltage of unloaded turbine.

26.9 Equalising Mode

To maintain some battery types healthy they have to be equalised approximately once a month. During that mode the controller creates higher than usually voltage on the battery during several hours. The matter of equalisation is batter addressed by specialised battery literature; not all batteries need equalisation; proper equalising technique is beyond the scope of this article.

Referring to a bankrupt PV company, a congressman condemned "ideologically driven pursuit of unproven energy sources" and advocated "market-based solutions". In their comments the readers pointed out that though the market mechanisms were important, the right and duty of defining the future of this planet had to belong to the people, not to short-sighted and heartless market.

The Turbine controller has also some features that were created for the sake of customer convenience. It can indicate the state of battery charge, the power coming from turbine, and the operational mode being executed. It is possible to manually initialise any of the described above modes, to switch between them, and to quit any of them if necessary. Recognising the value of hybrid system, Turbine controller also has a simple PV regulator making combining wind and solar possible without a dedicated (and expensive) PV controller.

26.10 Happy End

Figure 26.5 The Turbine in Nova Scotia.

Figure 26.6   The Turbine in Colorado (left) and in New Zealand (right).

With the new controller and most mechanical problems eliminated, Turbine enjoyed stable customer demand and good reputation. The path to that status was not straight, some mistakes were made on the way, some technical and business decisions were not simple and braving them was risky—but that was probably the only way the progress could be achieved: by working persistently, taking advantage of progress in the contiguous industries, respecting and listening to customers, working as a team and cooperating with each other, trusting people with ambitious tasks, uncovering problems and following up with solutions. The story of Turbine is not that remarkable taking into account its rated power (residential scale) or manufacturer (Company is definitely not a financial powerhouse) involved. But in that its strength lies, too. May it turn up to be the fortune of the whole residential scale wind industry: modest beginning, unassuming means, and hard work—a success at the end.

> *. . . Relating that which was Toward what will be.*
>
> (Velimir Khlebnikov, "There is that smell of honey-clover flowers")

About the author

Igor Avkshtol obtained his engineering degree from the Riga Technical University, Latvia, the degree of Candidate of Technical Science from Belarusian Technical University, Belarus, and the Doctor of Engineering from the Riga Technical University. He is a licensed professional engineer in the state of Ohio, USA.

He started working on electric drives as a research engineer at the Riga Technical University. Later he got interested in wind turbines and worked on grid-tie inverters. For a short time, he worked on a power electronics project at the Folkecenter for Renewable Energy, Denmark. When he moved to the United States, he was employed by Bergey Windpower Company and worked there for more than 10 years. There he designed and tested power electronics for mostly residential scale wind turbines as well as did customer support. Currently he is with Beacon Power LLC, Massachusetts, a manufacturer of flywheels for smart grid applications.

At present Igor's field of technical interests lies in the bi-directional AC-DC-AC converters and space vector modulation: harmonic mitigation, common mode voltage reduction, phase imbalance compensation, power factor correction, field weakening, and system stability in presence of significant grid impedance.

Chapter 27

Wind Power in China: Chasing a Dream that Creates Value

Qin Haiyan

Chinese Wind Energy Association (CWEA) and China General Certification Center (CGC), 11/F, Yiheng Mansion, No. 28, North 3rd Ring Road East, Beijing 100013, P. R. China

qinhy@cgc.org.cn

The development of the Chinese wind power sector has been the result of diligent work by many key companies, government ministries and individual people over a period of more than 30 years. As the director of China's General Certification Center (CGC), I am deeply respectful of the great leaders, executives and researchers who have built this great industry of the future. In this chapter, I will share a few of the many stories and perspectives about the development of wind power in China.

27.1   Wu Gang's Dream and Philosophy

During the spring festival of 2012, I received a text message from Wu Gang, Chairman and CEO of Goldwind, "Xiao Qin, I would like

Wind Power for the World: The Rise of Modern Wind Energy
Edited by Preben Maegaard, Anna Krenz and Wolfgang Palz
Copyright © 2013 Pan Stanford Publishing Pte. Ltd.
ISBN 978-981-4364-93-5 (Hardcover), 978-981-4364-94-2 (eBook)
www.panstanford.com

to express our respect for you, all on behalf of Goldwind, for there would be no wind power like what it is today in China without you. From the first version of Regulation in Tianchi, Xinjiang, to today's international recognition of CGC brand, how painful and strenuous it has been! Yet for China's wind power, you stayed, persisted and persevered. All thanks to you, China's wind power thrives!"

The affirmation and encouragement from Wu Gang is the best present I have received for the spring festival of 2012, which brought our team and me cordial gratification and joy!

The "Regulation" mentioned in Wu Gang's SMS refers to the project under China National High-Tech Research and Development Program (Program No. 863)—Specification of Wind Turbine Generating Units. Wu Gang and I were members of the compilation team. I was 29 years old that year, 12 years younger than Wu Gang, so it's been natural for him to call me "Xiao Qin" (meaning younger with "Xiao"). For several decades, he still calls me "Xiao Qin", which makes me feel close to him.

At that time, Wu Gang was the executive deputy general manager of Goldwind and 16 June 1998 was a memorable date in his career. On that day, Goldwind introduced the first 600 kW wind turbine made in China with German technology and was connected to the grid at no. 1 wind farm in Daban, Xinjiang, causing great attention from home and abroad in the wind power industry. Some commentators named it as "the beginning of domestic large wind turbines in China". Wu Gang, who usually keeps a low profile, could not resist the excitement and told me that "his dream of making wind turbines is more firm than the steel used to make them".

Wu Gang's dream to make wind turbines dates back to 16 March 1992, at 10 am, when he witnessed an accident caused by a failure in the braking system of a foreign brand. The situation was horrible. Wu Gang stepped up at the risk of his life, climbed to the nacelle at a height of 23 m, and used the jack to engage the braking system, exercising 90 degrees of deviation, which avoided a larger accident and bigger losses. With such experience, Wu Gang realised the central importance of quality, and dreamed of making a great wind turbine here in China.

At the end of 1999, while Goldwind was preparing for industrialisation of 600 kW wind turbines, it also launched

ISO 9001 certification, becoming the first manufacturer that passed this quality management standard among China's wind power equipment manufacturers. Wu Gang and I had more contacts due to work on certification and I paid more attention to Goldwind under his leadership.

In May 2002, Wu Gang became the chief executive of Goldwind, which was then building a modern wind turbine assembly base with an annual production capacity of 200 turbines of 600 kW–1 MW, aspiring to compete with international wind power equipment manufacturing giants with a strong commitment to build a corporate culture of learning.

One day in June, experts from one internationally established wind turbine manufacturer visited Goldwind, where Wu Gang convened senior executives to listen to their research and development. However, he was shocked, because what the experts meant was: since they have entered into wind turbine equipment manufacturing, is there a good reason for Goldwind to continue to exist?

Figure 27.1   Wu Gang (right) of Goldwind at the CWP in 2011.

Wu Gang and his team made special investigations and analysis before they came to a conclusion. He decided that, as long as Goldwind can create value for customers, it can survive, regardless of how powerful those international manufactures can be. This is how Wu Gang set up the philosophy of "Create Value for Customers" for Goldwind.

Truly, to achieve the philosophy of "Create Value for Customers", Goldwind undertook national science and technology projects such as "Program No. 863", "600 kW WTGs Industrialisation Technology", "750 kW WTGs Research" and "MW-scale WTGs and Key Components Research". All of these programs have proven that the philosophy could withstand tests.

27.2   Goldwind's Challenge

Looking at the development of Goldwind over recent years, Wu Gang once said "all these years, market and competition environment is changing, so Goldwind keeps adapting with strategy adjustment too". However, the core philosophy has remained.

In order to better create value for customers, Goldwind had experienced two transformations, with the first starting from MW-scale research, on the project of national "Program No. 863" in the subject of sustainable energy. At first, Goldwind's target was to make MW active stall-geared wind turbines, but after concept designs, load calculations, etc., they changed for a better direction of making other models. Then they received news from overseas about direct-driven permanent magnet wind turbine without the parts of high speed rotation and with synchronous connection to the grid *via* electronic components. At that time, power and electronic technology of big capacity was still immature. This challenge, they believed, with the rapid development in electronic technology, can be dealt with, especially with flexible and fast responding electronic devices replacing complex mechanic structures as the new trend.

At that time overseas research on direct-driven permanent magnet wind turbines was at a very early stage, with no sophisticated prototype to follow, which in turn consolidated the research direction of Goldwind. The choice of Goldwind to use direct-driven permanent magnet systems won recognition and approval from relevant national authorities and experts, and seriously shifted the MW-level technology design from active stall-geared wind turbines to direct-driven permanent magnet. This technology was jointly developed by Goldwind and Germany's Vensys, starting from early 2004 until June 2005, when research on 1.2 MW prototypes was completed and installed in the wind farm in Daban, Xinjiang, as the first domestic MW-scale turbine. During the next four years, Goldwind completed three upgrades on the capacity and realised the commercialised manufacturing of 1.5 MW and operation testing for 2.5 MW and 3.0 MW turbines, with commercialised production by orders unfolding.

However, the worldwide community will not have any idea about the hard choices made by Wu Gang and his team for the first transformation. Before 1.2 MW direct-driven permanent magnet, the major products of Goldwind were 600 kW and 750 kW, which were technologically transformed, further developed, innovated and integrated into industrialisation based on German technologies. Goldwind also won 10 patents, 13 patent applications and had different series of products of hubs at varied heights, rotors of varied diameters. Under such circumstances, research in MW-scale direct-driven permanent magnet wind turbine requires innovative approaches, with risks. In the face of such "huge bet and risk", Wu Gang was fearless.

At the end of 2007, Goldwind completed the transformation of its technology route and advanced with the second strategic transformation with a solid foundation, in order to shift from "wind power equipment manufacturer" to "wind power solution provider".

The second transformation of Goldwind started from the early spring of 2008, on the basis of its core technologies of direct-driven permanent magnet designs, control systems, converters, pitch mechanisms, etc., integrating its components of wind turbine manufacturing, wind power technology services and wind power project development, which have transformed into core competitiveness, thus providing wind power system solutions to customers at home and abroad, and creating optimal value for customers.

Based on the philosophy of "Create Value for Customers", Goldwind accelerated its pace of internationalisation. With Goldwind Wind Energy GmbH as its operation in Germany, it established an R&D centre and production base in September 2009, with the view to expanding its European market. In early 2010, Goldwind USA Inc. was set up in America, and at the end of 2009, a branch in Australia. "The internationalisation of the market for Goldwind is not simply selling products in local places, but also creating industry value chain from R&D, manufacturing and supply chain, etc.", with the low profile Wu Gang made a firm step in promoting the internationalisation of Goldwind.

Figure 27.2 The first Goldwind 1.5 MW direct-driven permanent magnet wind turbine.

Another memorable date for Wu Gang, since he values "world's Goldwind", is 8 October 2010. On this day, the Hong Kong Stock Exchange ratified transaction of 395,294,000 H shares issued by Goldwind, signifying the realisation of internationalising its capital. As of now, Goldwind has achieved "A + H" and became world's largest "market cap" among wind power equipment manufacturers.

27.3 Developing Certificates for China

Dating back to early 2000, when China expected its domestic wind turbines, standing in China's wind farms, to rival foreign ones, Goldwind, under the leadership of Wu Gang opened a gate of "Made in China" for this industry in China.

Back then, I was trying to contribute to the development of this industry in China too. In June 2001, I contacted Mr Sandy

Butterfield, the chief engineer of National Renewable Energy Laboratory in the United States (NREL) and the leader of standards and certification program, and hoped to receive some training in IEC certification procedures and bring relevant international standards to China.

As I remember, the training program went through several months both in NREL and in China, respectively. I followed through the whole training, where I began to dream of establishing China's wind energy certification system, through which China's wind energy industry can be provided with effective technical supports.

In early 2003, I joined China General Certification Center (CGC), and visited countries with developed wind energy like Germany or Denmark, for learning and technical exchanges. While improving ourselves, we would need to connect with international mainstream wind energy certification for technologies, in order to live up to the core philosophy of CGC: "Create Value Through Certification".

From the beginning of establishing wind energy certification scheme to the first certificate for wind energy in China, CGC spent over four years. On the one hand, we needed to put sufficient technological content in certification, in order to create value for customers; on the other hand, wind power products with certificates needed to satisfy certain standards, so that with such credibility embodied in products, an orderly market can be cultivated for sound development of the industry.

27.4   The Power of Sinovel

On 18 July 2007, CGC issued its first certificate to Sinovel, the first in China's wind energy industry, marking the progress made in quality supervision and technological advances in this industry in China.

Since 2010, due to the cooperation of wind turbine certification, my team and I have kept extensive communications with Han Junliang and the Sinovel team, which gave us more understanding and tremendous respect of Sinovel.

Han Junliang, with glasses on the shaped face and hair split in half, showing a temperament of freshness and scholarliness is

Chairman and General Manager of Sinovel. Born in Yancheng of Jiangsu Province, he graduated in 1987 from Taiyuan University of Science and Technology, majored in cranes. In 2003, when Chinese government started bidding for the first phase of inland wind power concession projects, he was already the director of Institute of Heavy Machinery Design in Dalian and also the General Manager of Dalian Heavy Mechanical & Electrical Equipment Engineering Co. Ltd. Those familiar with Han Junliang said he had an active mind, passion for work, was good at interpersonal skills and would never settle for a mere scholar behind the scene.

Han Junliang sensed the development of wind power and had no doubts that the government would soon enact policies to encourage domestic growth, so the 80% of foreign capital would soon be replaced and Dalian DHI-DCW Group Ltd. became an important equipment base in the Northeast, boasting of born advantages of providing upstream components.

In 2004, the 41-year old Han Junliang finally convinced investors with various backgrounds to fund RMB 70 million, and set up Sinovel, with a registered capital of RMB 100 million, jointly with Dalian Heavy Mechanical & Electrical Equipment Engineering Co. Ltd. Thereby, Han Junliang became the chairman and general manager of this company. He made commitments to "introduce the first 1.5 MW wind turbine technology from Germany and produce 100 turbines in the first year, 300 in the second and 500–600 in the third."

Not many people truly believed such commitments, particularly in a domestic market where 80% of manufacturing was then dominated by foreign brands; the newcomers would barely have any space, let alone confidence. However, Han Junliang's sensitive judgment of policies and predictability of wind power market fulfilled his aspirations.

In October 2004, Sinovel took over the manufacturing permit for FL1500 series of wind turbines of Germany's Führlander. This series is composed of nine parts and five core parts of heavyweight gearbox, yaw system, tower, hub, and main frames. As the company became proficient in manufacturing non-standard equipment, as well as designing and producing of transport mechanisms, like the launch pad of Shen Zhou VI ship, it is not a difficult task for Sinovel to manufacture wind power components.

Sinovel's timing was good in 2006, when domestic MW turbines were in strong demand. At that point in time, Sinovel beat any other manufacturer and produced the first 1.5 MW wind turbine in China. This model completed the whole local supporting industry chain in China and was put into production in June 2006.

Figure 27.3   The first Sinovel 1.5 MW double-fed wind turbine.

At the early stages of Sinovel, it trained the largest after-sale service team in China, in order to make up for any possible quality problems. This team had more than 1000 members at its peak. Han Junliang challenged them to locate and solve problems before owners experienced any failures. Sinovel paid high costs for this, but also maintained its reputation for a period of time, achieving improved quality and technology.

In 2008, the newly added installation capacity of Sinovel reached 1403 MW, ranking no. 1 in China and no. 7 in the world. In 2009, the newly added installation capacity of Sinovel reached 3510 MW, ranking no. 1 in China and no. 3 in the world.

Figure 27.4   Qin (left) and Han Junliang (second to right) at the CWP 2010.

In 2010, the newly added installation capacity of Sinovel reached 4 386 MW, ranking no. 1 in China and no. 2 in the world. On 13 January 2011, Sinovel successfully launched its initial public offering (IPO) on the Shanghai Stock Exchange.

If Sinovel was compared to the "Legend" in China's wind power manufacturing industry, Han Junliang would be the creator of such a Legend. Shareholders all agreed that daily operations, customer relations and government relations were all managed by Han Junliang, who would take care of details on his own and had turned to be a rare talent with vision and reliability.

27.5   Towards the Sea

What was commendable was in 2006, when Han Junliang ordered 1500 sets of main shafts from overseas and said "it will not be easy to buy these in future". Investors thought this to be too aggressive, yet felt reluctant to interfere. Nevertheless, the market staged a huge shake up in 2008 and sent the supply of main shafts into great scarcity, not available even with money. Han Junliang was proven insightful and right once again.

In 2009, Han Junliang was looking for sea terminals and planned to set up a port industrial zone in Changxing Island of Dalian of Liaoning, Tianjin and Yancheng of Jiangsu, etc. When investors asked why, he answered, the next priority would be offshore wind power, where equipment would be needed. In order to ship them, boats would be necessary, for which they needed to find sea terminals.

"Why not take the road transport?" an investor asked. Han Junliang explained that a 3 MW wind turbine had a diameter of 4.7 m at the base, yet with the 4.5 m high elevated overpass, it was not feasible to travel through, or to dismantle the overpass, Despite a long coast line in China, the number of sea terminals is limited, so it would be a good time to use them either for transport or future export at a low cost before competitors realised this, and made it difficult to get even at higher prices. His vision could even reach such extension and details; who else could have thought of the height of the overpass?! Han Junliang's prestige was thus consolidated.

In 2010 Sinovel completed China's first offshore wind power demonstration project deemed impossible by domestic enterprises—102 MW Shanghai Donghai Bridge Offshore Demonstration Project. At that time, both the Shanghai government and the National Development and Reform Commission (NDRC) expected domestic enterprises to undertake this project, which would save money and win honours for the country, but industry experts unanimously said it will be impossible, considering that domestic enterprises had just started installation of onshore MW turbines and would be doomed to fail in 3 MW offshore projects.

Nevertheless, Sinovel took its first step towards offshore wind turbines. In Han Junliang's view, this was another starting from scratch. In December 2007, Sinovel undertook projects and in September 2009, completed the first batch of 3 MW turbines connected to the grid. In February 2010, a successful installation of 34 offshore wind turbines was completed and started generation to the grid on 8 June 2010.

This set the tone for Sinovel to play an extraordinary role in China's offshore wind power industry, the curtain for which was gradually unveiled. On 18 May 2010, China's first batch of offshore wind power concession projects bidding was officially launched. The first four projects totalled 1 GW, equivalent to 10 Donghai Bridge wind farms in size. This is only the tip of an iceberg compared with massive sizes of installations submitted by local regions. The total installation capacity planned in Shanghai,

Jiangsu, Zhejiang, Shandong and Fujian by 2015 alone reached nearly 10 GW, together with targets of over 1 GW in other provinces.

As the early player in China's offshore wind power market, Sinovel embraced bigger market opportunities and challenges. In June 2011, Sinovel obtained complete IPR, had the biggest installation capacity and the globally leading technology of 6 MW offshore wind turbines. All this shows the advantages of Sinovel in future's offshore market competitiveness.

According to statistics of BTM, in 2010 already four Chinese enterprises entered the world top 10 in sales. In addition to Sinovel, three others, Goldwind, Dongfang Turbine, Guodian United Power ranked no. 4, no. 7 and no. 10, respectively. China MingYang, Sewind and XEMC Wind were ranked no. 11, no. 14, and no. 15, respectively, posing challenges to traditional manufacturers in Europe.

27.6   Success of Zhongfu Lianzhong

The improvement in China's wind turbine manufacturing has increased the demand for high quality components on the market. Zhang Dingjin, executive in charge of Zhongfu Lianzhong Composites Group, said when talking with me about certification for blades that what they gained through certification was not simply a certificate, but how the process itself assured Zhongfu Lianzhong's ability to make high quality blades, because the implementation of the certification required technical contents and stringent specifications to realise this.

This is the trust and hope from Zhongfu Lianzhong. I organised the team and it took us one and a half years of hard work to complete the thick-paged implementation rules of certification for wind turbine blades. All this was aimed at realising the top 3 values of CGC certification scheme:

(1) To promote the introduction, understanding and innovation of the imported technology of blades and testify the maturity and reliability of proprietary technologies;
(2) To assist producers to establish normative production process and improve the capability of product conformity control;
(3) To assist the producers to make key quality and technical information traceable during production process.

From the first blade in production in April 2006 to the biggest player in China's market in 2010, Zhongfu Lianzhong spent less than five years. In the view of Zhang Dingjin, what is more valuable than the title of "China's market share no. 1" is how it opens markets in Japan, India and, the United Kingdom, etc. with its high quality of blade manufacturing.

Figure 27.5 Zhongfu Lianzhong blade factory.

To conclude, I find mainly two reasons for Zhongfu Lianzhong's success:

First, open R&D in technologies. Zhongfu Lianzhong started manufacturing blades for 1.5 MW wind turbines. Blades have high requirements of performance of materials and under some circumstances, the features of materials have become technical barriers for blades manufacturing. "If Zhongfu Lianzhong entered into blade manufacturing, at least they would have advantages in such features with the company's years of experience in composite materials" Zhang Dingjin saw this business opportunity. What is the second reason of Lianzhong's success?

One day in March 2005, Wang Wenqi and Lu Zhunli, senior experts in China's wind power, came to Zhang Dingjin and asked him to introduce blade manufacturing technology from Germany's independent blade manufacturer NOI, in order to support the domestic manufacturing of blades for 1.5 MW wind turbines. Zhongfu Lianzhong completed all matters relating to technology introduction in China for NOI's 1.5 MW blades manufacturing in mid-June and started manufacturing the first on 16 April before mass production in half a year later, thus becoming the first enterprise to produce 1.5 MW blades.

Then there came a time when Zhang Dingjin looked at long-term development of the company, since blade manufacturing involves multiple technological sectors, with design capability, mould casting which determines the competitiveness of a company. This was the weakness of Zhongfu Lianzhong, which led Zhang Dingjin to think about obtaining blade technology from an existing manufacturer, which resulted in Zhongfu–NOI.

Therefore, Zhongfu Lianzhong underwent rounds of hard negotiations and successfully took over NOI in January 2007 and established SINOI in Germany as China's first blades manufacturer to set up an R&D centre in Europe. Zhang Dingjin believed that blade technology research and development is a systematic program and it takes a long time for a Chinese blade manufacturing enterprise to form its own professional technological R&D team. He said frankly that he did not believe a five-year journey for development of MW-scale blades in China could replace over two decades of such R&D in Europe. This was the reason for him to take over SINOI in Germany and position China's design team at an open platform, which not only produce blades for Zhongfu Liangzhong but also cooperate with other design teams all over the world in order to provide blade design for more wind turbine manufacturers and even to cooperate with other design companies for R&D and blade design.

Zhang Dingjin hoped to understand more problems in R&D and design, so that wind turbine manufacturers could have access to thorough international cooperation and more suitable blade solutions. Benefiting from such global open platform of blade design resources, by 2011 Zhongfu Lianzhong has already offered seven series of blades with 28 models for 1.25 MW, 1.5 MW, 2.0 MW, 3.0 MW, 3.6 MW, 5.0 MW and 6.0 MW wind turbines. Open platform enabled Zhongfu Lianzhong to use the best resources of blade design and to better realise value for customers.

However, an open platform does not lead to instant success, but to a realistic and innovative promotion of China's wind power technologies in line with the wind environment in China. Until 2012, Zhongfu Lianzhong has won five national patents of invention, released four enterprise standards, his blades for a 2 MW wind turbine won "China Standard for Innovation Contribution Award" in 2010.

Figure 27.6   Zhang Dingjin (left) at the ceremony for the first 5 MW blade production.

The second reason for Zhongfu Lianzhong's success was refined manufacturing. For certification, I visited the workshop of blades several times and noticed how much detail the staff paid attention to. This is a kind of capability not owned by many. It is precious that in March 2011 Zhongfu Lianzhong came back to fundamental research on blades even after five years of manufacturing. The senior engineer, Zhang Xiaoming explained, "On the one hand, the bigger the size of blades we make, the heavier they get, so we need to extract refined nutrition of making it from five years of experiences, in order to provide technical support for optimisation of blades manufacturing; on the other hand, those experiments and research projects halted due to limitations of experimental conditions. Now technological capabilities can be conducted for future technological reserve."

In terms of blade manufacturing techniques, manufacturers are not so much different from each other, and would mainly find this in refined work, which is the determinant of quality. Currently, the industry agrees that the difficulty of work lies in the control of manufacturing methods and processes. Without good control, quality problems could easily occur, such as cracks (failed performance of materials).

The main materials used for blade manufacturing are resin, adhesives and fibre, the combinations of which would create different effects in performance. The controlling parameters in refined work for different sizes of blades are also different, since they are generated from countless and often expensive experiments. Zhang Xiaoming told me, "Zhongfu Lianzhong tested almost all resins, adhesives and glass fibre from home and abroad, and knows clearly even about viscosity and infiltration of kinds of resins under various conditions. This homework might seem insignificant, but has laid a solid foundation for blade materials and technique optimisation. We have been paying attention to all the details in the manufacturing of blades."

Zhang Xiaoming also said that the manufacturing of big blades compared with smaller ones would simply require longer open periods for adhesives, as well as manual regulation of fluidity of resins on fibre. What is noteworthy is the factor of heat release on the quality for bigger blades. Even though many other countries have developed material systems for bigger blade manufacturing, we still need to make decisions on the selection of materials, matching report and technique details through experiments and data as support.

In the workshop of Zhongfu Lianzhong, I experienced the importance attached to the details of manufacturing by their staff. For example, the glass fibre rowing needs to be laid out evenly, as required by the blade design company. This work is related to the experience of the workers with the features of the materials. However, design companies do not have requirements in detail about these factors. Other blade enterprises might only require the woven method, density and fibre model while selecting fabrics, but Zhongfu Lianzhong would even identify the drying method, since the touch would be different as a result of different drying methods. The touch is another reference to tell the experience of the workers. Why would some workers fail to use the same experience on the same kind of fabric while laying out the blades? In fact, it relates to humidity in the workshop. It is not easy to identify with the naked eye how to lay out the fabric, so Zhongfu Lianzhong has a rule to start such work only when humidity is below a certain percentage.

How can laying of rowing and resin be evaluated? Zhongfu Lianzhong has his own inspection methods and standards. It requires a large amount of experimentation to obtain the best temperature for resins to shale the module under the best conditions.

Zhongfu Lianzhong paid attention to materials and featured experiments. The costs for experiments on materials alone exceeded RMB 10 million.

Refined manufacturing lays a solid foundation for high quality blades, but this alone cannot do the whole job. Zhongfu Lianzhong has detailed rating standards in all manufacturing bases for quality control, especially with onsite monitoring over key controlling points to ensure the process. On the one hand, the monitor would find common problems and locate the opportunity to improve detailed work; on the other hand, the workers would fully display their talents while enhancing awareness of quality.

27.7 Wind Power Pioneers

I am more of a junior standing on the shoulders of the giants, compared with the predecessors, senior experts and leaders in China's wind energy. I have been deeply inspired by their visions, spirit of tenacity, and persistence, and how their contributions shaped today's glory of China's wind power.

Chinese scientist Qian Xuesen devoted his attention and guidance to China's wind energy and promoted its development. On 10 September 1990, in the letter to Mr He Dexin, the president of Chinese Wind Energy Association, Qian Xuesen proposed an "electrical power program dominated with wind power" in China and emphasised "wind power generation should be nothing but big". He also said, "we should improve awareness first to embrace the socialist China in 21st century!"

Qian Xuesen proposed the following solutions for power supply in the 21st century:

- Have four million standardised 500 kW wind power generation units;
- Fully tap into China's large, medium and small hydropower resources;
- Use fuel gas peak value turbine generation units (but the gas has to come from places like petrochemical plants, free of pollution, instead of from coal. In addition, nuclear power plant can be considered as a supplement for insufficient supply).

This is a program for electric power dominated by wind energy. Of course, to realise this comprehensive planning is required as well as national support and research in wind energy regarding:

- How to standardise and reduce costs, while improving reliability;
- How to start using existing facilities and select for network building;
- How to address other issues such as mass production of equipment.

On 13 February 1994, in another letter to He Dexin, Qian Xuesen pointed out the issue of coordination between wind power plants and coal-fired power plants and believed the solution was to be found in the grid connection as the sole approach.

Qian Xuesen's guiding principles for China's wind energy development have never ceased to inspire Chinese talents in wind energy even until today.

At the end of the 1980s, Huang Yicheng was the Minister of then Energy Ministry of China and the member of standing committee of the 8th Chinese People's Political Consultative Conference (CPPCC), and called for related policies several times to develop wind power, "regard wind power as an important aspect of power generation", and suggested the government to issue a guiding ratio for the installation capacity for wind to coal-fired power plants, make a unified regulation of tariffs for wind power, reduce the VAT of wind power to the level of small hydro plants and provide favourable policies for wind power projects in bank loan interest and payment term, in order to promote the development of wind power in large scale.

When Xiao Gongren was working with the department of science and technology under China Water and Hydro Power Ministry, he undertook the role of organising wind power generation as part of the hydropower generation system. In 1978, the Ministry would research in wind turbine generators (WTGs) as the national key project, where Xiao Gongren was the head of panel of experts, in charge of organising R&D in the system. He said, "I did not feel a huge gap between domestic and European wind power, and I was watching Europe and other regions all over the world for wind power technologies for our own use."

From 1996 until 1998, Xiao Gongren had his busiest two years with the biggest load pressure. In March 1996, National Planning Commission launched "wind program", and Xiao Gongren was appointed as a member of the panel of experts and later the head, hosting the bidding for large-scale wind turbines in China. In 1997, the National Economics and Trade Commission arranged 70 MW of installed capacity in technological upgrading project "Two Enhance Program". The Ministry of Electric Power set up wind power projects, where Xiao Gongren was assigned the task to lead the panel of experts to provide technological support for wind farms. This project was completed and accepted at the end of 1998, with the capacity of 83.6 MW, signifying the demonstration phase for wind power generation in China entering into large scale development.

Named as the giant in China's wind power, Zhu Ruizhao explored the proven potential for wind energy resources in China to be around 160 GW, merely for resources available at 10 m height. "Apparently, with variations of height, this will change and at 50 m, the amount of resources could double". Such findings put light on wind power in China. Afterwards, Zhu Ruizhao made research in the zoning of wind energy in China and released the report of China's Wind Energy Resources Calculation and Zoning in 1985, elaborating on the wind energy resources, possibility of exploration and advice.

As the pioneer of wind farms, Wang Wenqi facilitated a historic event in China's wind power history. In 1980, he led some young people to explore wind power technology. In November 1986, they set up two large "windmills" at the lake of Chai Wopu and on the first day of 1987, put them officially into operation. After setting up the test station at Chai Wopu, Wang Wenqi had already started to prepare wind farms in the city of Daban, in Liaoning Province. With hardships overcome, the largest wind farm in size of installation back then in China and even in Asia was set up in October 1989. This fact exerted positive influences in wind power development in China.

As one of the pioneers of China's wind power, He Dexin initiated the establishment of the Chinese Wind Energy Association and China Wind Energy Technology Development Center as early as in 1980. He hosted national key projects on topics like technological breakthroughs in small wind turbines, research in aerodynamics of wind turbine aerofoil, concept designs of large wind turbine, research in wind farm planning methodology, development of wind turbine design software package, research in 200 kW generation units and certifications, etc., laying a good foundation for technological advances in WTGs, domestic production of large turbines and wind farms.

He Dexin made contributions to the theoretic evolution of aerodynamics for wind turbines by conducting comprehensive research in deviation features of wind turbines and 3D fluidity of blades in wind tunnels for the first time in the world, revealing the 3D fluidity and non-constant flow of blades, the results of which were used for validating design theories and model construction by scholars in China and abroad. In the meantime, he developed experimental techniques of wind tunnels for wind turbines, used improved wall pressure info matrix to successfully correct interference from the tunnel walls in the wind tunnel experiment for the first time in the whole world. These findings caught high values among counterparts. In over 30 years, the well-respected He Dexin is still active at the frontline of China's wind energy, where he contributes his experience and intelligence.

Shi Pengfei started his career in wind power from a leading role in Sino-Belgian governmental cooperative project. In 1986, he was involved in building the wind farm in Pingtan, Fujian Province, one of the earliest in China. With overseas study and experience in developing wind farms, he soon became an authoritative expert and won people's trust and respect with his dedicated work and frank expression.

Afterwards, Shi Pengfei took the position of director of new energy division under Water and Hydropower Planning and Design Institute of the Ministry of Electric Power and later was promoted to be deputy chief engineer, responsible for the early stage work of wind power projects in China.

Shi Pengfei left his footprints in almost each and every important wind farm and wind turbine manufacturing plant. With no exaggeration, no one would know better than he does about the true status of China's wind power industry. After retirement, there still is not much leisure for him and business trips are still normal. At the age of 70, he keeps riding a bicycle to work and home, two hours a day, 28 km, a habit he has had since high school.

27.8 Looking Ahead

Just like the predecessors that I deeply respect, I am fortunate to have witnessed the journey of growth of the wind power industry and to catch up with the wave of wind power development. As for China General Certification Center, this wave would not guarantee a thriving future, so CGC should continue to create higher values for customers with innovative certification technologies.

During the past decade, CGC has made positive contributions to the development of renewable energy in China by completing various tasks, including renewable energy projects of the 10th, 11th, and 12th Five-Year Plans, establishing a test lab for wind turbines and a national key lab for simulation, test and certification technology of wind and solar energy, and was the strategic partner of the newly-established China Renewable Energy Center. However, in face of the challenges of fossil energy exhaustion and climate change, CGC will need to work harder, to take more responsibilities to promote cooperation between China and the world, and to open up new development for clean energy. Only this way can we live up to the era we live in.

I am honoured that on 2 February 2012, Mr Sandy Butterfield, the chairman of IEC TC88, affirmed the contributions and accomplishments made by CGC in wind energy in his recommendation letter to Ms Magubane, director-general of the Department of Energy of South Africa. Sandy Butterfield said, "I see a well-established certification system operating in China, I see the CGC participating actively in the newly established IEC Certification Advisory Committee (CAC) which was established to facilitate international certification harmonisation. In short, I see a responsible certification organisation that is deeply engaged with the international certification process and deserves recognition for their accomplishments."

I feel honoured and obliged to Sandy Butterfield's affirmation on the work of CGC and his heartfelt assistance in the internationalisation of certification. CGC will always uphold the core philosophy of "Create Value for Customers" and make new contributions to the development of clean energy in China and in the world.

About the author

Qin Haiyan graduated from the Department of Power Engineering of Shanghai Jiaotong University in 1994 and got MBA from Renmin University of China in 2003. Since 1998, he has been engaged in renewable energy and has focused on policy and technology research in terms of wind energy, solar energy and other renewable energy development.

Since 2004, taking CGC and CWEA as two major platforms, he and his team have been working on wind power technology standards development and improvement, as well as the establishment of wind power equipment, solar water heater and PV production certification systems in China. All the efforts he and his team made are very favourable and supportive to the rapid and sound development of the industry in China.

Currently, Mr Qin is secretary general of Chinese Wind Energy Association (CWEA) and general director of China General Certification Center (CGC). He is also board member of PV GAP, China co-chair of ACORE, council member of China Renewable Energy Society and deputy secretary general of the Technical Committee of Wind Machinery Standardisation of China.

Chapter 28

Rising Wind Power Industry of XEMC

Zhou Jianxiong [a,b]

[a]XEMC, 302 Xiashesi Street, Xiangtan City, Hunan 411101, China
[b]Xiangtan Electrical Equipment Manufacturing Co. Ltd., No. 132, Tanshao Road,
Xiangtan, Hunan 411100, China

xtdq.vip@hotmail.com

28.1 Brief Introduction to Xiangtan Electric Manufacturing Company

Xiangtan Electric Manufacturing Co. Ltd. (XEMC) was established in 1936. Over seventy years, the company has developed into one of the most powerful enterprises in the electric industry of China regarding the overall technological superiority and product supporting capability. It is the cradle of electric industry of China.

XEMC has a national enterprise technology centre that ranks among the top 18 centres all over China, 1 803 engineering team members, and the capability of overall product design and development, and integration of complete equipment. The company possesses three core technologies, that is, ship electric propulsion,

Wind Power for the World: The Rise of Modern Wind Energy
Edited by Preben Maegaard, Anna Krenz and Wolfgang Palz
Copyright © 2013 Pan Stanford Publishing Pte. Ltd.
ISBN 978-981-4364-93-5 (Hardcover), 978-981-4364-94-2 (eBook)
www.panstanford.com

vehicle outfitting, and electrical assembly of motor. The company has successively developed 1 100 new products, which have filled many domestic production gaps. More than one hundred of them set records in China. In terms of technical strength and industrialisation, XEMC takes the lead in the electric industry of China.

Figure 28.1   The headquarters base of XEMC.

The company focuses on the businesses of wind-generating sets, large and medium AC motors, DC motors, large pumps, large electric-wheel dump cars, and metro rail and supporting products, which have a good sale all over the world, and are highly recognised and commended by clients. By the end of 2011, the group company has had 56 holding companies, 5 branch companies, and more than 1 2 000 employees.

XEMC has gained such development under close attention and strong support of the Chinese government. Jiang Zemin, Xi Jinping, Wen Jiabao, Jia Qinglin, Li Changchun and other party and state leaders have visited the company in person, which has inspired all employees of the company.

Figure 28.2   President Jiang Zemin inspecting wind power industry (left); Premier Wen Jiabao inspecting wind power industry (right).

28.2 Development Course and Superiority of XEMC in the Wind Power Industry

28.2.1 Long Development History

XEMC has begun the research on the wind power generation since the 1980s. It was one of the earliest enterprises that began to engage in the research on wind power generation in China. At the beginning of the 1990s, XEMC developed the first 315 kW and 600 kW asynchronous wind generators of China, both of which won the "Science and Technology Progress Award of China Machinery Industry".

In 2004, XEMC developed the first 1 300 kW wind turbine double-fed asynchronous generator in China. In 2007, XEMC successfully launched the first batch of complete 2 MW PMDD wind turbine generators in China. XEMC successfully developed the first 5 MW PMDD wind turbine generator installation in China in 2010, and, in 2011, the first 6 MW double-fed wind turbine generator and the first direct-driven electric excitation wind turbine generator in China.

Figure 28.3 Ceremony of XEMC's 5MW PMDD wind generator off the assembly line.

So far, XEMC has established its industrial bases in Xiangtan of Hunan province, Zhangzhou of Fujian province, Tongliao of Inner Mongolia, Wuwei of Gansu province, and Rotterdam of Netherlands, and now is able to manufacture 2 000 sets of complete direct-driven wind turbine generators, 2 500 direct-driven wind turbine generators, and 2 000 double-fed wind turbine generators. It is an industrial base of complete set of mega watt level wind power generation system approved by the National Development and Reform Commission of China, and the chairman unit of China Wind Equipment Society.

Figure 28.4 Wind turbine generators production base (left); Complete wind turbine generators production base of XEMC head-quarters (middle); Industrial base in Europe (right).

28.2.2 Solid Technical Strength

In the wind power industry, XEMC is famous for its leading technology and top-grade products. XEMC has three specialised wind power research institutions, that are, XEMC Wind Power Research Institute, XEMC Beijing Wind Power Academy and Holland Darwind Wind Turbine Academy. XEMC has more than 860 technicians engaged in the research on wind power, including 72 European researchers with top design level from XEMC Darwind BV. XEMC is the only national energy wind generator research and development (experimental) centre of national resources of China, and has the capability of development of first-class products, and autonomous innovation.

The wind driven products of XEMC are provided with complete proprietary intellectual property rights. Now, 116 applications for patent have been accepted, and 64 patents have been obtained.

XEMC is the unit to compile national standards for the wind turbine generators in China. Entrusted by Chinese government,

XEMC has successively hosted the compilation of such national standards as GB/T "Technical Conditions and Experimental Methods for the Grid-connected Asynchronous Wind Generators", GB/T "Permanent Magnet Synchronous Wind Generators—Part I: Technical Conditions", and GB/T "Permanent Magnet Synchronous Wind Generators—Part II: Experimental Methods".

Figure 28.5 National energy wind generator research & development (experimental) centre (left); Darwind Wind Generator Academy in Netherlands (right).

28.2.3 Leading Direct-Drive Types

XEMC specialises in the complete megawatt level direct-driven wind turbine generators and has advanced design concept. It removes speed-increasing gearbox that causes loud noise and is apt to be faulty, but adopts full power converter with a simple structure and high reliability. The converter transmits less high harmonic to the power grid, and generates high quality electricity, and has a great low-voltage ride through (LVRT) ability. The cost of maintenance of it is low. During the 20 years of design service life, the total cost of operation (TCO) was at least 15% lower than that of double-fed wind turbines with speed-increasing gearboxes.

28.2.4 Complete Product Series

Through years of high-speed development, the complete wind turbine products of XEMC have covered 5 kW–5 MW series, and wind turbine generator products 5 kW–5 MW PMDD series, and 1 MW–6 MW double-fed series. In terms of wind development and utilisation, XEMC has the integrated capability in the research and development, in manufacturing, test and commissioning of wind turbine generators, and in the measurement, design, development and operation of wind farm. It is one of few outstanding enterprises in the world that can provide integrated services and solutions to the wind power construction projects.

Figure 28.6 1 MW to 6 MW double-fed wind generator series (left); 1.5 MW to 5 MW direct-driven wind generator series (right).

Figure 28.7 5 kW to 200 kW direct-driven wind generator series.

28.2.5   Complete Test System

Quality is the life of the enterprise. All products of XEMC have passed the certification of ISO9000 system. Fan products have passed certification of China Classification Society and GL, and EC certification of the United Nations.

XEMC has set up a complete quality management system, and has advanced quality test equipment. It is the only state key laboratory of China for offshore wind power technique and test.

28.2.6   Convenient Marketing Channels

XEMC has more than 30 marketing institutes, a complete marketing network, and a large quantity of stable customer resources and partners all over the world, has a large number of XEMC wind turbines operating in the main wind farms in China. Meanwhile, XEMC has set up points of sales in North America, Europe and Southeast Asia, and sells fan products to European and American countries and regions, as well as Taiwan of China.

28.2.7   Outstanding Sales Performance

Since 2007, XEMC has totally manufactured more than 2 000 sets of 2 MW–5 MW direct-driven wind generators, which are operating stably on the offshore wind farm in the Netherlands and in the main wind farms all over China, for example, Zhangzhou of Fujian province, Zhuozi of Inner Mongolia, Zhangjiakou of Hebei province, Chenzhou of Hunan province, and Taiwan. Besides, XEMC has provided over 2 000 sets of double-fed wind generators to Beijing State Grid Corporation of China, Dongfang Turbine Co. Ltd., Shanghai Electric, Guangdong Mingyang Electric, etc.

Figure 28.8   5 MW wind turbines operating in the Netherlands (left); 2 MW wind turbines operating in Taizhong County, Taiwan (right).

28.2.8   Public Praise on the Market

According to statistics of owners of wind farms and quality inspection authority of China, all operation indexes and quality of products installed by XEMC are in the leading position in the industry. The advanced design concept, high quality and high performance of wind turbines have won XEMC the public praise in the market.

Figure 28.9 President of World Wind Energy Association Preben Maegaard's visit to the wind power industry of XEMC.

28.3 Rising Wind Power Industry of XEMC

XEMC has been carrying forward its advantage in the mechanical and electrical integration over the 70 years of development, taking the lead in the industry by virtue of its highly reliable complete direct-driven wind turbine generators, and is developing at a higher speed towards series large offshore wind turbine generators.

Figure 28.10 Patented offshore integrated power generation system of XEMC.

It keeps pushing technological innovation, improving the space layout of industry, and enhancing its professional advantage in the key parts and components, and is exploring international market steadily so as to become one of the largest manufacturers of complete wind turbine generators in the world, provide clean energy and make the world greener. On the new starting point of globalisation, the wind power industry of XEMC is ready for takeoff and will draw the curtain of wind power generation all over the world.

About the author

 **Zhou Jianxiong**, born in 1955, is a native of Changsha in Hunan Province, China. He holds a PhD, and is a professor-level senior engineer. Dr Jianxiong is president of Xiangtan Electric Manufacturing Group and president of Xiangtan Electric Machinery Co. Ltd. He has been long engaged in the scientific and technical innovation and corporate management in fields such as new energy and manufacture of advanced equipment, with enterprises under his leadership having been listed in top 500 manufacturing enterprises in China. Concurrently, Zhou Jianxiong has been serving as standing director of China Enterprise Confederation, director general of the Wind Power Generation Equipment Committee of China Electrical Equipment Industry Association, vice-president of China Urban Rail Transit Association and president of Electric Traction Equipment Committee of China Electrical Equipment Industry Association. He has been awarded many honours, such as Excellent Entrepreneur of China Excellent Database, Excellent Entrepreneur in National Machinery Industry, Award of Chinese Senior Professional Manager with Special Contribution, Top 10 Pioneers for New Energy in China, and the First Prize of Scientific and Technological Progress in Hunan Province.

Name Index

Subject Index

With financial support of